Banking on Climate Change

International Banking and Finance Law Series

VOLUME 24

The titles published in this series are listed at the end of this volume.

Banking on Climate Change

How Finance Actors and Transnational Regulatory Regimes Are Responding

Megan Bowman

Wolters Kluwer

Law & Business

Published by:
Kluwer Law International
PO Box 316
2400 AH Alphen aan den Rijn
The Netherlands
Website: www.kluwerlaw.com

Sold and distributed in North, Central and South America by:
Aspen Publishers, Inc.
7201 McKinney Circle
Frederick, MD 21704
United States of America
Email: customer.service@aspenpublishers.com

Sold and distributed in all other countries by:
Turpin Distribution Services Ltd
Stratton Business Park
Pegasus Drive, Biggleswade
Bedfordshire SG18 8TQ
United Kingdom
Email: kluwerlaw@turpin-distribution.com

Printed on acid-free paper.

ISBN 978-90-411-5223-7

Printed and Bound by CPI Group (UK) Ltd, Croydon, CR0 4YY.

Table of Contents

Preface and Acknowledgements   xi

List of Abbreviations   xv

List of Tables   xix

List of Figures   xxi

CHAPTER 1
Banking on Climate Change   1
§1.01   "It's the Economy, Stupid"   1
§1.02   Transnational Regulatory Responses to Climate Change   4
         [A]   The Reality and Rapidity of Climate Change   4
         [B]   Multi-jurisdictional Regulatory Regimes on Climate Change   5
                [1]   International Level   5
                [2]   National and Regional Climate Regulation   6
                       [a]   Regulation That Puts a Price on Carbon   6
                       [b]   Regulation That Incentivizes Energy Efficiency and
                              Investment in or Usage of Renewable Energy and
                              Clean Technology   7
                       [c]   Regulation That Mandates Compulsory Monitoring
                              and Reduction of Internal Corporate GHG Emissions   9
                       [d]   Regulation That Discourages or Prohibits Intensive
                              GHG-Emitting Projects   10
§1.03   Sustainable Development and Climate Finance   11
         [A]   Key Public and Multilateral Finance Entities   12
         [B]   Key Private Finance Actors   14
                [1]   Insurers and Reinsurers   16
                [2]   Institutional Investors   16

      [3]   Banks   17

      [4]   Entrepreneurial Finance Actors   17

§1.04   Scope and Plan of This Book   17

    [A]   Practical and Theoretical Context   18

    [B]   Empirical Case Study: Early-Moving Banks in Key Market
Economies   19

    [C]   Regulatory Recommendations   20

CHAPTER 2
A Unique Relationship: Private Finance Actors and Climate Change   21

§2.01   Overview   21

§2.02   The Money-Go-Round: Types and Functions of Private Sector
Financial Institutions   21

§2.03   Climate Risks and Opportunities for Private Finance Actors   26

    [A]   Relevance by Industry: Climate-Related Risks and Opportunities   27

      [1]   The Insurance Industry: Insurers and Reinsurers   27

      [2]   Institutional Investors   32

      [3]   Banks   37

    [B]   Relevance by Activity: Finance Sector Business Practices and
Climate Change   38

      [1]   Risk Assessment and Risk Management   38

        [a]   Credit and Investment Risks   40

        [b]   Reputation Risk   42

        [c]   Litigation Risk   47

      [2]   Investing, Lending & Financing   49

      [3]   New Market Entry and Innovative Product Design   55

§2.04   Banks as Corporate Change Agents?   57

    [A]   Banks as Creditors and Investors   59

    [B]   Banks as Advisers and Heads of Supply Chains   60

CHAPTER 3
Why Do Companies Go Green?   63

§3.01   Overview   63

§3.02   Theoretical Framework   63

§3.03   Meso Level: Organizational Field Level Drivers of Change   64

    [A]   CSR: Short History and Lack of Conceptual Clarity   64

    [B]   The Business Case for "Doing Good"   70

      [1]   The "Win-Win"/ "Win-Lose" Literature   71

      [2]   The "It Depends" Literature   75

§3.04   Macro Level: Socio-Cultural Drivers or "External" Factors   81

§3.05   Micro Level: Intra-organizational Drivers or "Internal" Factors   85

§3.06   Summation: Knowledge Gaps and Empirical Next Steps   89

CHAPTER 4
The Levers of Corporate Change: Case Study Evidence   91
§4.01   Overview   91
§4.02   Qualitative Methodology   92
         [A]   Sourcing the Banks and Respondents   92
         [B]   Data Analysis   94
§4.03   Drivers: Bottom-Up, Top-Down and Middle-Out   95
         [A]   Meaning and Role of Corporate Reputation as Driver   97
                [1]   Client Service Reputation   99
                [2]   Social Reputation   102
                       [a]   Type of Bank   103
                       [b]   Type of Employee   106
                       [c]   Regulatory Context   108
         [B]   Risk Mitigation as Driver   111
                [1]   Types of Regulation and Risk   111
                [2]   Leading Bank Approaches by Jurisdiction   113
                       [a]   Europe / UK   113
                       [b]   United States   114
                       [c]   Australia   116
§4.04   Non-driver: The "Care" Factor   118
         [A]   Manifestation of CSR   119
         [B]   Conceptualization of CSR   119
         [C]   Subordination of CSR   122
§4.05   Implications   125

CHAPTER 5
The Limits of Corporate Change: Case Study Evidence   129
§5.01   Overview   129
§5.02   Limitations of the Business Case as Driver   130
         [A]   The Requirement of a Business Case for Green Uptake   130
         [B]   The Countervailing Business Case for Non-Green Initiatives   131
§5.03   Banking Regulation, Corporate Law and Corporate Governance
         Norms   132
         [A]   Bank Regulation and Supervision   133
                [1]   United Kingdom   133
                [2]   Australia   135
                [3]   United States   136
         [B]   Corporate Law, Directors' Duties and Corporate Governance
                Norms   142
                [1]   United Kingdom   142
                [2]   Australia   144
                [3]   United States   146
                [4]   Summation   149
§5.04   Incremental versus Transformational Change   149

[A] "Organic" Mainstreaming via Isomorphism   149
[B] Going the Distance: Standard-Setting and Real Change   151
     [1] Green Activities as an Extension of Current Practice?   153
     [2] New Green Initiatives in Competition with Established
          Non-Green Activities   159
     [3] Built In versus Bolted On   160
     [4] Market Size   163
[C] Timeliness: The Imperatives of Expeditious and Radical Change 164
§5.05 Concluding Remarks   167

CHAPTER 6
Empirically Informed Regulation   169
§6.01 Summary of Empirical Findings   169
[A] A New Taxonomy of "Corporate Reputation" as Driver   170
[B] Climate Change as Risk or Opportunity? The Relevance of
     Regulatory Context   171
[C] CSR as a Non-driver   172
[D] The Limited Nature of Voluntarism for Mainstreaming   174
[E] Reflections   174
§6.02 Generalizability of Findings to Other Private Finance Actors 175
§6.03 Legal Implications of Findings: The Case for Regulatory Intervention 176
§6.04 Direct and Coercive Government Regulation   178
[A] Attributes and Benefits of Direct Coercion   178
[B] Examples of Direct and Coercive Climate Finance Regulation 179
     [1] Prescriptive and Proscriptive Legislation   179
     [2] Taxation   179
[C] Criticisms of Coercive Legislation   180
§6.05 Indirect and "Nudging" Government Regulation   183
[A] What's in a Nudge?   184
[B] Critiques of "Nudging" Regulation   189
     [1] Incentives Can Have Unintended and Unexpected
          Perverse Consequences   190
     [2] Governments May Not Know What Is "Best" or "Right"   191
§6.06 The Necessity of a Climate Finance "Regulatory Mix"   194

CHAPTER 7
Re-setting the Regulatory Sights   197
§7.01 Overview   197
§7.02 The Capital Asset Pricing Model and Risk/Return Requirements for
       Different Private Finance Actors   198
§7.03 Tax Equity Partnership Structures and Corporate Climate Finance 202
§7.04 Climate Bonds and Green Bonds   205
§7.05 Evergreen Buy-Out Strategies   209
§7.06 Concluding Remarks   214

CHAPTER 8
Looking to the Future: Chinese-Global Green Growth   215
§8.01   Overview   215
§8.02   China Goes Global   216
         [A]   Policy Background   216
         [B]   The Rise and Role of State-Owned Enterprises to Going Global   218
         [C]   Going Global Responsibly?   220
§8.03   China and Green Growth: Policy Framework   222
         [A]   Key Targets   222
         [B]   Policies That Encourage Green Investment Within China   223
                [1]   Incentives   223
                [2]   Taxation Measures   224
                [3]   National Emissions Trading Scheme   225
                [4]   Foreign Investment Catalogues   225
         [C]   Green Banking and Financing Policies   226
§8.04   Chinese Green Investment and Finance: Patterns, Levers and Limits   227
         [A]   Wind and Solar   228
         [B]   Financing Arrangements   229
         [C]   Foreign Regulatory Levers and Limits   230
§8.05   Future Directions   233
§8.06   Concluding Remarks   235

CHAPTER 9
Conclusion   237
§9.01   The Precipice and the Bridge   237
§9.02   Viewing Change through an Institutional Lens   239
§9.03   From Dream to Reality   242

Table of Cases   245
Table of Statues   247
Index   251

Preface and Acknowledgements

The climate finance field is extremely new. Architectural frameworks and key players are still emerging within an Escher-style landscape of fiscal, environmental and political mutability. But I think we are on the cusp of grasping the importance of private capital and public finance to facilitating the most significant and challenging global transition of our time: the transition to a low-carbon economy. I genuinely hope this book is a timely and useful contribution to that broader endeavor.

In recent years there have been rapid developments in all major economies around the world regarding financial markets and, quite separately, climate change. These developments have made the climate finance field highly relevant and frustratingly mercurial. Indeed a litany of important and seemingly unrelated world events occurred during the research and writing of *Banking on Climate Change*. Interview data were gathered in the aftermath of the 2009 global financial crisis and at a time when societal suspicion of banks had reached an all time high. The world was still suffering economically; collapse of the Eurozone loomed constantly with potential contagion of financial failings between EU member countries; and governments within the United States, Europe and Australia were formulating legal measures to avert future financial crises instigated by irresponsible bank practices. International climate negotiations had stalled, the then-UN Climate Chief had resigned, and nations were beginning to contemplate abandonment of the *Kyoto* endeavor entirely, only to reinvigorate it in 2014. The BP *Deepwater Horizon* oil spill occurred in American waters, which precipitated public scrutiny of the world's dependence on fossil fuels and resignation of BP's then-CEO. And in Australia, the Labor Government busily see-sawed not only between its leaders but also in its affection for a carbon price, breaking promises to eventually pass a carbon tax in 2011 that was, in an unprecedented move by a government that takes power with a tax *in situ*, subsequently repealed by a Coalition Government in 2014.

In short, during the past five years it has become very clear that now is a good time to understand the discrete topics of finance and climate change *together*.

This book embodies one of the first qualitative studies on corporate climate finance. To date, very little empirical work of this nature has been undertaken in the climate finance space so we haven't really understood what makes private finance actors "tick" let alone "go green" in real life. Banks in particular are multi-cellular and notoriously opaque creatures that do not usually welcome scrutiny. I have been very fortunate in gaining access to senior managers and invaluable information, which opened a window into bank decision-making and motivations vis-à-vis climate change. Many respondents would prefer to remain anonymous so I thank them collectively for their frankness, insights, and generosity in giving their time and knowledge to me. The empirical case study in this book – and the larger lessons learned thereby – could not exist without them.

I was fortunate enough to be selected for a number of international workshops where I could air my theses and receive insights from critical thinkers of exceptional quality. Workshops that were particularly influential on my thinking for this book include the C9/G8 Clean Energy symposium at Tsinghua University in 2011; the Cambridge University Regulation and Governance conference in 2012; the Harvard/Stanford International Junior Faculty Forums in 2012 and 2013; the Responsible Investing workshop at the University of British Columbia in 2013; and the Conference on Empirical Legal Studies at UC Berkeley School of Law in 2014. I thank participants at those fora, and other generous scholars outside of them, for their rich insights and encouragement. In particular, I give special thanks to Professors Eric Talley and Benjamin J. Richardson for their invaluable support and provocations, without which these chapters would be much the poorer; and for further nudges provided by Professors Bill Alford, Cass Sunstein, Bob Kagan, Lynne Dallas, Tima Bansal, Dan Farber, Forest Reinhardt, Venky Narayanamurti, Sam Fankhauser, Ethan Elkind and Steve Weissman.

I would also like to acknowledge the intellectual, financial and moral support of staff at the Australian National University's Regulatory Institutions Network (RegNet) where I undertook my PhD candidature and where this book began. RegNet is recognized internationally not only for its world-class research output but also for the warmth and generosity of its people. There are some people that deserve special mention for their genuine interest and inspiring role-modeling: Professors Neil Gunningham, Val and John Braithwaite, Veronica Taylor, Hilary Charlesworth, Terry Halliday, Howard Bamsey and Dr Kyla Tienhaara.

Many thanks also to those who gave professional stimulation and support along the way. In particular, my UNSW Law colleagues: Dr Rob Nicholls and Professor Justin O'Brien from the Centre for Law, Markets and Regulation; Centre interns Stephanie Cardy, Jacqui Vorreiter and Joe Shin for research assistance when I needed it most; Professors Brendan Edgeworth, Prue Vines, and Fleur Johns for additional (tor)mentoring; and Simone Degeling for faculty grants to Harvard and Berkeley Law Schools to discuss my work in progress. Outside of formal channels, I thank Professors Jaye Ellis and Andrew Clarke for their ongoing support and good advice; the international Network for Sustainable Financial Markets, particularly Cary Krosinsky and Raj Thamotheram; and for subject-specific discussions with Sean Kidney, Tim McDonald and Jim Osborne, Justin Ritchie, Dr. Wei Li, Professors Ruoying Chen, Nico Howson, Peter Drysdale, and the East Asian Bureau of Economic Research team.

I especially thank the Wolters Kluwer team for their professionalism, encouragement and patience: Professor Ross Buckley, Simon Bellamy, and Pritha at NewGen editing.

Finally, I give heartfelt thanks to my family and friends. To Jus and Dan (and their gorgeous girls), Katie and Vlad (and their beautiful offspring), Tamar, Thierry, Jono, Anna, Amanda – your unwavering love and support is a constant source of strength and joy for me. To all the dancers and musicians who inspire and touch me. To my beloved family Mum, Pierre, Marcus, Anjalee, Tristan, Penny, Lily, Gabriel, Wesley: we are so blessed to have each other. And to dear Papa: it was because of you that I always wanted to be a doctor.

List of Abbreviations

ADI	Authorized Deposit-Taking Institution
AGM	Annual General Meeting
APRA	Australian Prudential Regulation Authority
ASIC	Australian Securities and Investments Commission
BAU	Business-as-Usual
CalPERS	California Public Employees Retirement System
CAPM	Capital Asset Pricing Model
CBRC	China Banking Regulatory Commission
CCS	Carbon Capture and Storage
CDP	Carbon Disclosure Project
CER	Corporate Environmental Responsibility
CFP	Corporate Financial Performance
CFR	Code of Federal Regulations (US)
CIF	Climate Investment Fund
CO_2	Carbon Dioxide
COP	Conference of the Parties
CSP	Social and Environmental Performance
CSR	Corporate Social Responsibility
ENGO	Environmental Non-government Organization
EPA	Environment Protection Authority
EPC	Engineering, Procurement, Construction (Contracts)
ESG	Environmental, Social and Governance
ETS	Emissions Trading Scheme
EU	European Union
FAT	Financial Activities Tax

FCA	Financial Conduct Authority (UK)
FDI	Foreign Direct Investment
FDIC	Federal Deposit Insurance Corporation (US)
FIT	Feed-in-Tariff
FTT	Financial Transactions Tax
FRB	Federal Reserve Bank (US)
FSA	Financial Services Authority (UK)
FSOC	Financial Stability Oversight Council (US)
GCF	Green Climate Fund
GHG	Greenhouse Gas
GIB	Green Investment Bank (UK)
GFC	2008/2009 Global Financial Crisis
IFC	International Finance Corporation
IPCC	Inter-governmental Panel on Climate Change
IPO	Initial Public Offering
IT	Information Technology
KPI	Key Performance Indicator
KKR	Kohlberg Kravis Roberts
M&A	Merger and Acquisition
MOFCOM	Chinese Ministry of Commerce
MRET	Mandatory Renewable Energy Target
MTR	Mountain Top Removal
NGO	Non-government Organization
NDRC	National Development and Reform Commission (China)
OCC	Office of Comptroller of Currency (US)
ODI	Overseas Direct Investment
OECD	Organisation for Economic Co-operation and Development
OTS	Office of Thrift Supervision (US)
PE	Private Equity
POE	Privately Owned Enterprise
PPA	Power Purchase Agreement
PRA	Prudential Regulation Authority (UK)
PSI	Principles for Sustainable Insurance
PTC	Production Tax Credits
PV	(Solar) Photovoltaic
ITC	Investment Tax Credit
RAN	Rainforest Action Network

RBA	Reserve Bank of Australia
RE	Renewable Energy
RECs	Renewable Energy Certificates
RET	Renewable Energy Target
RPS	Renewable Portfolio Standard
SEC	United States Securities and Exchange Commission
SML	Security Market Line
SOE	State-Owned Enterprise
SRI	Socially Responsible Investment
SSE	State-Supported Enterprise
SSCN	Sustainable Supply Chain Network
TXU	Texas Utilities
UK	United Kingdom
UNEP	United Nations Environment Program
UNEPFI	United Nations Environment Program Finance Initiative
UNFCCC	United Nations Framework Convention on Climate Change
UNPRI	United Nations Principles for Responsible Investing
US	United States
USC	United States Code
VAT	Value-Added Tax
VC	Venture Capital
WBCSD	Business Council for Sustainable Development
WWF	World Wildlife Fund

List of Tables

Table 2.2   Classification and Activities of Private Sector Financial
             Institutions

Table 2.3   Climate Change and the Private Finance Sector: Mutual
             Impacts

Table 5.1   The US Financial Regulatory Framework After Dodd-Frank

Table 5.2   New US Financial Regulatory Institutions created by
             Dodd-Frank

List of Figures

Figure 2.1   The Financial System: Financial Markets, Intermediaries, and
              the Flow of Funds
Figure 4.1   A New Conception of Corporate Reputation
Figure 4.2   The Constitution of Social Reputation Versus Client Service
              Reputation
Figure 4.3   Modality and Effects: Social Reputation versus Client Service
              Reputation
Figure 7.1   Risk/Return Trade-Off: The CAPM and SML
Figure 7.2   Tax Equity Partnership Structure

CHAPTER 1

Banking on Climate Change

§1.01 "IT'S THE ECONOMY, STUPID"

The imperative of climate change mitigation is to urgently cap global warming at two degrees Celsius in order to prevent catastrophic global change. We know that the solution is expeditious reduction of global greenhouse gas emissions, particularly carbon dioxide.[1] Yet we do not seem to know how best to mobilize this solution.To date, scrutiny has been directed at the likelihood of an international agreement beyond the *Kyoto Protocol*[2] and the activities of high-emitting industries such as the fossil fuel sector. Far less consideration has been given to the facilitative role of financial intermediary actors. This is a critical oversight because addressing climate change will not only require behavioral and technological changes; it will also require lots of money.

Global greenhouse gas (GHG) emissions must peak and level no later than 2020 and then drop by half (of 1990 levels) by 2050 if we are to stay within the two degrees Celsius (2°C or 3.6°F) guardrail.[3] Massive financial mobilization in the order of US$15.2trillion of additional costs for both developed and developing nations will be required for global GHG emissions mitigation, which is an exponential increase on the current annual investment of US$160billion.[4] In addition, trillions of dollars will be

---

1. Intergovernmental Panel on Climate Change (IPCC) *Climate Change 2007: Synthesis Report. Contribution of Working Groups I, II and III to the Fourth Assessment Report of the Intergovernmental Panel on Climate Change* (IPCC, Geneva, 2007) (hereafter "IPCC 2007 Synthesis Report"), 36 (Figure 2.1). See also IPCC, *Fifth Assessment Report (AR5)* (IPCC, Geneva, 2014), available at http://www.ipcc.ch/report/ar5/.
2. *Kyoto Protocol to the Framework Convention on Climate Change* (1998) 37 ILM 22.
3. European Commission, *The EU Emissions Trading System (EU ETS)* (2013), 2, http://ec.europa.eu/clima/publications/docs/factsheet_ets_en.pdf(accessed Oct. 10, 2014).
4. Green Climate Fund, *Business Model Framework: Private Sector Facility, GCF/B.04/07* (Green Climate Fund, Jun. 12, 2013).

required to upgrade and expand energy and transport infrastructure to assist adaptation in developing nations;[5] and additional annual investment of nearly US$800billion will be required for electricity expansion, modern cooking fuels, energy efficiency, and renewable energy.[6]

In short, moving to a low-carbon global economy and increasing climate resilience will require significant capital outside of normal government channels and beyond business as usual. It will involve one of the largest market and economic transitions in modern society. Such a massive transition requires financial input and facilitation on an equally grand scale.

Public finance actors, such as the World Bank and the newly created Green Climate Fund, tend to take the spotlight here. Far less attention has been given to the potential of and processes for directly engaging private sector finance actors as positive societal change-agents. Specifically, transnational private sector financial institutions that are headquartered in developed countries are global economic gatekeepers and financial intermediaries, making them critical actors in the transition to a low-carbon global economy.[7] In addition to entrepreneurial actors such as venture capital funds and private equity firms, these "private finance actors" comprise conservative actors, namely insurers (including reinsurers), institutional investors (including pension funds) and banks. To date, their potential to assist the global shift to a low-carbon and climate-resilient economy has been largely overlooked by scholars and policy-makers.

Partly this oversight is because the private finance sector has not been recognized as relevant to climate change due to its minimal direct environmental impact; it does not engage in extractive or manufacturing activity or produce high-polluting services and products. Nonetheless, in recent years it has become apparent that the finance sector is impacted by climate change and has a central role in helping to address it.

For private finance actors, climate change presents new opportunities and risks.

Opportunities take the form of innovative climate-related products and services and new market entry and brokering roles that can create competitive edge and lucrative returns. For example, industry reports have estimated that carbon trading could reach US$3trillion by 2020 and will eventually total US$10trillion a year, which would make it the largest commodity market in the world.[8] Similarly, renewable energy and clean tech markets are gaining momentum with aggregate private sector "green" investment from 2007-2013 (inclusive) totaling US$5.3trillion globally.[9]

---

5. World Bank, *World Development Report 2010: Development and Climate Change* (The International Bank for Reconstruction and Development / The World Bank: Washington DC, 2010).
6. World Bank, *Sustainable Energy For All: Global Tracking Framework (v.3)* (The International Bank for Reconstruction and Development / The World Bank: Washington DC, 2013), http://documents.worldbank.org/curated/en/2013/05/17765643/global-tracking-framework-vol-3-3-main-report (accessed Feb. 7, 2014).
7. M. Bowman, "Development and Global Sustainability: The Case for Corporate Climate Finance" (2014) *Harvard College Review of Environment & Society* 12.
8. M. Carr, "China, Greenpeace Challenge Kyoto Carbon Trading," *Bloomberg.com* (Jun. 19, 2009), http://www.bloomberg.com/apps/news?pid = 20601080&sid = aLM4otYnvXHQ.
9. H. Henderson, R. Sanquiche, and T. Nash, *2014 Green Transition Scoreboard® Report: "Plenty of Water!"* (Ethical Markets Media, 2014), http://www.ethicalmarkets.com/wp-content/uploads/2014/03/GTS-report-water-focus-March-2014-3-29-14.pdf (accessed Oct. 10, 2014).

Opportunities often arise from government policies that lubricate market activity. Accordingly, in order to exploit these opportunities, private finance actors need to keep abreast of current policy and likely regulatory changes.

Concomitantly, climate change heralds increased risk for private finance actors. A new category of "climate risk" has emerged that extends the indirect environmental, social, governance (ESG) and material risks to which finance actors are vulnerable from clients and investments:

- Credit and investment risks arise from companies that are exposed to increased climate-related regulatory pressure and changes in market demand. Sectors such as real estate, agriculture, forestry and tourism are directly affected by wild weather; sectors such as oil, coal, heavy manufacturing and transport are competitively affected by government policies that price carbon.
- Reputation risk has become a key issue for private finance actors due to indirect socio-environmental impacts from financing environmentally risky or "dirty" borrowers and projects. In particular, naming and shaming efforts by non-government organizations (NGOs) have become sophisticated and transnational. Such campaigns can undermine the trust and reputation so crucial to financial operations.
- Financiers now also face litigation risk for their role in facilitating GHG emissions due to financial support for projects that contribute to climate change impacts on cities and civil society.

The key for finance actors is to explicitly internalize these risks into enhanced due diligence processes, and to learn good external stakeholder engagement. Moreover, as with regulatory opportunities, private finance actors must keep abreast of regulatory risk.

Importantly, there are also broader regulatory and societal implications beyond the business-related repercussions of climate change for private finance actors. Investors, insurers and banks are economic gatekeepers and financial intermediaries: they have access to large and multiple pools of money and the authority to move it around. Their *raison d'être* is to make intermediating decisions about where money (as an asset, debt or equity) comes from and where it flows to (via sourcing, allocation and advisory processes). They have a critical role in climate-related efforts because, as noted by Lord Stern, "reducing emissions and adjusting to climate change involves investment and risk."[10] Specifically, they can assist the transition to a low-carbon and climate resilient economy through their financing, lending, investing and advising activities, and their influence on the behavior of other corporate actors in a carbon-constrained world.

---

10. UNEPFI, *CEO Briefing: Carbon Crunch: Meeting the Cost* (United Nations Environment Program Finance Initiative: Geneva, 2007), 2.

§1.02 TRANSNATIONAL REGULATORY RESPONSES TO CLIMATE CHANGE

[A] The Reality and Rapidity of Climate Change

It is now known that our climate change trajectory is far beyond even the worst case scenario depicted by the Inter-governmental Panel on Climate Change (IPCC) and the Stern Review in 2007.[11] In 2009, with GHG emissions increasing three times faster than initially predicted, the Copenhagen Synthesis Report emphasized how we are moving beyond "the patterns of natural variability within which contemporary society and economy have developed and thrived."[12] In 2014, scientists nearly unanimously predicted that without urgent policy and multisectoral action the world would warm by 4°C above the preindustrial climate by the end of the century.[13] Such a rise would instigate unprecedented heat waves, droughts, flooding, cyclones and wildfires in developed nations as well as many of the world's poorest regions with serious impacts on infrastructure, ecosystems, human services, and human life.[14]

Although we know that the planet has experienced previous extremes in climate, the difference now is that it is happening in a much shorter time frame and is human-exacerbated. There is scientific consensus that the natural greenhouse effect is being exacerbated by anthropogenic activities such as burning of fossil fuels, deforestation of large portions of the planet, massive expansion of modern agricultural methods, rapid development of emerging economies and commensurate consumption of carbon-based products.[15] As Weiner notes: "[t]he story of life is punctuated by Ice Ages, volcanic winters, meteoritic collisions, mass dyings. And at the moment it is punctuated by us."[16]

For some of the worst risks of climate change to be avoided the average rise in global temperatures must be kept to no more than 2°C above pre-industrial levels. The reason is that once global warming reaches a certain point, carbon uptake from the land and oceans begins to lessen and even break down so that more carbon is released to the atmosphere, which causes additional warming so that more carbon is released to the atmosphere and so on in an unstoppable chain of positive feedback loops. For these reasons, addressing climate change requires urgent and multisectoral effort.

---

11. IPCC 2007 Synthesis Report, *supra* n. 1, at 45. N Stern, *The Economics of Climate Change: The Stern Review* (Cambridge University Press: Cambridge UK, 2007), chapter 6.
12. International Scientific Congress, *Climate Change: Global Risks, Challenges & Decisions – Synthesis Report* (Copenhagen 2009, March 10-12) (hereafter Copenhagen Synthesis Report), 6. See also: R. Garnaut, S. Howes, F. Jotzo and P. Sheehan, "Emissions in the Platinum Age: the Implications of Rapid Development for Climate-Change Mitigation" (2008) 24(2) *Oxford Review of Economic Policy* 377.
13. World Bank, *Turn Down the Heat: Why a 4°C Warmer World Must be Avoided* (The International Bank for Reconstruction and Development / The World Bank: Washington DC, 2012).
14. IPCC, *Climate Change 2014: Impacts, Adaptation, and Vulnerability: IPCC WGII AR5 Summary for Policymakers* (Intergovernmental Panel on Climate Change: Geneva, 2014).
15. IPCC 2007 Synthesis Report, above n. 1, at 30, 39.
16. Per Jonathan Weiner quoted in D. Suzuki, *The Sacred Balance: Rediscovering our Place in Nature* (Oxford University Press: Vancouver, 1997), 143.

[B]   Multi-jurisdictional Regulatory Regimes on Climate Change

[1]   International Level

The most obvious approach for tackling climate change is nation-state negotiation of an international treaty beyond the *Kyoto Protocol*. That Protocol was adopted in 1997 under the United Nations Framework Convention on Climate Change (UNFCCC) and entered into force in 2005 with its first commitment period ending in 2012. However, international negotiations to this end have been slow and uncertain. No binding individual or aggregate emissions reduction targets were agreed upon in the 2009 Copenhagen Accord,[17] the 2010 Cancun agreements[18] or subsequent decisions of the Conference of the Parties (COP) to the UNFCCC at Durban in 2011 and Doha in 2012.

In November 2013 at the 19th session of the UNFCCC COP in Warsaw, governments did agree to negotiate a new international climate treaty for adoption at the 21st COP in Paris 2015 with effect from 2020.[19] This forthcoming agreement is intended to replace the *Kyoto Protocol* by setting new binding national emissions reduction targets. Yet there are several obstacles to its timeliness and efficacy. The World Bank has reported that current emissions reduction pledges are not enough to prevent a 2°C temperature rise[20] so new pledges will need to be more ambitious, which raises political concerns within domestic borders for negotiating states. Further, any new agreement will need to include key emerging economies such as China, Brazil, India and Russia, and developed countries will need to provide technology, finance, and capacity-building support for developing countries to start on a clean-growth trajectory; yet there is a lack of clarity around the extent and application of UNFCCC "common but differentiated responsibilities" to this endeavor and also how technology-sharing may interact with extant intellectual property agreements. Moreover, the content is uncertain regarding reform of the Clean Development Mechanism[21] and/or inclusion of the United Nations collaborative program on Reducing Emissions from Deforestation and Forest Degradation in Developing Countries (REDD +).[22]

---

17. United Nations Framework Convention on Climate Change, *Report of the Conference of the Parties on its Fifteenth Session*, held in Copenhagen from Dec. 7 to 19, 2009, FCCC/CP/2009/11/Add.1 (Mar. 30, 2010), Decision 2/CP.15 (hereafter "Copenhagen Accord").
18. United Nations Framework Convention on Climate Change *Report of the Conference of the Parties serving as the meeting of the Parties to the Kyoto Protocol on its Sixth Session*, held in Cancun from Nov. 29 to Dec. 10, 2010, FCCC/KP/CMP/2010/12/Add.1 (Mar. 15, 2011), Decisions 1/CMP.6 and 2/CMP.6 and *Report of the Conference of the Parties on its Sixteenth Session*, held in Cancun from Nov. 29 to Dec. 10, 2010, FCCC/CP/2010/7/Add.1 (Mar. 15, 2011) Decision 1/CP.16 (hereafter "Cancun agreements").
19. United Nations Framework Convention on Climate Change, *New Key Documents From ADP Co-Chairs in Advance of the UNFCCC Negotiating Session in October* (UNFCCC, undated), http://unfccc.int/2860.php.
20. World Bank, *supra* n. 13.
21. United Nations Framework Convention on Climate Change, *Clean Development Mechanism (CDM)*, http://unfccc.int/kyoto_protocol/mechanisms/clean_development_mechanism/items/2718.php.
22. UN-REDD Programme, *About REDD +*, http://www.un-redd.org/AboutREDD/tabid/102614/Default.aspx.

Indeed the uncertain and halting progress of international climate negotiations has led some commentators to opine that "imperial dreams of a [top-down] global-level comprehensive solution to climate change" are no longer realistic.[23] Yet even if the need for global cooperation is assumed, constructive interim steps are still required. And any global agreement, if achieved, will require individual nations to take action, which in turn will require cooperation and action from sub-national constituents. The upshot is that, for climate change mitigation and adaptation to occur in a timely way, "subglobal efforts are not only desirable but indispensable."[24]

Accordingly, attention has turned to more plural and decentralized approaches. In addition to the climate-related actions of cities and civil society,[25] focus has shifted to national and regional climate change law and policy.

[2]   National and Regional Climate Regulation

National/regional climate change legislation and policy is still emerging and variable. This brings opportunities for financial actors in the form of new market entry; and it also brings increased risk from regulatory uncertainty. Indeed, national/regional climate regulation is highly subject to political vicissitude. It is worth noting here that regulatory uncertainty affects corporate behavior. How and to what extent provides the substance of later chapters.

Currently, national and regional climate regulation in key market economies comprises a narrow band of interventions which can be grouped into four categories.

[a]   Regulation That Puts a Price on Carbon

The predominant mechanisms to price carbon are taxation and emissions trading schemes (ETSs). While Scandinavian nations have the most extensive energy taxes,[26] the most active carbon market at the national/regional level is in Europe where the European Union Emissions Trading Scheme (EU ETS) began in 2005. In 2013 it moved into Phase III with more stringent emissions targets to keep on track for a 60-80 percent reduction by 2050. Despite its flaws, the EU ETS has reduced GHG emissions by an estimated 2.5 to 5 percent per year.[27]

Moreover, the UK ETS commenced under the Carbon Reduction Commitment (CRC) Energy Efficiency Scheme in April 2010 and was administratively simplified in May 2013.[28] It targets low energy-intensive users and covers both public and private sector organizations to affect 10 percent of total business sector emissions within the

---

23. E.W. Orts, "Climate Contracts" (2011) 29 *Virginia Environmental Law Journal* 197, 234.
24. D. A. Farber, "Carbon Leakage Versus Policy Diffusion: The Perils and Promise of Subglobal Climate Action" (2012) 13(2) *Chicago Journal of International Law* 359, 359.
25. See Orts, *supra* n. 23, at 224-233.
26. See A. Bruvoll and B.M. Larsen, "Greenhouse gas emissions in Norway: do carbon taxes work?" (2004) 32(4) *Energy Policy* 493.
27. E. Posner and D. Weisbach, *Climate Change Justice* (Princeton University Press: NJ, 2010), 68.
28. *The CRC Energy Efficiency Scheme Order 2013: Climate Change 2013* No. 1119, http://www.le gislation.gov.uk/uksi/2013/1119/made.

UK; it does not include emissions that have been captured by the EU ETS and Climate Change Agreements.[29]

There is equivocation in a number of other jurisdictions about pricing carbon at the federal level. Indeed, this category of intervention yields striking examples of regulatory uncertainty due to political vicissitude. For example:

– In the United States, the most robust legislative attempt to pass a federal carbon price was the *American Clean Energy and Security Act* (HR 2454) or Waxman-Markey Bill, which passed through the Lower House in 2009 but was defeated in the Senate the following year. Due to republican opposition, there is no national-level price on carbon despite the Obama Administration's original pledge to implement an ETS by 2016. In lieu, cap and trade regulation has been enacted at the sub-national level in California, Colorado, Florida, New Jersey and New York.

– Australia commenced a carbon tax in July 2012 with intent to shift to an ETS in 2015 under the *Clean Energy Act 2011* (Cth) which formed part of the former Labor Government's Clean Energy Legislative Package.[30] However, under a Coalition Government, the carbon price was abolished on July 17, 2014 with retroactive effect from July 1, 2014.[31]

[b]   *Regulation That Incentivizes Energy Efficiency and Investment in or Usage of Renewable Energy and Clean Technology*

Several mechanisms exist to incentivize renewable energy. Predominant mechanisms are feed-in tariffs (favored in Europe) and tax credits (favored in the United States).

Feed-in tariffs (FITs) offer subsidized rates to producers of electricity from renewable sources for power sold to the grid.[32] In other words, a government buys renewable energy at a set price for a long period of time (usually 15-20 years) based on the cost of generation, which means that returns are guaranteed.For example, Germany has a history of strong tariffs. Its *Renewable Energy Sources Act 2000* or

---

29. UK Government website, Department of Energy & Climate Change, *Policy: Reducing demand for energy from industry, businesses and the public sector* (last updated Jul. 29, 2014), https://www.gov.uk/government/policies/reducing-demand-for-energy-from-industry-businesses-and-the-public-sector--2/supporting-pages/crc-energy-efficiency-scheme.
30. The Clean Energy Legislative Package comprised: *Clean Energy Act 2011* (as amended by *the Clean Energy Legislation Amendment Act 2012*), *Clean Energy Regulator Act 2011, Climate Change Authority Act 2011,* and the *Clean Energy (Consequential Amendments) Act 2011.*
31. Repealing legislation comprised: *Clean Energy Legislation (Carbon Tax Repeal) Act 2014; Customs Tariff Amendment (Carbon Tax Repeal) Act 2014; Excise Tariff Amendment (Carbon Tax Repeal) Act 2013 2014; Ozone Protection and Synthetic Greenhouse Gas (Import Levy) Amendment (Carbon Tax Repeal) Act 2014; Ozone Protection and Synthetic Greenhouse Gas (Manufacture Levy) Amendment (Carbon Tax Repeal) Act 2014; True-up Shortfall Levy (Excise) (Carbon Tax Repeal) Act 2014; True-up Shortfall Levy (General) (Carbon Tax Repeal) Act 2014; Ozone Protection and Synthetic Greenhouse Gas (Import Levy) (Transitional Provisions) Act 2014.*
32. M. Mendonça, D. Jacobs, B.K. Sovacool, *Powering the Green Economy – the Feed-In Tariff Handbook* (Earthscan: Oxford UK, 2009).

Erneuerbare Energien Gesetz (EEG 2000) regulated the renewable electricity sector and applied FITs to a wide variety of energy sources, including not only wind and solar (photovoltaic and thermal) but also hydropower, landfill gas, sewage treatment, biomass, and geothermal (EEG section 3.3). The most substantial reform of the EEG entered into force on August 1, 2014 (EEG 2014) to address growing cost concerns by providing a revised system of fixed FITs and direct marketing mandates for electricity generated from renewable energy sources. The purpose of EEG 2014 reform was to "find a balance between cost effectiveness, environmental compatibility and security of supplies in the so-called energy policy triangle."[33]

Tax credits reward upfront investment in renewable power plants as well as back-end production of electricity from renewable energy. In the United States, federal-level tax credits have incentivized investment in renewable energy and clean tech via an energy investment tax credit (ITC) and a renewable energy production tax credit (PTC).[34] These tax credits reduce the federal income taxes of owners of renewable energy projects, either via capital investment in renewable energy projects in the case of ITCs, or the electrical output of renewable energy facilities pursuant to PTCs. ITCs have provided a 30 percent tax credit for upfront investments in solar, fuel cells and small wind energy projects; and 10 percent for all other eligible technologies. From January 1, 2017 the ITC will phase down to 10 percent of qualifying costs for all eligible renewable project expenditures.[35] In contrast, companies that generate electricity from wind, geothermal and biomass have benefited from the PTC, which provided a 2.3-cent per kilowatt-hour (kWh) incentive for the first ten years of a renewable energy facility's operation.[36] The PTC expired at the end of 2013 but captured eligible projects under construction before January 1, 2014.

Two other modalities of intervention are aimed at increasing renewable energy uptake. The first is the tender regime (or reverse auction mechanism) pursuant to which a government awards a long-term power purchase contract to the successful bidder for the supply of renewably-sourced electricity from a specific technology over a specified period of time.[37] The second modality is a sector-wide Renewable Energy Target (RET) or Renewable Portfolio Standard (RPS). These latter measures are

---

33. Norton Rose Fulbright, *Client Alert: The New German Renewable Energies Act 2014: An analysis of the current legislative proposal* (Apr. 17, 2014). http://www.nortonrosefulbright.com/know ledge/publications/115304/client-alert-the-new-german-renewable-energies-act-2014. See also T. Herbold, "German Renewable Energy Sources Act 2014 – Overview of the Most Important Changes," *GÖRG Partnerschaft von Rechtsanwälten* (Aug. 8, 2014), http://www.goerg.de/en/n ews/legal_updates/german_renewable_energy_sources_act_2014.40797.html.
34. *American Recovery and Reinvestment Act of 2009* (Pub. L. No. 111-5 §§1101-02, 123 Stat.319).
35. *Internal Revenue Code* 26 USC. §48(a)(2)(A)(ii). See also United States Department of Energy, "Business Energy Investment Tax Credit (ITC)," *United States Department of Energy* (Mar. 3, 2014), http://energy.gov/savings/business-energy-investment-tax-credit-itc.
36. J. Goodward and M. Gonzalez, *The Bottom Line on Renewable Energy Tax Credits* (World Resources Institute: Washington DC, 2010), http://www.wri.org/publication/bottom-line-rene wable-energy-tax-credits (accessed Jan. 31, 2014).
37. See e.g. Mendonça, Jacobs and Sovacool, *supra* n. 32; B. Mortenson, "International Experiences of Wind Energy" (2008) 2 *Environmental and Energy Law and Policy Journal* 179, 202.

quantity-driven and generation-based and require a minimum threshold of renewably-sourced electricity to be generated by a certain date. For example, the German EEG 2014 sets revised minimum requirements for the generation of electric power from renewable energy sources to 40 percent by 2025, 55 percent by 2030, and 80 percent by 2050. The consequence of RET and RPS measures is that electricity utility companies seek increasing supplies of renewable energy, which in turn incentivizes investment in renewable energy for larger-scale deployment. Utilities prove compliance with RET and RPS requirements through the use of renewable energy credits.[38]

Moreover, an innovation in this category of intervention is the creation of a government-supported bank or corporation to leverage private finance for energy efficient and low-carbon solutions. A prime example is the UK Green Investment Bank (GIB). The GIB was established under the *Companies Act 2006* (UK) in 2012 as a public company whose sole shareholder is the UK Government but which operates independently of it.[39] Its mission is to "accelerate the UK's transition to a greener, stronger economy" by mobilizing private sector capital for green projects on commercial terms across the UK.[40] The UK Government capitalized the GIB with £3.8billion of public funds in 2012.[41] By mid-2014, the GIB had mobilized £4.8billion from direct commitments of £1.3billion into energy efficiency, waste and bio-energy, and offshore wind projects within the UK.[42]

[c]   *Regulation That Mandates Compulsory Monitoring and Reduction of Internal Corporate GHG Emissions*

An example of compulsory monitoring is the Australian *Energy Efficiency Opportunities Act 2006* (Cth), pursuant to which corporate groups that use more than 0.5 petajoules of energy must assess and report publicly on their energy use and also implement identified energy saving opportunities. Similarly, the UK Carbon Reduction Commitment Annual Report Publication provides publicly-available information about the carbon dioxide (CO_2) emissions and renewable energy use of participating organizations.[43]

---

38. See F. Mormann, "Enhancing the Investor Appeal of Renewable Energy" (2012) 42 *Environmental Law* 681, 691-2.
39. Green Investment Bank, *Annual Report 2013*, 20, https://www.gov.uk/government/uploads/system/uploads/attachment_data/file/336552/green-investment-bank-annual-report-2013.pdf (accessed Aug. 7, 2014).
40. Green Investment Bank, *Annual Report 2014*, 14, http://www.greeninvestmentbank.com/media/25360/ar14-web-version-v2-final.pdf (accessed Aug. 7, 2014).
41. Green Investment Bank, *Annual Report 2013*, 20, https://www.gov.uk/government/uploads/system/uploads/attachment_data/file/336552/green-investment-bank-annual-report-2013.pdf (accessed Aug. 7, 2014).
42. Green Investment Bank, *Summary of Transactions*, 1, http://www.greeninvestmentbank.com/media/25380/gib_ar_transactions_250714.pdf (accessed Aug. 7, 2014).
43. See Environment Agency, "Guidance: CRC Energy Efficiency Scheme: Annual Report Publication," *gov.uk* (Mar. 18, 2014), https://www.gov.uk/crc-energy-efficiency-scheme-annual-report-publication#crc-performance-league-tables.

[d]   *Regulation That Discourages or Prohibits Intensive GHG-Emitting Projects*

Outright prohibitive regulation is yet to find favor with national governments. However, there are examples of regulatory discouragement of high-emitting projects and industries:

– In June 2014, the United States Environmental Protection Agency (EPA) proposed state-specific CO_2 emissions targets for power plants as well as guidelines for states when developing plans to achieve these targets.[44] The proposal aims to reduce CO_2 emissions nationally by 30 percent by 2030 (from 2005 levels). In so doing, it sets off a complex regulatory process in which the 50 states will each determine how to meet customized targets set by the EPA.[45] The EPA has authority to undertake this regulatory function for two related reasons: first, under the *Clean Air Act* (1970) 42 USC §§ 7401-112, the EPA is required to regulate pollution that threatens public health and welfare; secondly, in 2009 the EPA made a formal finding that GHGs endanger human health and safety in response to the Supreme Court decision in *Massachusetts v. EPA* (2007) that CO_2 could be classified as a pollutant.[46] By proposing these new measures as EPA regulations, the government has by-passed the need to introduce legislation to a Republican-dominant Senate. Nonetheless, the EPA proposal faces legal and political challenge from coal-producing states and industries.

– Moreover, as part of the United States' Climate Action Plan, the White House established new policy to limit public financing for coal plants abroad, which was adopted by the federal Export-Import Bank in December 2013.[47] The Danish, Swedish, Norwegian, Finnish and Icelandic governments similarly agreed to end their public financial support for new coal-fired power plants in foreign jurisdictions except in rare circumstances. The UK government has given qualified agreement: it will still consider "proposals for financing

---

44. Office of Air Quality Planning and Standards, Office of Atmospheric Programs and the Office of Policy of the United States Environmental Protection Agency, *Regulatory Impact Analysis for the Proposed Carbon Pollution Guidelines for Existing Power Plants and Emission Standards for Modified and Reconstructed Power Plants* (EPA-542/R-14-002, United States Environmental Protection Agency: Jun. 2, 2014).
45. United States Environmental Protection Agency, *Clean Power Plan Proposed Rule* (Aug. 5, 2014), http://www2.epa.gov/carbon-pollution-standards/clean-power-plan-proposed-rule.
46. *Massachusetts, et al., Petitioners v. Environmental Protection Agency, et al.* 549 US 497 (2007). See also United States Environmental Protection Agency, *Endangerment and Cause or Contribute Findings for Greenhouse Gases under Section 202(a) of the Clean Air Act*, http://www.epa. gov/climatechange/endangerment; upheld by the DC Circuit in *Coalition for Responsible Regulation, Inc v. Environmental Protection Agency*, 684 F3d 102, 113 (DC Cir 2012).
47. Export-Import Bank of the United States, *Press Release: Export-Import Bank Board Adopts Revised Environmental Guidelines to Reduce Greenhouse Gas Emissions* (Dec. 12, 2013), http://www.exim.gov/newsandevents/releases/2013/EXPORT-IMPORT-BANK-BOARD-ADOPTS-REVISED-ENVIRONMENTAL-GUIDELINES-TO-REDUCE-GREENHOUSE-GAS-EMISSIONS.cfm.

coal-fired power plants in the world's poorest countries where no other economically feasible alternative exists" on a case-by-case basis.[48]

§1.03   SUSTAINABLE DEVELOPMENT AND CLIMATE FINANCE

At the 2013 World Economic Forum, conversation amongst delegates focused on resilience to economic risk post-global financial crisis (GFC). Ms. Christine Lagarde, Managing Director of the International Monetary Fund, also cast a spotlight on the notion of climate risk, describing climate change as "by far the greatest economic challenge of the 21st century."[49] This statement is one of the strongest acknowledgements by a global financial leader that the world's economic and environmental systems are inextricably intertwined.

Indeed, the role of financial capital in addressing climate change becomes clear by examining its relevance to sustainable development and environmental issues more generally.

Financial support for projects and technological innovation will almost always have environmental effects of some kind whether adverse or beneficial. Wholesale decisions regarding future development often arise in the finance sector and, as such, this is where future pressures on the environment begin. As Benjamin J. Richardson notes: "[i]f sustainable development is understood to imply, among other things, maintenance of natural and human-made capital for posterity, the role of capital markets must be recognized as pivotal to this goal."[50]

To this end, an inquiry into the design of a sustainable financial system was initiated in January 2014 by the United Nations Environment Programme (UNEP). This Inquiry presents the case for "linking the development of the financial system with green and inclusive policy objectives":

> The recent financial crisis, reinforced by the failure of today's global economy to deliver the jobs needed and steward the natural environment, has eroded trust in the financial system's capacity to serve its intended beneficiaries and the long-term interests of the real economy. Aligning the financial system to the needs of a green and inclusive economy is a pre-condition for achieving sustainable development, complementing policy and private action in the real economy.[51]

48. Department of Energy & Climate Change and The Rt Hon Edward Davey MP, *Written statement to Parliament: UK position on public financing of coal plants overseas* (Nov. 21, 2013), https://www.gov.uk/government/speeches/uk-position-on-public-financing-of-coal-plants-overseas.
49. S. Scouler, "Opinion: Banking on Climate Change," *Australian Banking and Finance (Online)* (Apr. 1, 2013), http://www.australianbankingfinance.com/banking/opinion--banking-on-climate-change/.
50. B.J. Richardson, "Sustainable Finance: Environmental Law and Financial Institutions" in BJ Richardson and S Wood (eds) *Environmental Law for Sustainability* (Hart Publishing: Oxford UK, 2006), 309-310.
51. UNEP Inquiry: Design of a Sustainable Financial System, *Aligning the Financial System with Sustainable Development: An Invitation* (UNEP, Geneva, 2014), 2.

After wide consultation, the Inquiry will promote policy design options in 2015 for step change in "better aligning the financial system to the needs of sustainable development."[52]

So what exactly is sustainable development? The answer is simple in form but complex in substance. The seminal report of the World Commission on Environment and Development in 1987 (known as the Brundtland Report), adopted subsequently by the United Nations General Assembly, provides the standard formulation of the term as development that "meets the needs of the present without compromising the ability of future generations to meet their own needs."[53] At its heart, the concept of sustainable development was created due to the realization that economic growth and environmental degradation are interrelated within a globalized world. Specifically, the definition acknowledges that "new" environmental problems, such as climate change, are global concerns that require concerted multisectoral effort and innovation to solve.

Thus, in recognition that "[e]conomics and ecology bind us in ever-tightening networks,"[54] the role of finance actors to the sustainability endeavor is both undeniable and critical.

[A]   Key Public and Multilateral Finance Entities

Since the 2007 Bali Action Plan, international action on climate finance has focused on multilateral and bilateral government interventions regarding the provision of financial aid by developed countries to developing countries to facilitate GHG mitigation and build their resilience to climate impacts. These efforts are facilitated by the World Bank Group together with the G20 and the United Nations. In addition to the World Bank, key players in international public finance are the International Finance Corporation (IFC), which has become a primary issuer of green bonds and "one of the world's largest financiers of wind and solar power for emerging markets";[55] the International Monetary Fund (IMF), which focuses on the macroeconomic challenges of climate change and provides advice for developing nations on fiscal policies to mitigate it;[56] and six multilateral/regional development banks, which together in 2012 delivered US$27billion to developing nations to assist with climate change mitigation and adaptation:

- For example, the African Development Bank established the Africa Climate Change Fund in April 2014 with a contribution of €4.725million for an initial

---

52. *Ibid.*, 3.
53. *Report of the World Commission on Environment and Development: Our Common Future*, transmitted to the General Assembly as an Annex to document A/42/427 – *Development and International Co-operation: Environment* (1987) (hereafter the "Brundtland Report"). Adopted by UN General Assembly resolution A/RES/42/187.
54. *Ibid.*, at para. 4.
55. International Finance Corporation, "Climate Business at IFC," *ifc.org* (undated), http://www.ifc.org/wps/wcm/connect/Topics_Ext_Content/IFC_External_Corporate_Site/CB_Home/.
56. See International Monetary Fund, *Factsheet: Climate, Environment, and the IMF* (Mar. 18, 2014), http://www.imf.org/external/np/exr/facts/enviro.htm.

three-year period from Deutsche Gesellschaft für Internationale Zusammenarbeit on behalf of the German Federal Ministry for Economic Cooperation and Development. The Fund is described as a "bilateral thematic trust fund" and through a series of grants it aims to:

> enhance the capacity of African countries to improve their national institutional governance for direct and international access to climate finance, and...develop transformational policies, programs and projects for climate resilience and low carbon growth, in alignment with UNFCCC decisions.[57]

– Moreover, Climate Investment Funds (CIFs) are managed by the World Bank and implemented jointly with regional development banks. CIFs leverage support from developed nations and buy-down the costs of low-carbon technologies in developing countries. Specifically, CIFs operate in 48 developing countries and with US$7.2billion capital investment have leveraged an additional US$43billion in low emissions and climate resilient development.[58] One of the first CIFs established in 2008 was the Clean Technology Fund. Its aim is to scale up, deploy and transfer renewable energy, energy efficiency and cleaner transport technologies for country-initiated low-carbon projects. This Fund disburses financing as grants, risk-mitigation instruments, and highly concessional loans. For every dollar disbursed from public sources it has mobilized an estimated US$8billion in co-financing from private sources, governments, and multilateral/regional financial institutions (being a highly successful leverage ratio of 1:8).[59] In 2013 the Fund committee approved US$150billion for two "dedicated private sector programs" designed to engage the private sector in public initiatives consistent with country priorities.[60]

Importantly, the Green Climate Fund (GCF) was established in 2010 as an operating entity of the UNFCCC's financial mechanism. The GCF's purpose is:

> to promote, within the context of sustainable development, the paradigm shift towards low-emission and climate-resilient development pathways by providing support to developing countries to help limit or reduce their greenhouse gas emissions and to adapt to the unavoidable impacts of climate change.[61]

---

57. African Development Bank Group, "Africa Climate Change Fund Launches First Call for Proposals," *afdb.org* (Jul. 4, 2014), http://www.afdb.org/en/news-and-events/article/africa-cl imate-change-fund-launches-first-call-for-proposals-13353/.
58. World Bank, "Climate Change Report Warns of Dramatically Warmer World This Century," *worldbank.org*, (Nov. 18, 2012), http://www.worldbank.org/en/news/feature/2012/11/18/Cl imate-change-report-warns-dramatically-warmer-world-this-century.
59. World Economic Forum, *The Green Investment Report: The Ways and Means to Unlock Private Finance for Green Growth* (World Economic Forum: Geneva, 2013), 21.
60. Climate Investment Funds, "Rooted in Learning, Growing with Results," *2013 CIF Annual Report* (The World Bank Group: Washington DC, 2013), 9.
61. Green Climate Fund, *Press Release: Green Climate Fund Board takes key decisions on operations and makes progress on "Essential Eight"* (Feb. 22, 2014).

According to the *Cancun Agreements* the GCF's main role is to channel new public financial resources from developed nations, in addition to the existing pledge of US$100 billion per year by 2020, to affect private and public finance for mitigation and adaptation in the most vulnerable developing countries.[62] The GCF is a work in progress. Essential operating requirements such as how it will receive, manage, and disburse funds were established in May and October 2014.[63] Capitalization of the GCF is evolving with UNFCCC Executive Secretary Ms. Christiana Figueres estimating that at least US$10billion is required.[64]

Clearly, multilateral and public financing efforts are vital. In particular, the GCF is a most welcome and timely global initiative. However, there are at least two initial concerns. First, looking at the sums of money cited in the Introduction together with the limited availability of public funds, investments at scale require private sector funding. To this end, the GCF employs a Private Sector Facility (PSF) to promote the participation of local private sector actors in developing countries, particularly "small and medium-sized enterprises and local financial intermediaries."[65] Foreign private sector entities (like Google or Coca Cola) can provide funds through the GCF's External Affairs division, alongside public contributions. Yet this raises the second concern: that a vital opportunity to directly engage private *finance* actors will be missed under these arrangements. Neither the PSF nor the External Affairs (donations) division will capture or engage multinational and transnational financial intermediaries, such as a large American pension fund or a European bank.

Furthermore, the efforts of private finance actors are required to assist mitigation and adaptation efforts not only in developing nations but also in developed nations. Specifically, developed nations such as the United States, Europe and Australia are responsible for much of the world's GHG emissions. Thus, economic and energy transitions in these jurisdictions are as important as those in developing nations if we are to move to a low-carbon and climate resilient *global* economy.

[B]   Key Private Finance Actors

The need for private finance actor engagement in sustainability was acknowledged by 30 banks that worked with the UNEP to create the *Statement by Banks on the Environment and Sustainable Development* prior to the 1992 Rio Earth Summit. Three years later the number of signatories had grown to 74 banks, and in May 1997 the statement was revised to become the *UNEP Statement by Financial Institutions on the Environment & Sustainable Development*, which embraced a broader membership

---

62. Cancun Agreements, *supra* n. 18, at Part IV A.
63. Green Climate Fund, "Decisions of the Board – Seventh Meeting of the Board, 18-21 May 2014," *Board meeting document, GCF/B.07/11* (Jun. 19, 2014).
64. International Institute for Sustainable Development Reporting Services, "GCF Board Paves Way for Mobilising Resources," *International Institute for Sustainable Development* (May 21, 2014), http://climate-l.iisd.org/news/gcf-board-paves-way-for-mobilizing-resources/.
65. Green Climate Fund, *Business Model Framework: Private Sector Facility, GCF/B.04/07* (Jun. 12, 2013), 1.

of "financial institutions."[66] Since 2011 it has been re-titled the *UNEP Statement of Commitment by Financial Institutions (FI) on Sustainable Development* (hereafter UNEP Finance Statement on Sustainability) and there are now over 200 signatories.[67]

Whilst the text has been changed in several places from the original statement, the UNEP Finance Statement on Sustainability retains the original sentiment that private finance actor signatories commit to: sustainable development in their business practices;[68] developing products and services that promote sustainable development;[69] publishing statements of their own environmental and social policies and their implementation;[70] and, through the sharing of information and knowledge, encouraging customers and other financial institutions[71] to strengthen their own capacity to promote sustainable development.

Specifically, signatories "regard financial institutions as an important contributor towards sustainable development."[72] In so doing, signatories acknowledge that "economic development needs to be compatible with human welfare and a healthy environment" and they are "committed to working cooperatively [with government, business, individuals] toward common sustainability goals" (preamble).

All members of the UNEP Finance Initiative (UNEPFI) must adhere to the UNEP Finance Statement on Sustainability, which is described as the backbone of UNEPFI.[73] Importantly, the UNEP Finance Statement on Sustainability expressly recognizes climate change in its sustainability commitments.[74]

That brings us to discussion of how private finance actors can help to address climate change. To this end, private finance actors have at least three roles.[75] First, they are capital providers that supply or facilitate finance for GHG intense projects as well as clean technology initiatives. Secondly, they act as valuers that price risks and predict company profitability and investment preferences in an increasingly carbon-constrained world. Thirdly, they are lenders and shareholders that exercise influence over corporate management and climate-related corporate governance. Actions and decisions taken (or not taken) by finance sector industries help to mitigate climate change or to exacerbate it.

The main groupings of private finance actors relevant to climate change are the insurance (including reinsurance) industry, the institutional investment (including pension fund) industry, the banking industry, and entrepreneurial finance actors (including venture capital funds and private equity firms). Brief synopses of each are provided below prior to detailed discussion in later chapters.

---

66. S Schmidheiny and F Zorraquin, *Financing Change: The Financial Community, Eco-Efficiency and Sustainable Development* (MIT Press: Cambridge MA, 1996), 106.
67. See http://www.unepfi.org/.
68. *UNEP Statement of Commitment by Financial Institutions (FI) on Sustainable Development* (hereafter "UNEP Finance Statement on Sustainability"), items 1.1, 1.3, 1.4, 2.1-2.6.
69. *Ibid.*, item 2.7.
70. *Ibid.*, item 3.1.
71. *Ibid.*, items 3.2, 3.5.
72. *Ibid.*, item 1.3.
73. See http://www.unepfi.org/statements/index.html.
74. UNEP Finance Statement on Sustainability, *supra* n. 68, item 1.5.
75. Richardson, *supra* n. 50, 309.

[1]   Insurers and Reinsurers

Insurers and reinsurers have been the canary in the climate change coal mine. They are among the first to be impacted by climate change due to risks they accept as underwriters and investors of premiums. Since the early 1990s leading insurers around the world have spoken openly of the threat of bankruptcy due to unpredicted and unmanageable catastrophic losses from wild weather. Although it is difficult to directly link any particular wild weather event to human-induced climate change, the sudden rise in frequency and severity of such disasters caused many insurers in the 1990s to consider the role of human activity and to predict the impacts of a warming world with uncanny accuracy. Extreme loss events have since translated into extreme financial losses for insurers and reinsurers.

Accordingly, the insurance industry has a proactive role to play in *ex ante* climate/disaster mitigation as well as conventional *ex post* financial mitigation and recovery. In addition to significant financial self-interest in ESG issues and climate risk mitigation, the insurance industry has three interrelated roles relevant to addressing climate change. First, it has a unique knowledge base with the ability to provide information on climate change impacts and to motivate appropriate responses from other private sector actors and also policy-makers. Second, the insurance industry has a collaborative and advocacy role to work with and actively lobby governments to ensure construction quality and sound land use planning. Third, in their role as institutional investors, insurers need to consider and address climate risk in their investment portfolios.

[2]   Institutional Investors

The institutional investment chain comprises institutional investors (or asset owners) such as pension funds and life insurance funds, and also asset/investment managers such as banks and insurance companies. Institutional investors have a fiduciary duty to act in the best interests of their members. In Anglo-American jurisdictions this manifests as unswerving dedication to beneficiaries' financial interests in terms of profit maximization. Climate change creates both risks and opportunities for institutional investors through their long-term asset class returns. Many funds invest in property, agriculture, utilities, transport and network infrastructure, which are all vulnerable to climate change impacts. If investment value decreases due to ESG or climate-related factors, then that is an issue to which a fiduciary must attend.

As such, institutional investors are being urged to consider the long-term implications of climate change on investment decisions by using "fiduciary capitalism" or "socially responsible investment" strategies. Moreover, there is evidence that funds can pool their expertise and leverage their influence as universal owners for positive change at two levels: first, at a sector- or economy-wide level via voluntary codes of conduct such as the UN Principles for Responsible Investment; and secondly, at a company-level via shareholder activism.

[3]   Banks

Banks are typically organized around key clusters of activity being retail banking, private banking, investment banking, asset management, and sometimes also insurance and leasing. Like insurers and institutional investors, banks are exposed to climate-related credit, investment, reputation and litigation risks through their clients and investments. Yet climate change also presents financial and reputational opportunities for them. In their capacity as asset managers, banks can steer investments into low-carbon companies; as traders, they provide liquidity in the carbon and clean tech markets; and as participants in the capital markets, banks can raise debt and equity capital to fund renewable energy/clean tech companies and projects. Moreover, in their role as advisers, banks can help corporate clients to survive and thrive in a carbon-constrained economy. Arguably, addressing climate change through their daily business practices and facilitating a low-carbon global economy through their corporate networks comprise a redemptive opportunity for the banking industry post-GFC.

[4]   Entrepreneurial Finance Actors

Different sources of finance come into play at different stages of low-carbon project/tech development and deployment. The later stages of manufacturing scale-up and commercial roll-out tend to be financed by public equity markets as well as debt and project finance. Prior to that, technology development is generally financed by venture capital funds and private equity firms. These actors have much higher risk tolerance than conservative finance actors such as banks and institutional investors. Venture capital funds are keen to invest in new technology and new markets; private equity firms are interested in companies and projects with more mature technology. Accordingly, the role of these actors in addressing climate change arises from their focus on early or growth stage technology companies and projects that conservative finance actors are reticent to support.

§1.04   SCOPE AND PLAN OF THIS BOOK

Banking on Climate Change is aimed at practitioners, policymakers and civil society. This approach is quite deliberate even though these three audiences are heterogeneous. They are often not across each other's domains or even speaking the same language when it comes to finance and/or climate change. This book seeks to remedy this communicative lacuna in order to move forward on the issue of climate finance. Moreover, different chapters will appeal to different audiences at different times. For example, policy-makers might be most interested in my regulatory recommendations toward the end of the book; practitioners may be more interested in legal regimes and industry responses to climate change at the front end. All readers will want to immerse themselves in the empirical case study, which comprises the heart of the book.

Regarding the scope of *Banking on Climate Change*, there are a few things to note from the outset.

First, "regulation" is broadly conceived within these pages. It includes not only law and policy in the form of government interventions, but also industry self-regulation and social institutions or norms. In certain contexts, specificity is required to pinpoint which regulatory modality is in play and to what end; at these times I am quite explicit.

Secondly, in recognition of the urgent need to halt and reduce GHG emissions in order to stay within the 2°C guardrail this book focuses more on mitigation of climate change than adaptation to it. In brief, the difference is that mitigation focuses on reducing GHG emissions in order to slow and stop climate change, whereas adaptation focuses on increasing resilience of human communities and circumstances in order to withstand and minimize the inevitable impacts of climate change. Nonetheless, the importance of mobilizing finance for adaptation measures in both developed and developing countries is acknowledged throughout this book. Building climate resilience is vital as we approach and look set to pass the 2°C guardrail.

Thirdly, focus is placed on private finance actors and relevant law and policy in key market economies. Specifically, finance sector activities and regulatory environments are compared and contrasted between the United States, Europe and Australia. Nonetheless, this book acknowledges that a low-carbon *global* economy cannot occur without key emerging economies particularly China. Hence the final chapter, titled "Looking to the Future," is dedicated to an analysis of Chinese "state capital" in green investment and finance.

Finally, this book is divided into three parts or themes:

(A)   Practical and theoretical context
(B)   Empirical case study
(C)   Regulatory recommendations.

[A]   Practical and Theoretical Context

The front end of the book details the different types of private sector financial institutions and their function in a market economy, covering a broad suite of financial actors, including finance companies, mutual funds, pension funds, insurance companies, and banks. In Chapter 2 these financial institutions are classified into two broad categories of depository and non-depository, and grouped according to activities and services. This chapter then reveals the risks and opportunities of climate change for finance actors, and introduces the concept of "corporate climate finance," being the deployment of finance and investment by private finance actors into climate-related spaces. It provides practical examples of the voluntary responses to climate-related risks and opportunities that early-moving private finance actors are taking. It hones in on the banking industry's "network change potential" to facilitate climate change mitigation via decision-making processes and product/market innovations that influence the GHG emitting behaviors of other corporate actors.

Chapter 3 provides theoretical background to the question of why companies go green. The literature is conceptualized using micro (intra-organizational), meso (inter-organizational) and macro (socio-cultural) lenses of focus to shed some light on why firms voluntarily adopt climate-related practices. This chapter posits that all three levels can provide theoretical insights about the internal and external influences on corporate environmental behavior. Yet it concludes that there is little real understanding of what motivates private finance actors in particular to adopt climate-related strategies despite the plethora of theory about why companies do and should "go green."

[B] Empirical Case Study: Early-Moving Banks in Key Market Economies

Why banks for a case study? The dearth of empirical investigation regarding the intersection between government interventions and the "real life" practices in banking, finance and law has created a communicatory and comprehension deficit between these spheres. From my discussions with industry, there is a real desire to know what other industry players are doing (especially if they are doing well) and how policy-makers are likely to regulate in this space. From my discussions with regulators, there is a real need to understand how private finance actors "tick" and to access their expertise without succumbing to it.

Thus, I use the banking industry as an empirical case study in order to provide practical lessons for both practitioners and policy-makers regarding: the components and limitations of the business case; social vs. corporate reputation; and the relevance of corporate social responsibility (CSR) to firms in real life.

The case study captures empirical data from bank personnel and third-party respondents in the United States, Europe, and Australia. Interviews with senior management bankers and analysts regarding organizational structures and managerial decision-making processes within banks have allowed new depth of perspective. My hope is that findings from the data assist practitioners, policy-makers and civil society to move the field of corporate climate finance forward.

The first of two empirical chapters is presented in Chapter 4 to answer the question of why early-moving banks are voluntarily adopting climate-related practices. In other words, why are these finance actors doing anything about climate change at all when not legally compelled to do so? The data confirm that climate-related initiatives of early moving banks are driven by business case logic. In so doing, they reveal a deeper and more complex conceptual understanding of corporate reputation as a powerful "soft" motivator of bank behavior and clarify the role of regulatory contexts (which include government and civil society interventions) that shape bank decision-making of climate change as a risk or opportunity.Importantly, the data challenge the CSR literature that posits that ethical conceptions of "doing the right thing" drive corporate greening. In so doing the case study reveals the limited ability of CSR to mobilize substantive corporate change in real life.

The second empirical chapter is presented in Chapter 5. This chapter considers the limitations or barriers to climate-friendly bank action as demonstrated via analysis of: (a) company law, directors' duties and the corporate governance norm of short-term shareholder wealth maximization; (b) banks' obsession with enhancing and protecting client service reputation and profits; and (c) the current limited size of clean energy markets. This chapter reveals that mainstreaming of "enlightened" climate-related practices through purely voluntary means will be modest and incremental, not rapid and transformational. It concludes that the network change potential of the banking industry (as posited in Chapter 2) cannot be realized without something more.

[C]   Regulatory Recommendations

Chapter 6 provides empirically informed recommendations for effective corporate climate finance regulation. These recommendations target actual points of leverage with private finance actors as extrapolated from the empirical findings of the banking case study. Specifically, this book recommends government intervention and canvasses both direct (coercive) finance regulation and indirect (nudging) climate regulation to this end.

Given the centrality of the business case to financial activity, Chapter 7 broadens the regulatory sights beyond conservative or low risk-tolerant financial actors such as banks and pension funds to include capital providers with greater risk tolerances such as venture capital funds and private equity firms. In particular it presents the logic behind capital asset pricing (or risk/return) theory across a range of financial actors and analyzes three entrepreneurial and innovative green investment modalities, being tax equity partnership structures; climate and green bonds; and Evergreen buy-out strategies.

Chapter 8 further expands the regulatory focus beyond climate-related investment activity in market economies to the investing activities of Chinese state-backed enterprises in renewable energy and clean tech markets domestically and abroad. In so doing, it raises questions about the way forward in a truly global energy transition.

Banking on Climate Change concludes in Chapter 9 with final remarks about reframing the climate issue from an institutional perspective and the likely direction of corporate climate finance into the future.

CHAPTER 2

A Unique Relationship: Private Finance Actors and Climate Change

§2.01 OVERVIEW

The Introduction pinpointed two insights. First, that climate change has ever-increasing relevance to the business activities of private finance actors. Secondly, that the private finance sector has a critical role in facilitating the reduction of GHG emissions and the transition to a low-carbon global economy. The purpose of this chapter is to consider those insights more deeply.

As a first step, this chapter outlines the various types of private finance actors, their activities and their role in a market economy. This covers a broad suite of financial actors, including finance companies, mutual funds, pension funds, insurance companies, and banks. The chapter then reveals the risks and opportunities of climate change for conservative finance actors, being insurers and reinsurers, institutional investors and banks; and introduces the concept of "corporate climate finance" being the deployment of finance and investment by private finance actors into climate-related spaces. It gives practical examples of the voluntary responses that these actors are taking to climate-related risks and opportunities. In so doing, it raises questions of why and to what extent early-moving finance actors are motivated to engage with climate change, which are addressed squarely via empirical investigation in subsequent chapters.

§2.02 THE MONEY-GO-ROUND: TYPES AND FUNCTIONS OF PRIVATE SECTOR FINANCIAL INSTITUTIONS

In general, the role of the financial system is to permit and promote the flow and efficient allocation of funds throughout an economy. Financial systems comprise financial markets and institutions.

Financial markets have five main functions:[76]

(1) facilitate the flow of funds through an economy;
(2) provide the transfer of funds (settlement of transactions) via the payments system;
(3) facilitate risk transfer and management;
(4) provide sufficient information to enable participants to make informed investment decisions; and
(5) minimize the effects of perverse incentives that arise in financial contracts, such as information asymmetry, agency problems and moral hazard, which can undermine an efficient and equitable financial system.

Financial institutions are the lead actors that permit and perpetuate properly functioning financial markets worldwide. As depicted in Figure 2.1, they do so primarily by transferring funds from savers or lenders (also known as suppliers of funds or surplus spending units) to borrowers (demanders of funds or deficit spending units) through means of indirect financing, more commonly known as financial intermediation.[77] Examples of financial institutions or intermediaries include commercial banks, insurance companies, pension funds, credit unions, and finance companies. Investment banks straddle both direct and indirect financing; their expertise includes facilitating private placements of a security issue to a single institutional investor, and helping to bring newly-created financial claims known as initial public offerings (IPOs) to market.[78]

---

76. D.S. Kidwell, M. Brimble, A. Basu, L. Lenten and D. Thomson, *Financial Markets, Institutions & Money* (John Wiley: Milton Australia, 2011), 12.
77. *Ibid.*, 13-16.
78. *Ibid.*, 14-15.

Figure 2.1   *The Financial System: Financial Markets, Intermediaries,*
and the Flow of Funds

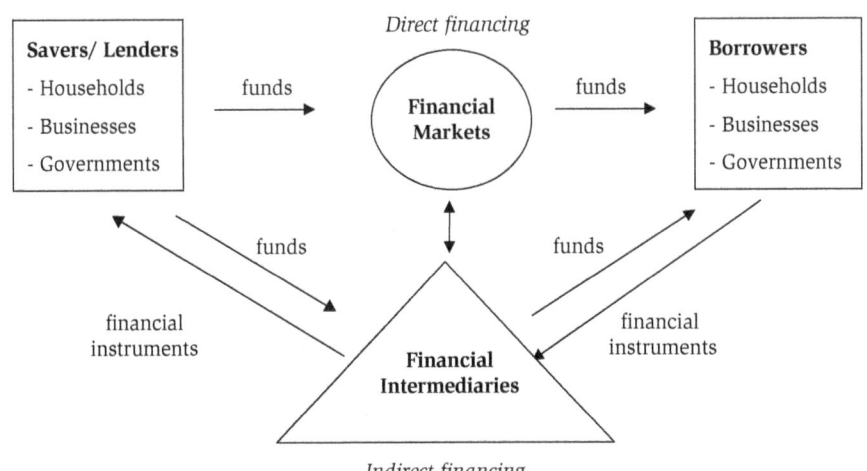

Source: Adapted from D.Kidwell, M. Brimble, A. Basu, L. Lenten and D. Thomson, *Financial Markets, Institutions & Money* (John Wiley: Milton Australia, 2011), 14.

Financial intermediaries bring together and link savings and investments (money as an asset) and lenders and borrowers (debt and equity capitalization).[79] Doing so promotes an increased flow of funds in an economy because savers/lenders (households, businesses and governments) can structure their savings according to their needs in order to create wealth, which then provides an increased pool of investment capital that can be used by borrowers (again, households, businesses and governments) to increase the productive capacity of an economy and so generate economic growth.[80]

In order to carry out these functions, multiple types of private finance actors exist. The types, classifications and activities of key private finance actors are summarized in Table 2.2. An important caveat is that areas of activity are not mutually exclusive. Indeed, many large private finance actors are involved in multiple areas concomitantly. For example, "universal banks" – such as HSBC, Deutsche Bank andCredit Suisse – are typically organized around key clusters of activity being retail banking, investment banking, asset management, and sometimes also insurance and leasing. By contrast, an entity such as Morgan Stanley is a hybrid commercial and investment bank that does not engage in retail banking.

In categorizing different types and activities of private finance actors, it is useful to distinguish between two broad groupings, being *depository* and *non-depository*.[81]

---

79. M. Jeucken, *Sustainable Finance and Banking: The Financial Sector and the Future of the Planet* (Earthscan: Oxford UK, 2001), 56-7.
80. Kidwell et al., *supra* n. 76, at 6-7.
81. *Ibid.*, 19-27. Jeucken, *supra* n. 79, at 53-54, 62.

Depository institutions "create" money by lending out large sums of the funds entrusted to them in the form of loans and mortgages. There are several main types of depository institution:

(1) Commercial banks, retail banks and private banks are the predominant types of depository institution. These banks use funds to extend loans to large corporate clients and/or wealthy private individuals and/or civil society ("mom and pop") customers. They distribute profits back to shareholders.
(2) Mutual savings banks specialize in attracting savings and extending home financing, and distribute surplus profits to customers.
(3) Cooperative banks and credit unions are owned by their members and focus on providing credit to customers and retaining any profits.

By contrast, *non-depository institutions* finance their activities through client fees and by issuing shares, bonds and other securities. Examples of public non-depository institutions are multilateral banks and governmental financial institutions. Private sector non-depository institutions fit into three categories:

(1) Securities market institutions, which are those actors involved in capital market transactions and advisory services with large corporate clients, institutional investors, and governments. These actors include investment banks, brokers, and traders.
(2) Investment institutions, which invest in higher-risk securities and loans than commercial and retail banks usually find acceptable. These institutions include mutual funds, and also finance companies (including venture capital funds).
(3) Contractual savings institutions, which comprise insurers, reinsurers and pension funds.

Table 2.2 Classification and Activities of Private Sector Financial Institutions

Services	Depository Institutions			Non-depository Institutions				
				Securities Market Institutions	Investment Institutions		Contractual Savings Institutions	
	Commercial Banks	Retail Banks	Private Banks	Investment Banks	Mutual Funds	Finance Companies	Pension/ Superannuation Funds	Insurance Companies
Transaction accounts	x	x						
Savings	x	x	x		x		x	x
Consumer (civil society) lending	x	x	x			x		
Corporate lending	x		x			x		
Project finance	x		x	x		x		
Mortgages	x	x	x			x		
Commodities trading	x		x	x	x			
Asset investment/ management	x		x	x	x		x	x
Insurance	x							x

Source: Adapted from M. Jeucken, Sustainable Finance and Banking: The Financial Sector and the Future of the Planet (Earthscan: Oxford UK, 2001), 53-54.

By appreciating the key functions and various types of private finance actors as financial intermediaries, we can grasp their qualitative effect on economic growth by shaping not only the development but also the direction of the economy.[82] In short, they *transform* money in global markets. They do this not only by scale and risk, but also by time (short versus long term investments), by space (a lender on Wall Street allocating money to a borrower in Indonesia) and even temporally (facilitating money transference between generations). If money makes the world go round, then private finance actors help to make the money-go-round.

§2.03   CLIMATE RISKS AND OPPORTUNITIES FOR PRIVATE FINANCE ACTORS

So where does global warming fit into this discussion? How are private finance actors and climate change relevant to each other?

Table 2.3 summarizes the relationship between private finance actors and climate change. It depicts risks and opportunities that climate change presents for finance actors, the most significant of which are indirect and relate to their unique business practices in risk mitigation, lending, financing, investing, insuring and brokering. It also identifies the influence of private finance actors on the business practices and GHG emissions of other corporate actors via capital resource allocation, supply chain standards and product /market innovations.

Table 2.3   Climate Change and the Private Finance Sector: Mutual Impacts

	Climate Change Impacts on Private Finance Actors	Private Finance Actors' Impacts on Climate Change
Direct Impacts	Asset damage	Internal GHG emissions
Indirect Impacts	Via corporate and insurance clients & investment targets:	Via corporate clients, suppliers & investment targets:
Risk-creating	– Sectors vulnerable to wild weather. – High GHG emitting companies/projects vulnerable to changing government regulation, market preferences & NGO campaigns.	– 'Climate risk' due diligence. – Financing, lending, investing decisions – GHG intense targets vs. renewables & clean tech. – 'Green' supply chain management standards.

---

82. Jeucken, *supra* n. 79, at 52, 55.

	Climate Change Impacts on Private Finance Actors	Private Finance Actors' Impacts on Climate Change
Opportunity-creating	– Innovative climate-related products and services. – New market entry & brokering opportunities.	– Innovative climate-related products and services. – New climate-related markets.

Source: Adapted from M. Bowman, "The Role of the Banking Industry in Facilitating Climate Change Mitigation and the Transition to a Low-Carbon Global Economy" (2010) 27 *Environmental and Planning Law Journal* 448, 452.

The nexus between private finance actors and climate change has been publicly acknowledged by both climate- and industry-related bodies in recent years. In 2001 the IPCC suggested that the finance sector "is sensitive to climate change and offers an integrator of effects on other sectors."[83] In 2009 the Carbon Disclosure Project stated that "the Financials sector holds great promise in the transition to a low-carbon economy and will be a driving force in helping its clients prepare for, and manage, the risks and opportunities presented by climate change."[84]

Private finance actors create and benefit from opportunities that flow to their own industries via climate-related business efforts such as carbon trading, financing low-carbon and renewable energy, weather derivatives, catastrophe bonds, and micro-finance.[85] Simultaneously, however, "the new economics of a carbon-constrained world" brings challenges and risks for the private finance sector due to environmental and regulatory changes, reputational and legal pressures, and competitive concerns.[86]

The remainder of this chapter focuses on the role and activities of conservative finance industries, namely insurance, institutional investment and banking, due to their ubiquity in financial markets. Higher risk-tolerant financial entities, such as private equity firms and venture capital funds, are examined later in the book.

[A] Relevance by Industry: Climate-Related Risks and Opportunities

[1] The Insurance Industry: Insurers and Reinsurers

Insurers are among the first to be impacted by climate change due to risks they accept as underwriters and investors of premiums.[87] Reinsurers are similarly impacted

---

83. IPCC Working Group II, *Climate Change 2001: Impacts, Adaptation and Vulnerability*, which forms part of the IPCC Third Assessment Report, specifically chapter 8 "Insurance and Other Financial Services": available at http://www.ipcc.ch/ipccreports/tar/wg2/index.php?idp = 322.
84. Carbon Disclosure Project (CDP), *Carbon Disclosure Project 2009: Global 500 Report* (CDP: London, 2009), 94.
85. UNEPFI, *CEO Briefing on Climate Change* (UNEPFI, 2002), 3-4.
86. A. Dlugolecki and S. Lafeld, *Climate Change and the Financial Sector: An Agenda for Action* (Worldwide Fund for Nature and Allianz: Gland, Switzerland, 2005), 10.
87. A. Dlugolecki, "Climate Change and the Insurance Sector" (2008) 33 *Geneva Papers on Risk and Insurance: Issues and Practice* 71. See also M. Paterson, "Risky Business: Insurance Companies in Global Warming Politics" (2001) 1(4) *Global Environmental Politics* 18.

because they insure the insurers, providing "a risk management tool for insurance companies to reduce the volatility in their portfolios and improve their financial performance and security."[88] The insurance industry bears the lion's share of climate risk by absorbing financial shocks from natural disasters exacerbated by global warming. Specifically, severe local weather events impact upon primary insurers due to an increasing attritional loss burden; and large-scale extreme events cause reinsurance losses.[89] Concomitantly, however, helping governments and the private sector to address climate change presents business and reputational opportunities for the insurance industry.

As early as 1992 leading insurers around the world were speaking openly of the potentially fatal financial threat to their industry from climate change due to unpredicted and unmanageable catastrophic losses from wild weather.[90] Insurance companies pay for much of the damage caused by improbable events such as 100-year storms that occur three or four times in a decade; taking heavy losses from such events in the late 1980s and early 1990s, insurers' primary vulnerability to climate change became apparent. Christopher Flavin explains that for many insurance companies Hurricane Andrew in Florida prompted a "profound reappraisal of their business" in 1992.[91] While Hurricane Andrew was the third most powerful hurricane to hit the United States in the 20th Century, total losses were estimated at US$25billion, which equated to the combined losses of the three most costly weather events in the country previously. Specifically: the Prudential Insurance Company paid out US$1.1billion, Allstate paid US$2.5billion and State Farm paid US$3.5billion in claims; and within months, eight insurance companies serving Florida had collapsed with many others threatening to cease business unless they could be protected from such extremity in the future. Reinsurers received a concomitant wake-up call after being exposed to the combined effects of unprecedented windstorms, firestorms and flooding in North America and Northern Europe. Lloyd's of London, for example, suffered losses of US$4.4billion in 1990 and 1991 from wild weather claims with the result that many syndicate members became bankrupt and approximately 8,000 members resigned. In 1993 Frank Nutter, President of the Reinsurance Association of America, openly claimed that climate change "could bankrupt the [entire] industry."[92]

Thus, insurers and reinsurers have been the canary in the climate change coal mine. Although it is difficult to directly link any particular wild weather event to

---

88. *Testimony of Franklin W. Nutter, President, Reinsurance Association of America: Climate Change: It's Happening Now* (before the United States Senate Committee on Environment and Public Works, Jul. 18, 2013), 2, http://www.epw.senate.gov/public/index.cfm?FuseAction = F iles.View&FileStore_id = f86b767e-7a71-48b4-8eef-7bd9ad1d3884 (accessed May 18, 2014).
89. F. Niehörster, *Warming of the Oceans and Implications for the (Re)insurance Industry* (Geneva Association: Geneva/Basel, June 2013) at https://www.genevaassociation.org/media/616661/ga2013-warming_of_the_oceans.pdf (accessed May 18, 2014), 19.
90. S.C. Jagers and J. Stripple "Climate Governance Beyond the State" (2003) 9 *Global Governance* 385, 390.
91. C. Flavin, "Storm Warnings: Climate Change Hits the Insurance Industry" (1994) 7(6) *World Watch Magazine*, http://www.smartcommunities.ncat.org/articles/world-watch-storm-warnings.shtml (accessed May 14, 2014).
92. *Ibid.*

human-induced climate change, the sudden rise in frequency and severity of such disasters caused many insurers to consider the role of human activity early on. Indeed, prior to publication of the IPCC reports that confirmed anthropogenic climate change and its likely consequences, insurers were predicting the impacts of a warming world with uncanny accuracy:

> In an age when many people live in air-conditioned homes and eat fresh food trucked in from farms located thousands of kilometers away, it is easy to lose awareness of the degree to which we are dependent on a narrowly prescribed range of climatic conditions... The chief concern about global warming...is not the increase in average temperatures, but the possibility that in the course of heating up, the atmospheric and oceanic systems that regulate the world's weather could be suddenly and dramatically disrupted. Areas that now receive ample rainfall might become deserts, regions now safe from catastrophic wind storms and floods could suddenly be vulnerable, and oceanic currents that now moderate both marine and continental climates might unexpectedly shift course.[93]

Fast forward to present day and insurance trends corroborate those earlier predictions.According to global reinsurer Munich Re, the number of worldwide natural catastrophes, including storms, heat waves and forest fires rose from around 300 a year in 1980 to 880 in 2013 with the ten-year average from 2003-2012 sitting at 790.[94] Those loss events have translated into financial losses for insurers. In 2013 alone, Munich Re estimated worldwide insured losses at US$31billion.

Yet that number is only one quarter of overall direct financial losses from global wild weather catastrophes, which totaled US$125billion in 2013 alone.[95] Not surprisingly then, climate change impacts, particularly due to ocean warming, could make some high-risk areas of the world "uninsurable" according to the Geneva Association, which is overseen by executives from some of the world's largest insurance firms. The Geneva Association contends that the coupling of risk transfer and risk mitigation is an essential step to avoiding market failure and addressing the insurability challenges of rising risk levels.[96]

In other words, the insurance industry has a proactive role to play in *ex ante* climate/disaster mitigation as well as conventional *ex post* financial mitigation and recovery.

This role is recognized in the UNEPFI Principles for Sustainable Insurance (PSI), a soft law instrument launched at the Rio + 20 Summit in Brazil in June 2012. The UNEPFI advocates a notion of "sustainable insurance" as a multifarious endeavor that "aims to reduce risk, develop innovative solutions, improve business performance, and contribute to environmental, social and economic sustainability."[97] Against this background, the PSI comprise four aspirational principles with accompanying potential actions: (1) embedding ESG issues in insurance decision-making; (2) working with

93. *Ibid.*
94. Munich RE, *2013 Natural Catastrophe Year in Review* (Munich RE: Munich, Jan. 7, 2014), 17-18.
95. *Ibid.*, 17-18.
96. Niehörster, *supra* n. 89, at 20.
97. Principles for Sustainable Insurance & UNEP Finance Initiative, *The Principles* (undated), http://www.unepfi.org/psi/the-principles/.

clients and business partners to manage ESG risk and develop solutions; (3) working with governments, regulators and other key stakeholders to promote widespread action on ESG issues; and (4) publicly disclosing progress in implementing the PSI.[98] The take-up of the PSI has been promising. Within the first two years, 42 insurance and reinsurance companies worldwide had become signatories including industry heavy-weights Munich Re, Swiss Re and AXA. A further 27 entities that conduct activities relevant to the insurance industry (such as insurance regulatory or supervisory authorities, and insurance institutes and associations) had become "supporting institutions" to the PSI during the same timeframe.

Yet the PSI are clearly a work in progress. The UNEPFI explicitly describes these principles as legally non-binding and aspirational only, with any signatory "free to decide which actions it deems appropriate to implement the Principles."[99] The only sanction for egregious non-compliance is potential de-listing subsequent to notification and discussion with the UNEPFI. Moreover, and this is an important caveat, any implementing actions taken by signatories are "subject to applicable laws, rules and regulations and duties owed to shareholders and policyholders."[100] For insurers that are also institutional investors, this caveat preserves fiduciary duties owed under relevant corporate law and obeisance to the corporate governance norm of shareholder wealth maximization, which may limit truly sustainable practices as examined later in this book.

Nonetheless, UN Under-Secretary-General and Executive Director of the UNEP, Mr Achim Steiner, gave a message of positive obligation at the first year anniversary of the PSI:

> It is our hope that in the years to come, this initiative will be an example for how we can...allow an industry to evolve in the context of the public policy arena, but also to provide new, innovative markets and products and services that the world needs more than ever.[101]

In summary, in addition to significant financial self-interest in ESG issues and climate risk mitigation, the insurance industry has three interrelated roles relevant to addressing climate change:

(1) It has a unique knowledge base with the ability to provide information on climate change impacts and to motivate appropriate responses from other private sector actors and also policy-makers. Importantly, unlike most other decision-makers, insurers are not rendered inert by uncertainty; indeed, for the insurance industry, "the idea that one should only assign dollar values to

---

98. UNEP Finance Initiative, *PSI Principles for Sustainable Insurance* (UNEPFI: Geneva, 2012), 7, http://www.unepfi.org/psi/wp-content/uploads/2012/06/PSI-document1.pdf.
99. *Ibid.*
100. *Ibid.*
101. Principles for Sustainable Insurance & UNEP Finance Initiative, "From Rio + 20 to Beijing 2013: Principles for Sustainable Insurance Initiative Reaches Next Milestone," *unepfi.org* (November 2013), http://www.unepfi.org/fileadmin/communications/PSI_in_Beijing_2013 .pdf.

things that are certain is nonsensical."[102] This has critical implications for addressing climate change, the existence of which is now a certainty but the effects of which will be variable and ambiguous across regions. Leading insurers have developed statistics and competencies in these areas through the creation of extreme weather and loss databases and also catastrophe models. These granular-level data have beneficial flow-on effects for the private sector by enabling businesses to assess and manage their climate-related risks and to make viable long-term investment decisions. The UNEPFI has found that without such information, the private sector is likely to fail to develop appropriate risk management and climate adaptation responses.[103]

(2) The insurance industry has a collaborative and advocacy role. Specifically, it can work with and actively lobby governments to ensure construction quality and sound land use planning.[104] From a climate change mitigation perspective, insurers may encourage governments to tighten the energy efficiency codes on buildings, such as requiring weather stripping or double-glazed windows, which can both save energy and reduce the potential for short-term weather damage. For adaptation, insurers can increase investment in flood defenses, and advocate for stronger building codes in high-risk locations to mitigate the fallout of extreme weather hazards (such as fire, wind and water damage).[105] This advocacy role was first exhibited by the insurance industry in July 1996 when a large delegation of approximately 60 insurers and reinsurers attended the UNFCCC COP in Geneva and signed a statement calling on governments to implement measures for substantial GHG emissions reductions. This advocacy role is further evidenced in the work of the Geneva Association and UNEPFI.

(3) Finally, insurers are also significant institutional investors. For example, at the end of 2012 insurers held assets worldwide worth US$26.8trillion, which comprised nearly one third of the assets from all conventional funds (pension funds, mutual funds and insurance).[106] Specifically, life-, property- and casualty-insurers have financial liabilities maturing over several decades so they must possess not only short-term liquidity to meet early surrender of

---

102. C. Flavin, "Update Climate Change and Storm Damage: The Insurance Costs Keep Rising" (1997) 10(1) *World Watch Magazine,* http://www.smartcommunities.ncat.org/articles/world -watch-storm-update.shtml (accessed May 14, 2014).

103. UNEPFI and SBI, *Advancing adaptation through climate information services: Results of a global survey on the information requirements of the financial sector* (United Nations Environment Program Finance Initiative and the Sustainable Business Institute: Geneva, 2011), 19.

104. See eg: Dlugolecki and S. Lafeld, *supra* n. 86, at27-28, 31; E. Mills, "A Global Review of Insurance Industry Responses to Climate Change" (2009) 34(3) *Geneva Papers on Risk and Insurance: Issues and Practice* 323; C. Herweijer, N. Ranger and R.E.T. Ward, "Adaptation to Climate Change: Threats and Opportunities for the Insurance Industry" (2009) 34(3) *Geneva Papers on Risk and Insurance: Issues and Practice* 360.

105. Niehörster, *supra* n. 89, at 21.

106. CityUK, *Fund management: September 2013* (TheCityUK: London, 2013), 6, http://www.thec ityuk.com/research/our-work/reports-list/fund-management-2013/.

policies but also long-term investment horizons similar to pension funds.[107] Indeed, approximately 90 percent of UK insurance investment funds come from long-term insurance policies whereby premiums paid over many years are invested by insurance institutions in order to meet their liability at maturity.[108] Thus, in their concomitant role as institutional investors, insurance companies need to consider and address climate risk not only in their underwriting policies but also in their investment portfolios.

[2]   *Institutional Investors*

Institutional investors comprise asset owners such as pension funds and insurance companies; they are assisted by asset/investment managers such as banks and (again) insurance companies. Climate change creates both risks and opportunities for institutional investors through their long-term asset class returns. For example, a report by the financial consulting firm, Mercer LLC, found that investments in low-carbon technologies could accumulate to US$5trillion by 2030 but that, concomitantly, the costs of physical damage to assets could accumulate to US$4trillion.[109]

Due to these concerns, climate-aware investor groups have coalesced in most market economies. The European Institutional Investors Group on Climate Change is based in the UK and currently represents assets of around €7.5trillion; and the North American Investor Network on Climate Risk supports over 100 institutional investors with assets exceeding US$10trillion.[110] In Australia and New Zealand, the Investor Group on Climate Change represents institutional investors with nearly AU$1trillion of funds under management.[111] All of the institutional investors in these coalitions have funds invested in assets and markets that are impacted by climate change, namely property (residential and commercial), transport infrastructure (roads, bridges, airports), utilities and network infrastructure, and agriculture.

Institutional investors have a fiduciary duty to act in the best interests of members. Specific legal obligations in a fiduciary relationship will vary with the type of finance industry and its jurisdiction.[112] Yet the traditional core of fiduciary law in

---

107. B.J. Richardson, *Fiduciary Law and Responsible Investing: In Nature's Trust* (Routledge: Oxon, 2013), 200.
108. TheCityUK, *supra* n. 106, at 4.
109. Mercer, *Climate Change Scenarios – Implications for Strategic Asset Allocation* (Mercer LLC, Carbon Trust and International Finance Corporation, 2011) (hereafter "Mercer 2011"); Mercer, *Through the Looking Glass – How Investors are Applying the Results of the Climate Change Scenarios Study* (Mercer LLC: London, 2012) hereafter "Mercer 2012").
110. Ceres, *Press Release: Global Climate Change Investor Groups Publish Report on Investor Practices Relating to Climate Change* (Ceres: Boston MA, 2013), http://www.ceres.org/press/press-rele ases/global-climate-change-investor-groups-publish-report-on-investor-practices-relating-to-c limate-change.
111. See Investor Group on Climate Change (IGCC), *IGCC submission to Review of the Renewable Energy Target Scheme*, dated Sep. 17, 2012 (IGCC: Sydney, 2012).
112. Richardson, *supra* n. 107, at 112-152, 153-225.

Anglo-American jurisdictions comprises the following duties:[113] act loyally in the interests of beneficiaries/principals; treat beneficiaries even-handedly, impartially and with care, skill and prudence; act in accordance with prevailing law (such as pension fund or superannuation legislation)[114] and governing instruments (such as trust deeds). The over-arching duty is unswerving dedication to the interests of beneficiaries, particularly their financial interests in terms of profit maximization.[115]

Due to its extensive time frame, climate change creates "sticky" or long-term investment impacts. So it is particularly relevant for long-term institutional investors such as pension funds and life insurance funds. Institutional investors have been urged in recent years to consider the longer-term ESG implications on and by investments through "responsible" or "impact" investment strategies.[116]

In 2014, John Rogers, President and CEO of the global Chartered Financial Analyst Institute put forward the notion of "fiduciary capitalism" as the future of finance.[117] The term itself is not new: James Hawley and Andrew Williams coined it in 2000 when proposing the "universal owner" thesis whereby large institutional asset owners are sufficiently diversified and pervasive to be self-interested in the long-term health of the entire economy.[118] Self-interest extends to curbing social and environmental externalities that become internalized by firms in a portfolio, thus reducing portfolio value and financial returns for beneficiaries. Rogers defines fiduciary capitalism as *"long-term-oriented institutional investors* [pension funds, endowments, foundations, and sovereign wealth funds] shaping behavior in the financial markets and the broader economy"; and contrasts it with "finance capitalism" which, largely through the work of banks, asset managers and brokerage firms, achieved tremendous and unsustainable growth in market capitalization from the 1980s through to the 2008/2009 financial crisis.[119] Rogers argues that due to legal duties of care and loyalty that "place the needs of their beneficiaries above all other considerations," fiduciary actors are best-placed to encourage long-term thinking, foster deeper governance engagement with investee management and policy-makers, and seek to minimize negative externalities while rewarding positive ones in capital markets. This thesis has important implications for responsible investing:

---

113. *Ibid.*, 101-102.
114. E.g. The *Employee Retirement Income Security Act of 1974* (*ERISA*) (United States); The *Pensions Act 1995* (UK); *Superannuation Industry (Supervision) Act 1993 (Cth)* (Australia).
115. T. Hebb, *No Small Change: Pension Funds and Corporate Engagement* (Cornell University Press: Ithaca, NY, 2008); Richardson, *supra* n. 107, at 103. See also R. Davis, *Democratizing Pension Funds: Corporate Governance and Accountability* (UBC Press: Vancouver, 2009).
116. See eg: B.J. Richardson, *Socially Responsible Investment Law: Regulating the Unseen Polluters* (Oxford University Press: New York, 2008); Freshfields Bruckhaus Deringer, *A Legal Framework for the Integration of Environmental, Social and Governance Issues into Institutional Investment* (UNEPFI, 2005), 100; Parliamentary Joint Committee on Corporations and Financial Services, *Corporate Responsibility: Managing Risk and Creating Value* (Commonwealth of Australia: Canberra Australia, 2006), 74; R. Addis, J. McLeod and A. Raine, *IMPACT –Australia: Investment for social and economic benefit* (Australian Government and JBWere, 2013).
117. J. Rogers, "A New Era of Fiduciary Capitalism? Let's Hope So" (2014) 70(3) *Financial Analysts Journal* May/June, 6.
118. J. Hawley and A. Williams, *The Rise of Fiduciary Capitalism* (University of Pennsylvania Press: Philadelphia, 2000).
119. Rogers, *supra* n. 117, at 6-7 (emphasis in original).

> Under fiduciary capitalism, these corporate management teams will have longer-term owners of their shares, who will hold them accountable for long-term performance and for the true environmental and social costs of their activities. Truly patient capital, which considers the total costs and benefits of a strategy over many years, will drive talent and innovation in finance, rather than the "I won't be here, you won't be here, so who cares" stereotype that has plagued the industry for years.[120]

The idea is that if investment value decreases due to ESG or climate-related issues, then that is an issue to which a fiduciary must attend.

For example, the UN Principles for Responsible Investment (UNPRI) self-recognize that the principles are "firmly within the bounds of investors' fiduciary duties" because "ESG issues can affect investment performance" and therefore need to be considered in order to deliver superior risk-adjusted returns.[121]

Launched in April 2006, the UNPRI were developed by a group of the world's largest institutional investors in conjunction with the UNEPFI and UN Global Compact initiatives. Like the PSI, the UNPRI are a global, voluntary and aspirational framework supported by the UN and focusing on the risks and opportunities associated with ESG issues. They apply to the institutional investment industry, which covers a broad spectrum of institutions such as insurance companies, pension funds, government reserve funds, foundations, endowments, depository organizations, and investment management companies. The UNPRI comprise six principles and an accompanying "menu of possible actions for incorporating ESG issues into investment practices across asset classes."[122] The principles are: (1) incorporate ESG issues into investment analysis and decision-making processes; (2) incorporate ESG issues into ownership policies and practices and be active owners; (3) seek appropriate disclosure on ESG issues from investees; (4) promote acceptance and implementation of the UNPRI within the investment industry; (5) work together to enhance effective implementation of the UNPRI; and (6) report on activities and progress towards implementation. From approximately 70 founding signatories in 2006, the UNPRI take-up has accelerated to over 1200 signatories comprising assets owners, investment managers and service providers with a total of US$45trillion under management in 2014.[123]

As with the PSI, the UNPRI do not provide benchmarks against which signatories' efforts can be measured and/or held accountable, and there are no legal sanctions for non- or under-compliance. The only minimum requirement for remaining a signatory is participation in the reporting framework which requires signatories to complete an annual survey regarding their progress in integrating the UNPRI into investment practices. This may seem modest; but in a positive sign that the UNPRI Secretariat is tightening disclosure standards, it de-listed five signatories that had failed to complete

---

120. *Ibid.*, 9.
121. PRI, *FAQs* (UNPRI, undated), http://www.unpri.org/about-pri/faqs/.
122. PRI, *About the PRI Initiative* (UNPRI, undated), http://www.unpri.org/about-pri/about-pri/.
123. *Ibid.*

the survey in 2009 and it implemented mandatory public reporting for all signatories in 2013, the first cycle of which closed on March 31, 2014.[124]

Moreover, the UNPRI are explicit that their application is to be consistent with fiduciary duties, which may potentially constrain investors' attention to ESG factors including climate change. As Richardson notes, "much uncertainty and controversy persists regarding the legality of RI [responsible investing] from a fiduciary perspective."[125] Indeed, as much as fiduciary duty is touted as the lever for ethical investment by some, it is concomitantly perceived by others as "the most pertinent barrier" to doing so.[126] Complexity in this space is created by not only variable interpretations of fiduciary finance law but also variable understandings of the term "responsible investing."[127]

Specifically, there is concern that the duty of loyalty and profit maximization may clash with the desire to advance socio-environmental causes especially if investment in pro-social causes or divestment from anti-social ones creates short-term profit loss.[128] Dermine writes that we live in "a world in which financial markets reward short-term reported profits"[129] yet, as noted by Salzmann et al., "the economic value of more sustainable business strategies...only materializes in the long term."[130] Indeed, while pension funds have a long-term focus, their fund managers may not. And an immediate business case for climate mitigation endeavors by investors may not be apparent to fund managers due to their short-term bias.[131] Nonetheless, investee firms have reported a perceived growing scrutiny of their sustainability credentials by equity analysts,[132] which in turn is prompting their consideration of socio-environmental factors early in the investment allocation process.

Overall, however, Richardson accurately identifies only four situations in which responsible investment can be made by Anglo-American institutional investors within fiduciary parameters:

---

124. See PRI, *Reporting and Assessment* (UNPRI, undated), http://www.unpri.org/areas-of-work/reporting-and-assessment/; PRI, *2013-14 Public RI Transparency Reports* (UNPRI, 2014), http://www.unpri.org/areas-of-work/reporting-and-assessment/reporting-outputs/individual-2013-14/; and PRI, *The New PRI Reporting Framework* (UNPRI, 2012), http://www.unpri.org/viewer/?file = files/2012.04.30%20New%20Reporting%20Framework%20at%20a%20glance.pdf.
125. Richardson, *supra* n. 107, at 102.
126. R. Koo, "Ethical Finance: Can Ethical Objectives Be Achieved Through Financial Investments?" (2008) 26 *Company and Securities Law Journal* 127, 136.
127. Richardson, *supra* n. 107, at 103-104.
128. See e.g., D. Hayton, "English Fiduciary Standards and Trust Law" (1999) 32 *Vanderbilt Journal of Transnational Law* 555; D. Tennent, "Ethical Investment in Superannuation Funds; Can it Occur Without Breaching Traditional Trust Principles?" (2008) 17 *Waikato Law Review* 98.
129. J. Dermine, "Bank Corporate Governance, Beyond the Global Banking Crisis" (2013) *Financial Markets, Institutions & Instruments* 259-281, 268.
130. O. Salzmann, A. Ionescu-Somers and U. Steger, "The Business Case for Corporate Sustainability: Literature Review and Research Options" (2005) 23(1) *European Management Journal* 27, 33.
131. A. Harmes, "The Limits of Carbon Disclosure: Theorizing the Business case for Investor Environmentalism" (2011) 11(2) *Global Environmental Politics* 98.
132. Ernst & Young, *Six Growing Trends in Corporate Sustainability* (Ernst & Young and GreenBiz Group: New York, 2012), 12, 27.

1. When ESG issues are judged to be financially material to investment performance. This approach fulfils the duty to invest prudently.

2. When alternative investments or investment portfolios are equally financially prudent, then ethical considerations may be the "tie breaker."

3. When a trust deed, investment prospectus or other governing instrument provides a mandate for RI [responsible investment], as for an endowment fund that is established to fulfill philanthropic goals.

4. When beneficiaries consent. In a pension plan, while beneficiaries might agree to an RI policy, these funds tend to be subject to legislative duties that restrict trustees' latitude to follow non-financial criteria.[133]

Clearly, there is a narrow bandwidth for institutional investors to engage in socio-environmental investment.

As such, some commentators seek to obviate limitations created by direct investment for greener outcomes by arguing that institutional investors can *indirectly* influence corporations to these ends. That is, institutional investors can leverage their influence over investee corporate governance to encourage good environmental management and GHG mitigation strategies.[134] This argument is based largely on the enormous wealth holdings of institutional investors and their status as universal owners with diverse stock portfolios. For example, the largest 1,000 institutional investors worldwide accounted for US$25trillion of global equity market value in 2013, which was more than half the total value of that market.[135] Rogers points to the sheer financial size of institutional investors' wealth holdings and states that "[w]hen these institutions act together, as they increasingly do in matters of corporate governance and market structure, they can shape the financial markets into the form they desire."[136]

Nonetheless, it remains uncertain whether universal owners can take account of beneficiaries' non-economic interests, let alone those of anyone else (consumers, future generations, the poor or unemployed).[137] Moreover, as Richardson highlights, even if they can, the health of the economy may not be synonymous with, or even "an adequate proxy" for, "the long-term health of the biosphere."[138]

Yet there is evidence that funds can pool their expertise and mobilize their leverage for positive change. Specifically, this is occurring at two levels: first, at a sector- or economy-wide level via voluntary codes of conduct such as the UNPRI described above; and secondly, at a company-level via shareholder activism, as described later in this chapter.

---

133. Richardson, *supra* n. 107, at 104-105.
134. See e.g. C St Anne, "Super Funds Have Role in Climate Change," *InvestorDaily* (Mar. 27, 2007); Rogers, *supra* n. 116.
135. L. Kennedy, "Top 1000 Pension Funds: The Thrifty Thousands," *Investment & Pensions Europe* (September 2013).
136. Rogers, *supra* n. 117, at 8.
137. See e.g. M. Patry and M. Poitevin, "Why Institutional Investors Are Not Better Shareholders," in RJ Daniels and R Morck (eds.), *Corporate Decision-Making in Canada* (University of Calgary Press: Calgary, 1995), 341.
138. Richardson, *supra* n. 107, at 81.

Finally, like the insurance industry, institutional investors have an important advisory and advocacy role vis-à-vis climate-related regulation and governance frameworks. This role is taken very seriously by climate-aware investor groups such as the European Institutional Investors Group on Climate Change, the North American Investor Network on Climate Risk, and the Australia/New Zealand Investor Group on Climate Change. Together, these groups have formed the *Global Investor Coalition on Climate Change* to "provide a global platform for dialogue between and amongst investors and governments on international policy and investment practice related to climate change."[139] Importantly, the Coalition's stated aim embraces the universal owner thesis, dedication to individual fiduciary duty, and a public-private partnership role to: "help reduce the risks from climate change to investors and the global economy by simultaneously encouraging improved governmental climate policies and facilitating increased low carbon investment."[140]

These groups acknowledge that private capital and public regulation are together indispensable for effective and systemic climate finance. I will explore the type and mix of legal and regulatory tools for this task in Chapters 6 and 7.

[3]   Banks

The relevance of banks to climate change is summarized aptly by Deutsche Bank:

> As *an asset manager* we can steer investments into low-carbon companies, as *a trader* we provide liquidity in the carbon market, and as *a capital markets participant* we can raise debt and equity capital to fund clean tech companies and projects and provide solutions to all clients who face the inevitable impact of climate change.[141]

Yet, of all the finance actors discussed thus far, banks appear the least likely candidate for a gold medal in good corporate citizenship for doing "the right thing." Although rarely loved, banks have been increasingly perceived as bad actors since the advent of the GFC, which precipitated sweeping legislative changes to banking regulation in the United States, Europe and Australia, and increased social pressure around the world for responsible banking practices.

Moreover, the GFC impacted upon the viability of the banking industry, particularly in North America and Europe, with the result that even too-big-to-fail commercial and investment banks have since merged, been acquired, or collapsed entirely. Given the adverse impact of the GFC on certain banks' financial stability and public confidence in the industry generally, are banks able (let alone willing) to engage in climate-related efforts?

---

139. Global Investor Coalition on Climate Change, *Global Investor Coalition on Climate Change* (2014), http://globalinvestorcoalition.org/.

140. Global Investor Coalition on Climate Change, *Global Investor Action Plan on Climate Change: 2013-2015* (undated), http://globalinvestorcoalition.org/global-investor-action-plan-on-climate-change-2013-2015/.

141. Deutsche Bank, *Deutsche Bank's Commitment to Addressing Climate Change* (Deutsche Bank: Frankfurt am Main, undated), 2 (emphasis added).

The Carbon Disclosure Project (CDP) provides a ready gauge of industry concerns and opportunities on this issue. Since the GFC, CDP reports reveal that high-risk areas such as project finance and emerging market investment had been impacted by the economic downturn; however, climate change remains near the top of bank stakeholders' agendas for the long-term, and is perceived by banks as presenting lucrative opportunities in carbon trading and renewable energy markets.[142] Moreover, banks that are cautious due to the GFC are now more likely to engage in ESG due diligence in order to mitigate climate-related risks.

Ironically, the global climate crisis may present an opportunity for banks to "make good."

[B]   Relevance by Activity: Finance Sector Business Practices and Climate Change

The preceding discussion depicted how climate change brings both direct and indirect risks and opportunities for private finance industries. Moreover, it made clear that the unique business practices of private finance actors indirectly impact upon climate change mitigation and adaptation , for better or worse.

But how do these realizations play out in practice? The remainder of this chapter considers the relationship between groupings of financial practices and climate change, specifically:[143]

- Pricing risk and predicting company profitability and investment preferences in an increasingly carbon-constrained world via due diligence processes that include attention to "climate risk."
- Providing capital for GHG intense projects as well as clean technology solutions. The role of private finance actors in mitigating climate and environmental degradation through financing decisions is recognized in a suite of international soft law codes.
- Brokering and innovative product design in lucrative global markets for carbon, renewable energy and clean tech.

[1]   Risk Assessment and Risk Management

Climate risk makes private finance actors vulnerable due to indirect credit, investment, reputation and litigation risks inherent in their corporate clients, projects, and investments.

---

142. See Carbon Disclosure Project, *supra* n. 84, at 90-94; Carbon Disclosure Project (CDP), *CDP Global 500 Report 2011: Accelerating Low Carbon Growth* (CDP, London, 2011), 42-43, http://www.cdproject.net. See also W. Sun, J. Stewart and D. Pollard (eds), *Reframing Corporate Social Responsibility: Lessons from the Global Financial Crisis, Critical Studies in Corporate Responsibility, Governance and Sustainability, v.1* (Emerald Group Publishing Limited: Bingley UK, 2010).
143. Adapted from M. Bowman, "The Role of the Banking Industry in Facilitating Climate Change Mitigation and the Transition to a Low-Carbon Global Economy" (2010) 27 *Environmental and Planning Law Journal* 448, 452.

Thus, as awareness of the impacts of climate change increases, finance actors are broadening their risk assessment and management procedures to include "climate risk," which includes regulatory risk and material risks. Regulatory risk comprises uncertainty around climate-related policy such as the exact rules under which a carbon market might be implemented or whether a financial incentive for renewable energy will be retroactively revoked. It is largely outside the control of corporations and therefore the most difficult risk for them to mitigate. Conversely, material risks, which include credit, investment, reputation and litigation risks, can be more readily managed and mitigated.

Reinsurers bear the greatest financial stake in appropriate assessment of climate risk. Accordingly, they have the most obvious and rather critical role of translating and communicating the interdependencies of climate risk assessment and pricing.[144]

Institutional investors (including insurers) and banks need to account for climate risk in their due diligence procedures for clients, projects and investments. This is not a big leap for them however. Since the 1980s they have had to consider environmental risk in lending, investing and financing decisions due to legal obligations that created financial liability for contaminated land. For example, application of the *Comprehensive Environmental Response, Compensation and Liability Act 1980* (US) Pub L No 96-510, 94 Stat 2767 (CERCLA or Superfund) in the United States triggered a number of high-profile court cases in which some banks were held liable for cleanup costs of contaminated land even though they did not "own" the land for the purposes of the Act.[145] Other OECD countries similarly introduced tough environmental legislation that mandates strict liability for contamination; however there are usually exceptions for secured creditors not involved in corporate operations.[146]

Regarding the more complex matrix of climate risk, Mercer LLC concluded in 2012 that: "Traditional models for strategic asset allocation cannot adequately capture the effects of climate change."[147] This is because conventional risk modeling relies on historical quantitative data, which are proving unreliable predictors of future climate change effects given the uncertainty of political and natural responses to climate change. Thus, climate risk assessments need to look forward and include qualitative risk factors such as technology impacts (which will increase returns), physical impacts (which will increase risk) and impacts from delayed or uncoordinated climate policy (which will likely decrease returns and increase risk).[148]

Consequently, institutional investors need to take the following actions to manage climate risks and capture climate-related opportunities:[149]

- embed climate risk into the asset allocation process in order to identify potential sources of risk and opportunity for portfolios;

---

144. *Testimony of Franklin W. Nutter, supra* n. 88.
145. See J.R. Burcat, "Environmental Liability of Creditors: Open Season on Banks, Creditors, and Other Deep Pockets (1986) 103 *Banking Law Journal* 509.
146. See Richardson *supra* n. 116, at 74, 351-352.
147. Mercer 2012, *supra* n. 109, at 6.
148. *Ibid.*; Mercer 2011, *supra* n. 109.
149. Mercer 2012, *supra* n. 109, at 6, 12.

   – "kick the tires" of existing investments across asset classes to assess their climate resiliency, and develop an "early warning" system by monitoring key climate change developments over time; and
   – allocate investment to assets that have a higher sensitivity to climate-related risks and opportunities in order to improve the resilience of a portfolio mix. These include long-term assets such as infrastructure, private equity, and real estate.

It becomes clear that the main question to be asked by private finance actors is whether they have adequately revised their climate risk assessment based on the categories of information below. If not, they may end up with a situation where their exposure to climate risk is greater than their risk appetite.

[a]   Credit and Investment Risks

Two categories of client and investment create credit risk being "the risk of non-payment or non-timely payment of the principal or interest of a loan by the borrower to the lender."[150] First, wild weather directly affects the operations and physical assets of organizations in climate-vulnerable sectors such as real estate, agriculture, forestry, tourism, and even the oil industry (given that many refineries in Louisiana were shut down due to the impacts of Hurricane Katrina in 2005).[151] In these cases, extreme weather events may increase client default risk from underestimated costs of mitigation and adaptation measures as well as physical damage to corporate assets.

   Secondly, intensive GHG-emitting corporations in sectors such as oil, gas, coal, heavy manufacturing and transport are competitively affected by government regulation that prices carbon. Standard & Poor's March 2013 report *What A Carbon-Constrained Future Could Mean For Oil Companies' Creditworthiness* analyzed the potential impact on oil sands and large oil companies from climate mitigation policies. In its "stress scenario" analysis, the report predicted increased pressure on cash flows, projects and dividends for some companies.[152] Regarding investment risk, HSBC's 2012 report titled *Coal and Carbon, Stranded Assets: Assessing the Risk* estimated that carbon constraints could impact discounted cash flow valuations of coal assets by as much as 44 percent from 2020.[153] Moreover, companies that are competitively disadvantaged in the market due to regulatory and reputational pressures may produce

---

150. P. Mudde and A. Abadie, *From Principle to Action: an Analysis of the Financial Sector's Approach to Addressing Climate Change* (Sustainable Finance Ltd: London, 2008), 27. See also Dlugolecki and S. Lafeld, *supra* n. 86, 33.
151. B.J. Richardson, "Climate Finance and its Governance: Moving to a Low Carbon Economy Through Socially Responsible Financing?" (2009) 58(3) *International & Comparative Law Quarterly* 597, 604.
152. Standard & Poor's Rating Services, *Ratings Direct: What A Carbon-Constrained Future Could Mean For Oil Companies' Creditworthiness* (Standard & Poor's Financial Services LLC, 2013).
153. HSBC, *Coal and carbon – Stranded assets: assessing the risk*, (HSBC, London, 2012). See also N. Robins, *Global High-Impact Risks for Banking & Investment: Navigating disruptive change*, September 25 (HSBC: London, 2013), 13.

lower investment and shareholder returns.[154] The key for finance actors is to explicitly internalize climate change into credit and investment risk assessments.

Moreover, climate-related credit and investment risks are increasingly reflected in market conduct regulation. In February 2010, the United States Securities and Exchange Commission (SEC) released the world's first guidance on climate change disclosures that public companies should provide in financial filings titled *Commission Guidance Regarding Disclosure Related to Climate Change*.[155] Interestingly, the move was motivated by repeated petitions from institutional investors in the United States that represented US$1.5trillion in assets with support from additional international investors representing US$6.5trillion in assets. The Guidance affects not only American companies but also international companies that operate in the United States as well as overseas supply chains.[156] It does not create new legal requirements but extends existing SEC disclosure requirements to include material regulatory risks and opportunities, physical risks and water shortages due to climate change. As with traditional types of risk, if a climate-related risk poses a material liability to a company, it must be disclosed.

Subsequently, in Canada, *CSA Staff Notice 51-333 Environmental Reporting Guidance* was issued in October 2010 by the Canadian Securities Administrators to provide guidance on compliance with existing environmental disclosure requirements under securities legislation and *National Instrument 51-102 Continuous Disclosure Obligations* (NI 51-102). Under these regulations, securities issuers are required to disclose material environmental matters that bear upon physical, litigation, regulatory, reputation and business model risks to the issuer. While the notice focuses on environmental disclosure, it encompasses climate risk. For example, relevant physical risks include "changing weather patterns and water availability" and relevant regulatory risks comprise "carbon pricing systems, carbon limits and trading systems, energy efficiency standards and building codes."[157]

These regulatory developments reflect a changing market awareness of climate risk with attendant consequences for corporations operating in a carbon-constrained economy. However, in many ways the corporate sector is yet to catch up with these regulatory developments, which creates heightened risk for financiers. In February 2014, Ceres found that publicly traded corporations in the United States are producing deficient climate disclosures despite the formal Guidance.[158] Analyzing S&P 500 company reporting on climate disclosure from 2000 to 2013 and more than 40,000 SEC

---

154. D. Austin and A. Sauer, *Changing Oil: Emerging Environmental Risks and Shareholder Value in the Oil and Gas Industry* (World Resources Institute: Washington DC, 2002).
155. Securities and Exchange Commission, 17 CFR Parts 211, 231 and 241 [Release Nos. 33-9106; 34-61469; FR-82] *Commission Guidance Regarding Disclosure Related to Climate Change* (Feb. 8, 2010), http://www.sec.gov/rules/interp/2010/33-9106.pdf.
156. J. Coburn, "SEC Climate Guidance Could Increase Supply Chain Scrutiny," *GreenBiz.com* (Jul. 26, 2010).
157. CSA/ACVM, *CSA Staff Notice 51-333: Environmental Reporting Guidance* (Oct. 27, 2010), 8, http://www.osc.gov.on.ca/documents/en/Securities-Category5/csa_20101027_51-333_envir onmental-reporting.pdf.
158. Ceres, *Cool Response: The SEC and Corporate Climate Change Reporting* (Ceres: Boston MA, 2014), at http://www.ceres.org/files/investor-files/sec-guidance-fact-sheet.

letters sent to companies for the same period, Ceres found that the majority of financial reporting on climate change was largely superficial and non-compliant with SEC requirements. Where a filing was not in compliance, the SEC discussed the issue in a comment letter sent to the company. However, according to the study, the SEC did not send any climate-related comment letters in 2013 and sent only three in 2012 (down from 49 in 2010-2011) despite the low quality of corporate reporting on climate risk. These findings echo earlier concerns of Canadian investors that environmental information provided by issuers lacked sufficient granularity and reliability to be meaningful to them.[159]

Given the importance of climate disclosure to investor decision-making and risk management and mitigation, these reporting and enforcement inadequacies are deeply concerning. It places a greater burden on private finance actors to conduct enhanced due diligence when assessing climate risks from potential clients and investments. Moreover, disclosure inadequacies are prompting shareholders to proactively seek information from private finance actors via securities regulators, as discussed in the next section.

[b]   Reputation Risk

Risk to reputation or brand is the second type of indirect climate risk for private finance actors. Put simply: "Reputation is crucial for financial institutions."[160] Explaining why this is so, Leo Johnson of the *Wall Street Journal* uses the example of banks:

> Forget credit risk or legal risk. Let's look at the brand. What holds a great bank together? What is the one emotion you as the depositor must feel toward your bank? In a word, trust. When you give a bank your savings, you are putting your life in their hands. What's more, they are unverifiable; you can't kick the tires.[161]

Reputation as a trustworthy and responsible institution is germane to a private finance actor's business success. This is acknowledged by the finance sector itself. For example, the PSI preamble states that the "insurance industry's core business is to understand, manage and carry risk. We depend on the trust people place in our industry to fulfil its obligations."[162] Similarly, fiduciary duties, which lie at the heart of an institutional investor's investment mandate, embody the strictest legal relations between a trustee and beneficiaries.

Trust and reputation are most starkly appreciated when they are lost. The importance of public trust in the private finance sector came to the fore in the aftermath of the GFC. The sector still contends with that trust deficit, which is fueled by public perception as much as historical events. For example, in 2014 the banking industry took equal last position with the media industry as the least-trusted industry globally

---

159. CSA/ACVM, *supra* n. 157, at 4, at http://www.osc.gov.on.ca/documents/en/Securities-Categ ory5/csa_20101027_51-333_environmental-reporting.pdf.
160. per John Tobin, Head of Sustainability Affairs, Credit Suisse, quoted in D. Enskog, "International Banks Address Climate Concerns," *Online Publications* (Oct. 13, 2008).
161. L. Johnson, "Open Season," *Wall Street Journal online* (Jul. 5, 2007).
162. Principles for Sustainable Insurance & UNEP Finance Initiative, *supra* n. 97.

according to the public relations company Edelman.[163] Importantly, this lack of trust existed despite positive turnarounds in the banking industry's stock and business performance.

In maintaining and regaining public trust, a key policy issue for the private finance sector has now become risk to reputation through association with environmentally risky or "dirty" companies and projects.[164] A particular source of reputational risk is public "naming and shaming" campaigns by NGOs in response to financing of companies or activities that perpetrate environmental harm. This is a relatively new area of concern for private finance actors. Traditionally, they have been invisible and not considered a frontline environmental threat.[165] As such, NGO campaigns traditionally targeted only polluters, a memorable example being the 1995 Greenpeace campaign against Shell UK for its planned disposal of the Brent Spar oil tanker at sea.

However, in the late 1990s NGOs started focusing attention on financial institutions in the private sector, particularly for their financing of large oil, gas and dam projects in emerging economies. O'Sullivan and O'Dwyer document a growing realization by NGOs at that time that "private [sector] banks had been able to operate in relative anonymity" while holding "the real power" over unsustainable projects and companies.[166] In particular, private sector banks had begun financing projects previously redlined by multilateral financial institutions on the basis of socio-environmental concerns, such as the Three Gorges Dam in China.[167]

This realization prompted not only extensive campaigns by individual NGOs but also, since 2003, coordinated NGO campaigns under the auspices of the Collevecchio Declaration and BankTrack. The Collevecchio Declaration is a coalition of NGOs including Friends of the Earth, the Rainforest Action Network (RAN), and the World Wide Fund for Nature (WWF), which calls upon financial institutions to incorporate commitments to the following six principles in all financial operations: sustainability; "do no harm"; responsibility; accountability; transparency; and sustainable markets and governance.[168] BankTrack is a global network that: tracks the socio-environmental operations of private sector banks; acts as a clearing house for reports and press

---

163. Edelman, *2014 Edelman Trust Barometer Executive Summary*, 5, http://www.edelman.com/insights/intellectual-property/2014-edelman-trust-barometer/about-trust/executive-summary/.
164. Eg: P. Watchman, *Banking on Responsibility. Part 1 of Freshfields Bruckhaus Deringer Equator Principles Survey 2005: The Banks* (Freshfields Bruckhaus Deringer: London, 2005); Oekom Research, "Global Sustainability Rating of Financial Services Providers: Australia Gets Gold, USA is Eliminated," *Industry News* (Oekom Research: Munich, 2006); G Cooper, "Room for Improvement on Banks' Green Risks" (2002) (October) *Environmental Finance* 7; M. Lundgren and B. Catasús, "The Banks' Impact On the Natural Environment – On the Space Between 'What Is' and 'What If'" (2009) 9 *Business Strategy and the Environment* 186.
165. Richardson, *supra* n. 116.
166. N. O'Sullivan and B. O'Dwyer, "Stakeholder Perspectives on a Financial Sector Legitimation Process: the Case of NGOs and the Equator Principles" (2009) 22(4) *Accounting, Auditing & Accountability Journal* 553, 562.
167. J.M. Conley and C.A. Williams, "Global Banks as Global Sustainability Regulators?: The Equator Principles" (2011) 33(4) *Law & Policy* 542, 543-544.
168. BankTrack, *Collevecchio Declaration: the Role and Responsibility of Financial Institutions* (Banktrack: Nijmegen, Netherlands, 2003).

releases by NGOs on same; and seeks to influence responsible banking practice.[169] It was formally established in 2005 subsequent to the Collevecchio Declaration and comprises both individuals and large NGOs such as Greenpeace and Friends of the Earth.

Accordingly, private finance actors, particularly banks, have become visible in the past decade due to heightened NGO scrutiny of their financing decisions that create indirect socio-environmental impacts.

Most recently, scrutiny has extended to their climate-related financing activities.

An example is the 2007 campaign mounted by RAN against financiers of Texas Utilities (TXU).[170] TXU planned to build 11 "cleaner" coal-fired plants; the estimated annual increase in CO_2 emissions due to the power plants was 78million tons. Citigroup, Morgan Stanley and Merrill Lynch were mandated to arrange financing for construction. They were targeted by RAN, which issued formal letters to 54 financial institutions requesting that they withhold finance for the project. The approval process by the Texan Governor was also litigated and the proposal became a magnet for deleterious media and public attention. Eventually, TXU was bought out by Kohlberg Kravis Roberts and Texas Pacific Group, pursuant to which TXU agreed to build only three coal-fired power plants, to invest in "green" energy technology, and to invest US$400million into new initiatives such as alternative energy sources and technologies. Apparently the deal helped to prevent the release of 56 million tons of CO_2 emissions each year.

Importantly, NGO campaigns can seriously hurt a finance actor's brand by publicly undermining the trust and reputation so crucial to its operations.

Local offices and Annual General Meetings (AGMs) of banks have become prime targets for NGO campaigners in order to gain maximum traction with senior management and civil society. For example, HSBC was forced to shut its flagship London branch on October 10, 2013 due to public protests against its investment in fossil fuels. At Barclays' AGM in London on April 24, 2014, the World Development Movement accused the bank "of fuelling climate change and destroying people's lives and the environment by financing the global coal industry."[171] In Australia, ANZ Bank's Chairman ended a series of questions about the bank's funding of coal at its 2013 AGM in Brisbane, which only fuelled an article in *Business Spectator* that "the board of ANZ is totally out of touch with community feeling" on the issue of climate change.[172]

Institutional investors are similarly not immune. The fossil fuel divestment movement has been gaining momentum since its inception in 2011. Its aim is to persuade investors to sell out of oil, gas and coal companies due to concerns about climate change and other polluting emissions. From 2011 to 2014 the movement successfully persuaded 11 universities and other investment institutions to commit to divestment. In a highly public campaign reported in the *Financial Times* in May 2014,

---

169. See http://www.banktrack.org/show/pages/about_banktrack.
170. See Mudde and Abadie, *supra* n. 150, at 14.
171. Banktrack, "Press Release: Barclays AGM: Protestors to Target Barclays for Bankrolling Coal," *Banktrack.org* (Apr. 23, 2014).
172. I. Lowe, "ANZ Bank is Blowing the Carbon Budget," *businessspectator.com.au* (Dec. 20, 2013).

students blockaded Harvard University's administrative offices seeking to persuade the University's US$33billion endowment to sell off its fossil fuel investments.[173]

In addition to causing public embarrassment, reputational campaigns can have real impacts on a finance actor's business by chilling customers and shareholders.

For example, customers of Australia's four major banks threatened to close their accounts in protest against funding of coal and gas projects in May 2014. The action was organized by activist investment group Market Forces and climate campaigner 350.org, and began a year earlier when hundreds of customers sent letters to the National Australia Bank, Commonwealth Bank of Australia, Westpac and ANZ "putting them on notice they would lose customers unless they committed to ruling out future loans to coal and gas projects."[174] Despite the relatively small number of customers involved, the organizers contended that their action threatened up to AU$120million in investment.

In addition to NGOs, shareholders are also paying increasing attention to climate-related matters. FundVotes reported that from 2004 to 2011 the number of resolutions in the categories of "climate change" and "environment" that went to a vote in the United States increased only slightly from 36 to 44.[175] However, the percent of shareholder *support* for resolutions rose dramatically from 0 percent in 2004 to: 30 percent in 2013 on the topic of GHG emissions disclosure; 22 percent in 2013 on the topic of GHG emission reduction efforts and goals; 42 percent in 2013 on sustainability reporting; and 34 percent in 2013 on hydraulic fracturing.[176] Certainly, disclosure is a key issue for shareholders. The Investor Network on Climate Risk reported that North American companies received 50 percent more shareholder resolutions in 2010 that sought increased disclosure and action on climate-related risks than in 2009. Investors withdrew half of those resolutions because company management took steps to address the demands.[177]

Some shareholder requests have elicited support from securities regulators. For example, in February 2013 the SEC directed PNC Financial Services Group Inc. to consider a shareholder request to assess "the greenhouse gas emissions resulting from its lending portfolio and its exposure to climate change risk in its lending, investing, and financing activities."[178] In making its request, shareholder Boston Common Asset Management LLC stated that:

> A bank's financed emissions can dwarf its other climate impacts and expose it to reputational and operational risks... Investors should seek additional disclosure from PNC regarding its climate change related programs as well as assurance that

---

173. E. Crooks, "Fossil Fuel Protesters Blockade Offices at Harvard University," *ft.com* (May 1, 2014).

174. A.Macdonald-Smith, "Bank Accounts Closed in Fossil Fuels Protest," *smh.com.au* (May 2, 2014).

175. FundVotes, *Shareholder Resolutions: Average Shareholder Support by Sub-Category*, http://www.fundvotes.com/resolutionsbysubcategory_countavg.php.

176. *Ibid.*

177. Editorial, "Investor Activism on Climate Shows No Sign of Slowing," *GreenBiz.com* (Jul. 8, 2010).

178. M. Benton, *Proxy Memo: PNC Financial AGM*, dated Mar. 18, 2013 (Boston Common Asset Management).

it understands the implications climate change may have for their business... *PNC has not shared with investors the ways in which it has sought to analyze, assess, or strategically manage its exposure to climate change risk.* Broad vague public statements are an insufficient basis for analysis and benchmarking of investment opportunities.[179]

PNC had earlier contended that the United States' *Securities Exchange Act of 1934* allowed companies to exclude shareholder proposals dealing with "ordinary business operations" from proxy statements. However, in its first display of push-back against a company on the climate issue, the SEC responded succinctly that "We are unable to concur in your view... In arriving at this position, we note that the proposal focuses on *the significant policy issue of climate change.*"[180] Undoubtedly, active investors will continue to make similar requests of financial firms, which must then include those requests in their proxy statements and put them to a shareholder vote.[181]

The SEC stance elevates climate change to the same level as traditional material risks; it indicates that finance actors need to add GHG emissions and liabilities to their extant information list with a view to disclosure where materially relevant. Although it is a potentially onerous task for private finance actors to provide disclosures on climate-related risks that are inherent in large portfolios, it is within their capacity to do so as they must already gather a great deal of information about the nature of their assets and investments.

A particular concern to financiers of environmentally risky projects is shareholder and local community push-back, which can create adverse market and public reactions to a brand akin to NGO campaigns. Indeed, there are prominent examples of NGO, community and shareholder activism converging on this issue. For example, in response to NGO questioning at Deutsche Bank's 2014 AGM in Frankfurt, Deutsche Bank Co-Chair Mr. Juergen Fitschen stated that the bank would not become involved with the expansion of one of the world's largest planned coal ports at Abbot Point in Australia's world heritage-listed Great Barrier Reef – even though it had federal and state government approval to do so. While Mr. Fitschen indicated multiple reasons for this decision such as regulatory and environmental impact uncertainty, the public pressure by NGOs undoubtedly played a key role. Specifically, a wide coalition of NGOs throughout Europe and Australia had campaigned intensively for weeks leading up to the decision, which included placing a full page advertisement in the *European Financial Times*, hundreds of emails to the bank, and a 200,000 strong online petition calling on Deutsche Bank to not fund the project.[182] Other finance actors, including HSBC and Royal Bank of Scotland, also dropped support for the proposal.

---

179. *Ibid.* (emphasis in original).
180. Letter from A. Kim, United States Securities and Exchange Commission, Division of Corporate Finance to George P. Long, III, The PNC Financial Services Group, Inc., dated Feb. 13, 2013 (emphasis added), http://environblog.jenner.com/files/a-letter.pdf.
181. See K. Goldberg, "Banks Dragged Into Climate Change Fight With SEC Order," *law360.com* (Mar. 1, 2013), http://www.law360.com/environmental/articles/419950/banks-dragged-into-climate-change-fight-with-sec-order.
182. Banktrack, "Press Release: Deutsche Bank Says No to Financing of Coal Harbor on the Great Barrier Reef," *banktrack.org* (May 22, 2014).

Similarly, in January 2014, Goldman Sachs announced the disposal of its remaining equity investment in Carrix, the parent company of Pacific International Terminals and SSA Marine which were responsible for the proposed coal export terminal Gateway Pacific Terminal at Cherry Point in the United States. That project aimed to be the largest such terminal in North America, exporting up to 48 million tons of coal to Asian markets each year. Apart from the climate-related consequences, NGOs and local communities had campaigned against its potential detrimental effect on the "rich biodiversity and unique cultural legacy found in the region."[183] Yet the exact motivation for Goldman Sachs' exit remains unclear. NGOs at the time explained it as a response to reputational damage inflicted by campaigning and that the bank had made a positive commitment to "walking their sustainability talk."[184] Yet it is just as probable that the move was based on profit calculations given that project proponents had dropped half of the proposed terminals in the Pacific Northwest in the previous two years.[185]

Investigating and understanding private finance actors' climate-related motivations is key to moving forward on corporate climate finance for both policy-makers and practitioners. That inquiry forms the basis of the empirical exposition in this book.

[c]   Litigation Risk

Finally, financiers now face litigation risk for their perceived role in facilitating GHG emissions.

From a legal perspective, the issue of causation is problematic in climate-related litigation. Traditionally, environmental liability attaches to the entity that directly causes damage, best illustrated by the "polluter pays principle" pursuant to which the "dirty" entity bears responsibility for reparations. Yet climate change is a global and ubiquitous problem, the sources of which are multifarious. Regulators and complainants are looking beyond polluters for succor, as they did in the United States with CERCLA and are doing now through climate litigation.

Climate-related common law actions in public nuisance were aimed first at power and utilities companies and automobile manufacturers in the United States. For example, in *Lockyer v. General Motors* six vehicle manufacturers were sued in tort by the State of California for allegedly causing public nuisance and injury to it due to GHG emissions from the manufacturers' vehicles.[186] However, federal courts in the United States have been reluctant to recognize these claims due to: (a) the difficulty of attributing blame among multiple emitters; and (b) the declaration by the Supreme Court on June 20, 2011 that federal legislation to regulate emissions displaces federal

---

183. Rainforest Action Network, "Goldman Sachs sacks coal terminal investment," *banktrack.org* (Jan. 7, 2014).
184. *Ibid.*
185. *Ibid.*
186. *Lockyer v. General Motors* No 3: 06. Civ 05755 (ND California, filed Sep. 20, 2006).

common law claims.[187] This has not deterred plaintiffs who are instead filing claims in state courts attempting to sue utilities for increased insurance premiums due to climate change. For example, the class action complaint *Comer v. Murphy Oil USA Inc.* filed on May 27, 2011 in the Southern District Court of Mississippi alleged the following: "Insurance companies have begun to incorporate the risks associated with global warming into their premiums, which has dramatically increased Plaintiffs' insurance costs. Thus, Defendants are directly responsible for the increased risk [of wild weather damage] to Plaintiffs' property as well as the concomitant increases in costs to insure Plaintiffs' property."[188] As such, claims for damages are based on increased risk rather than actual harm, with attendant implications for private finance actors.

In 2002 the Overseas Private Investment Corporation (OPIC) and the Export-Import Bank of the United States (Ex-Im) were sued by Greenpeace, Friends of the Earth, the City Council of Oakland, and four more American cities.[189] The lawsuit alleged that the defendants had breached requirements under the *National Environmental Policy Act* by providing over US$32billion in financing and insurance to fossil fuel projects during 1990-2003 without conducting environmental impact assessments of whether the projects contributed to global warming or adversely affected the environment. The Plaintiffs further claimed that the Defendants were liable for the adverse effects of cumulative GHG emissions from approved but unassessed projects, being the equivalent of nearly one third of annual emissions in the United States in 2003. On August 23, 2005, Judge Jeffrey S. White of the US District Court for the Northern District of California denied the Defendants' motion challenging the Plaintiffs' standing to bring their claims and allowed the lawsuit to proceed. His Honor held that there was sufficient evidence "to demonstrate it is reasonably probable that emissions from projects supported by OPIC and Ex-Im" threatened the Plaintiffs' "concrete interests."[190]

The case finally settled on February 6, 2009 after an intermediary ruling by the court on cross motions for summary judgment in March 2007 that had encouraged the parties to engage in settlement discussions.[191] Under the settlement agreement, Ex-Im agreed to create an organization-wide carbon policy and to take CO_2 emissions into

---

187. *American Electric Power Company, Inc., v. Connecticut*, 564 U.S._ (2011). See also L. Hurley and G. Nelson, "Climate: High Court Blocks States' Lawsuit Over Coal Plant Emissions," www.eenews.net (Jun. 20, 2011).
188. *Comer v. Murphy Oil USA Inc.* (case 1:11-cv-00220-LG -RHW), filed on May 27, 2011 in the Southern District Court of Mississippi, para. 38.
189. *Friends of the Earth, Inc., Greenpeace, Inc., City of Boulder, Colorado, City of Oakland, California, City of Arcata, California, and City of Santa Monica, California (Plaintiffs) v. Peter Watson in His Official Capacity as President and Chief Executive Officer of the Overseas Private Investment Corporation, and Philip Merrill in his Official Capacity as President and Chairman of the Export-Import Bank of the United States (Defendants)* Civ. Bo. C 02 4106 JSW, US District Court Northern District of California, San Francisco Division, Second Amended Complaint, http://www.greenpeace.org/usa/Global/usa/report/2009/2/plaintiff-s-complaint.pdf.
190. US District Court for the Northern District of California, 23rd August 2005, summary judgment of Judge Jeffrey S. White, Case No. C 02-04106 JSW.
191. See *FRIENDS OF THE EARTH, INC., et al., Plaintiffs, v. ROBERT MOSBACHER, JR., et al., Defendants* No. C 02-04106 JSW UNITED STATES DISTRICT COURT FOR THE NORTHERN DISTRICT OF CALIFORNIA 488 F. Supp. 2d 889; 2007 US Dist. LEXIS 24268 (Mar. 30, 2007, Decided; Mar. 30, 2007, Filed).

account in evaluating fossil fuel projects. OPIC agreed to aim to reduce GHG emissions associated with projects by 20 percent over ten years. Both agencies committed to increasing their financing for renewable energy by US$500million. The undertakings, binding only on the parties to the settlement, have been described by American attorneys as "unquestionably substantial" that "go beyond any prior settlement by the federal government."[192]

Although the Defendants in that case were public finance entities, the litigation and subsequent settlement create a dangerous precedent for private finance actors, especially those operating in the United States. Providing an analysis of the extent of corporate reliance on external finance, Richardson writes that "financial sponsorship is intimately etched into the "cause" of corporate activities harming the environment."[193] If this notion of cause can be translated into a legal concept of causation then private finance actors are an increasingly obvious target for GHG-related litigation.

Concern for climate change and its impacts on corporate viability is clear. As such, private finance actors that support carbon-intense projects and companies are increasingly vulnerable to credit, investment, reputation and litigation risks. Early-moving actors recognize this connection and "are addressing climate change as a risk management issue as they would other credit, operational and reputation issues."[194]

[2]   *Investing, Lending & Financing*

Flowing on from the private finance sector's role in risk assessment and mitigation is its role as investor, financier and lender.

As noted earlier, there has been increasing pressure from NGOs and civil society groups for divestment of coal, oil and gas from portfolios of institutional investors, which includes large insurers and pension funds. In a move that signals longer-term impacts, London Stock Exchange-owned FTSE Group together with Blackrock (the world's largest fund manager) launched the new FTSE Developed ex Fossil Fuels Index Series on April 29, 2014 in order to track the performance of non-fossil fuel stocks. The index specifically excludes companies involved in: (a) ownership, exploration and/or extraction of fossil fuels; (b) investment in oil, gas, coal mining, exploration and production; and (c) deriving revenues from such activities or with proven reserve exposure to fossil fuels. FTSE explained that the indices are necessary for market participants "increasingly looking to manage carbon exposure in their investments, and reduce write-off or downward revaluation risks associated with stranded assets," which it defined as "fossil fuels deposits, including oil, gas and coal, that must remain unburned or in the ground in order for the world to avoid the worst impacts of climate

---

192. Arnold & Porter LLP, "Client AdvisoryMarch 2009: Settlement by Federal Agencies Accepts Obligations tro Take Global Warming into Account in Supporting Overseas Projects," *arnold-porter.com* (March 2009), 4, http://www.arnoldporter.com/resources/documents/CA_Settle mentByFederalAgenciesAcceptsObligationsToTakeGlobal_032309.pdf.
193. Richardson, *supra* n. 116, at 346.
194. D.G. Cogan, *Corporate Governance and Climate Change: The Banking Sector* (Ceres, San Francisco, 2008), 2.

change."[195] More specifically, FTSE chief executive, Mr. Mark Makepeace, noted an increasing demand from investors "for indices that reflect their overall business culture and values."[196]

Moreover, institutional investors as shareholders have a key role to play in the management of any company in which they own stock. Shareholder engagement or activism arises as a partial solution to the agency problem between company owners and managers whereby the potential for misaligned interests can result in managers not acting in the best interests of owners. Thus, rather than being an exercise in micromanagement, shareholder engagement focuses on areas that can increase the long-term value of firms or address specific operational concerns.[197] Recent examples of shareholder activism have focused on executive remuneration and management liability for stock downturn during the GFC, with variable efficacy. For example, Citigroup shareholders in the United States unsuccessfully sued bank directors in 2007-2009 for financial losses suffered as a result of decisions to invest in sub-prime mortgage-related securities.[198] In contrast, in 2003 the shareholders of GlaxoSmith-Kline in the UK successfully voted down a proposed payment of £22million to outgoing CEO Mr. Jean-Paul Garnier.

Until recently, institutional shareholders had not systematically exerted influence regarding socio-environmental corporate concerns for a variety of reasons, including: lack of knowledge and cohesion on ethical issues; monitoring costs; collective action barriers; and aversion to increased political visibility.[199] Yet with the growing concern over climate risk, institutional investor activism is growing in the ESG space. One example is the aforementioned Boston Common Asset Management request to PNC Financial Services Group Inc. for disclosure on investment-related climate change risk.[200] Another active shareholder is the California Public Employees' Retirement System (CalPERS), which often uses the media to name and shame recalcitrant companies and has even excluded entire developing countries in the face of systemic problems.[201] Its corporate influence is significant due to the sheer scale of its investment portfolio and the potential negative impacts of divestiture.[202]

---

195. FTSE, *FTSE Factsheet: FTSE Developed ex Fossil Fuels Index Series* (May 30, 2014), 1. http:// www. ftse. com/ Indices/ FTSE_ Developed_ ex_ Fossil_ Fuels_ Index_ Series/ index. jsp.
196. A.Macdonald-Smith, *supra* n. 174.
197. R. Barker, "Ownership Structure and Shareholder Engagement: Reflections on the Role of Institutional Shareholders in the Financial Crisis," in W. Sunn et al. (eds.) *Corporate Governance and the Global Financial Crisis: International Perspectives* (Cambridge University Press: Cambridge MA, 2011), 144, 151-2.
198. *re Citigroup Inc. S'holder Derivative Litig.*, 964 A.2d 106 (Del. Ch. 2009), 124; *Gantler v. Stephens* 965 A2d 695, 705-06 (Del 2009). For detailed discussion of these cases see Chapter 5 *infra*.
199. See e.g. W.T. Proffitt and A. Spicer, "Shaping the Shareholder Activism Agenda: Institutional Investors and Global Social Issues," 2006 4(2) *Strategic Organization* 165; J. Parkinson, *Corporate Power and Responsibility: Issues in the Theory of Company Law* (Clarendon Press: Gloucestershire, 1995), 168-69; P. Myners, *Institutional Investment in the United Kingdom: A Review* (H.M. Treasury: London, 2001); B.S. Black, "Shareholder Activism and Corporate Governance in the United States," in P. Newman (ed.), *The New Palgrave Dictionary of Economics and the Law* (Palgrave, 1998).
200. Benton, *supra* n. 178.
201. Hebb, *supra* n. 115.
202. Kidwell et al., *supra* n. 76, at 729.

Regarding lending and financing, private finance actors (particularly banks) decide whether or not to finance major projects or to lend to corporate clients and, if so, on what terms. This aspect of financial practice embodies a gate-keeper function regarding which projects "get the green light." Financiers and lenders are in a position to decide whether to provide full finance, conditional finance or an outright refusal to finance major projects. These projects might facilitate GHG pollution or, conversely, innovative solutions in renewable energy, low-carbon technology, energy efficiency and/or green infrastructure. Thus, for better or worse, finance decisions impact upon the timely transition to a low-carbon economy.

Specifically, the relevance of corporate loans and project finance to climate change is recognized in three international voluntary codes: the Equator Principles, the Carbon Principles, and the Climate Principles.

The Equator Principles are the longest-standing of the three codes. They were created in 2003, reviewed and updated in 2006, then subsequently reviewed and updated in June 2013 to form "EPIII," which came into force on January 1, 2014.[203] The Equator Principles currently cover project finance only, which comprises approximately 2 percent of an average bank's financing activities. The ambit of the Equator Principles may be expanded in future given that the underlying IFC Social and Environmental Performance Standard was expanded in May 2011 to cover financial advisory products.[204] Nonetheless, project finance is an important target for climate change efforts as it is a frequent modality for energy deals in coal, electricity, oil and gas.[205] The Equator Principles require that financiers screen all projects worth more than US$10million (down from US$50million in 2003) for potential ESG impacts in accordance with standards set by the IFC. The Equator Principles were designed to have effect primarily in developing countries where competent environmental regulation may be lacking; however some banks apply the principles to all projects regardless of location or monetary size. Signatories pledge to provide loans only to borrowers who conform to the principles. Screening leads to a classification into one of three environmental risk categories: high (A), medium (B) or low (C). Higher-risk borrowers must: complete a comprehensive Environmental and Social Impact Assessment and disclose summaries of those Assessments; disclose details of GHG emissions for intensive projects; and comply with a risk mitigating Action Plan, or be declared in default. Signatories to EPIII are also required to assist clients to quantify, monitor and manage their CO_2 emissions if the project is CO_2 intensive (over 100,000 tons CO_2 equivalent per year). Over 70 financial institutions around the world (predominantly banks) have adopted the Equator Principles, covering over 95 percent of the global project finance market.[206]

The Carbon Principles were adopted in February 2008 by six financial institutions (Bank of America, Citi, Credit Suisse, JPMorgan Chase, Morgan Stanley and Wells

---

203. Equator Principles Association, *Equator Principles III* (2013), http://www.equator-principles. com/resources/equator_principles_III.pdf (viewed Feb. 5, 2013).
204. See e.g. Conley and Williams, *supra* n. 167, at 546.
205. BankTrack, *Close the Gap: Benchmarking Credit Policies of International Banks* (Banktrack: Nijmegen, Netherlands, 2010), 103.
206. Richardson, *supra* n. 107, at 90.

Fargo) in consultation with power companies and environmental groups.[207] The Carbon Principles set procedural standards for advisers and lenders to power companies in the United States. They were created to address potential investment risks due to uncertain regional and national climate change policy.[208] Signatories encourage their clients to invest in energy efficiency initiatives, renewable energy and low-carbon technologies, and advocate for regulation that encourages such investment. Signatories also acknowledge that "conventional energy generation" will be required in addition to renewable and efficient energy and they agree to encourage government policy that facilitates carbon capture and storage. Despite initial expectations that other financial institutions would adopt the Carbon Principles, the number of signatories remains at the original six.

The Climate Principles were finalized in December 2008 as a product of the Climate Group, an NGO, in dialogue with approximately 20 financial institutions. Seven financial organizations originally adopted these Principles, namely BNP Paribas, Credit Agricole, F&C Asset Management, HSBC, Standard Chartered, Swiss Re and Munich Re. The Climate Principles were the broadest of the three voluntary codes, applying to all financial services including insurance, corporate and investment banking, project finance and asset management.[209] They guided signatories on areas such as: advising clients on climate risks and GHG emissions reduction technologies; requesting clients quantify and disclose GHG emissions associated with high-emitting projects (over 100,000 tons of CO_2); financing low-carbon technologies and projects; and assessing the climate consequences of investments.[210] However, by 2012 the Climate Principles had become defunct with several signatories pulling out on the basis that they could do similar engagement elsewhere.[211] The Climate Group no longer has a specific finance work stream.

A detailed analysis of the efficacy of these voluntary codes to achieve environmental and sustainability objectives is beyond the scope and intent of this book. Suffice to say that it is a contested field amongst NGOs, academics, and industry participants.[212] What is relevant here is that none of these soft law instruments stipulate an

---

207. The Carbon Principles, "Carbon Principles Statement of Intent," *carbonprinciples.org*, http://www.carbonprinciples.org/documents/Carbon%20Principles.pdf (accessed Jan. 15, 2014).
208. The Carbon Principles, "Press Release: Leading Wall Street Banks Establish The Carbon Principles: Guidelines to Strengthen Environmental and Economic Risk Management in the Financing and Construction of Electricity Generation," *carbonprinciples.org* (Feb. 4, 2008).
209. The Climate Group, "The Climate Principles: A Framework for the Finance Sector," *climate-group.org*, http://www.theclimategroup.org/_assets/files/The-Climate-Principles-English.pdf, last updated Dec. 2, 2008 (accessed Jan. 15, 2014).
210. *Ibid.*, commitment 2.0.
211. As told to the author by an NGO CEO on Mar. 3, 2012.
212. See eg: Conley and Williams, *supra* n. 167; O'Sullivan and O'Dwyer, *supra* n. 166; R. Macve and X. Chen, "The 'Equator Principles': A Success for Voluntary Codes?" (2010) 23 *Accounting, Auditing & Accountability Journal* 890; K. Miles, "Targeting Financiers: Can Voluntary Codes of Conduct for the Investment and Financing Sectors Achieve Environmental and Sustainability Objectives?," in K. Deketelaere, J.E. Milne, L.A. Kreiser and H. Ashiabor (eds), *Critical Issues in Environmental Taxation – Volume V* (Oxford University Press, London, 2008); B.J. Richardson, "The EPs: The Voluntary Approach to Environmentally Sustainable Finance" (2005) 14(11) *European Environmental Law Review* 280; P. Watchman and C. July, "A New Environment" (2006) (Feb) *Legal Week: Insolvency, Banking and Finance* 24; BankTrack, *Bold Steps*

absolute prohibition on financial support for high GHG polluting projects. NGOs had advocated for "categorical exclusion of the financing of climate destructive projects such as coal mining, coal power plants and oil exploration projects" from EPIII;[213] but no such prohibition was included in the final text. Thus, debate exists as to whether private finance actors *should* refuse to support GHG intense companies and projects.

Advocates of non-funding, non-investment and divestment argue that the socio-environmental risks to humanity of coal, oil and gas projects cannot be mitigated by corporate proponents or financiers and therefore ought not to be supported to any degree.[214] Some commentators go further to assert a positive obligation on finance actors to facilitate broader beneficial change. For example, Flavin writes that insurance companies and investors "could spur the development of a less threatening energy system" by "dump[ing] some of their stocks in oil and coal companies, or actively invest[ing] some of their funds in new, less carbon-intensive energy technologies."[215]

While divestment arguments might seem radical, they are not new. Examples include the South African apartheid divestment campaign in the 1970s-80s;[216] and the elimination of tobacco companies from the investment portfolios of Harvard University, the City University of New York, and a number of American health insurers in 1990. In this latter case, tobacco stock had comprised approximately US$3.5million of Harvard University's $60million portfolio. Leading up to the divestment, students at the Harvard public health school had pressured Harvard management through radio advertisements that chided then-Harvard President Mr. Derek Bok for failing to take leadership on the issue. Subsequent to divestment, Mr. Bok stated that Harvard's decision "was motivated by a desire not to be associated as a shareholder with companies engaged in significant sales of products that create a substantial and unjustified risk of harm to other human beings."[217] Analogous arguments are being made by NGOs and civil society groups to prompt fossil fuel divestiture.

In contrast to investors and insurers, banks provide a good case study for counter-arguments on this issue. Watchman surmises that banks take an environmental "mitigation of harm" as opposed to a "do no harm" approach to their lending activities, whereby they attempt to reduce or manage harm as opposed to avoiding

---

Forward: Towards Equator Principles that Deliver to People and the Planet (BankTrack: Nijmegen, Netherlands, 2010); BankTrack (2009) *Meek Principles for a Tough Climate: Why the Carbon and Climate Principles Will Not Stop the Melting of the Ice* (Banktrack: Nijmegen, Netherlands, 2009); Rainforest Action Network (RAN), *The Principle Matter: Banks, Climate & the Carbon Principles* (RAN: San Francisco, 2011). See also: PricewaterhouseCoopers LLP, *Climate Principles Progress Review* (The Climate Group: London, 2010).

213. J. Frijns, "Equator Principles III better be good: Launch New Equator Principles Again Postponed; Expectations Rise on What New Principles Will Deliver" *BankTrack.org* (Mar. 29, 2012).

214. *Ibid.*

215. Flavin, *supra* n. 102.

216. See e.g., D. Pederson, "Divestment of Securities in Companies Doing Business in South Africa: Conflicting Moral and Legal Imperatives?" (1986) 1 *Inquiry and Analysis* 1; J.C. Dobris, "Argument in Favour of fiduciary divestment of "South African" Securities" (1986) 65 *Nebraska Law Review* 209.

217. T. Lewin, "Harvard and CUNY Shedding Stocks in Tobacco," *New York Times Archives* (May 24, 1990), http://www.nytimes.com/1990/05/24/us/harvard-and-cuny-shedding-stocks-in-tobacco.html.

projects *per se.*[218] Coulson debates this generalized approach.[219] Some banks do veto (or, in bank parlance, "redline") certain investments, such as the ABN AMRO Forest Investment Policy which went further than IFC policies to exclude loans for projects in old-growth forest; and Credit Suisse's outright veto of mountain top removal coal mining, which is a legal but highly controversial method of mining coal as it creates irreparable socio-environmental harm by "blowing the tops off of mountains and dumping the debris into nearby streams and valleys."[220] Other banks veto investments in certain industries only, such as nuclear energy. Yet other banks such as Barclays, HSBC and Westpac are actively opposed to redlining and purport instead to work with potential borrowers to improve their environmental performance and thus reduce harm.

Some argue that banks baulk at denying loans on environmental criteria because to do so would make them "surrogate environmental regulators."[221] This concern is evidenced in a 2007 UNEPFI report:

> Calls are being made in some quarters for financial institutions to act more aggressively by simply cutting off finance to high carbon emission activities. However these calls do not account for the broader economic and business context in which financial institutions operate. They are in effect asking the financial institutions to take a hand in determining economic policy. Financial institutions make the investment decisions, but it is national governments that set the rules for the markets in which they invest and operate.[222]

Of course, this discussion of "to fund or not to fund" goes beyond a governance legitimacy conundrum. The fact is that the fossil fuel sector is still a lucrative investment and clientele base for private finance actors; it is not currently in their business interests to redline this sector and avoid it entirely.[223] Indeed, some finance actors that claim to be leading on finance and investment for clean technology and renewable energy also publicly state that they will continue to work with fossil fuel and GHG intense clients. For example, Westpac has declared that: "[o]ver the medium term fossil fuel energy generation will continue to play a major role in the economies where we operate ... [so we] will continue to support clients in the fossil fuel based energy sector for the immediate term" even while it simultaneously commits to facilitating renewable and efficient energy.[224] This rather instrumentalist approach is consistent with Richardson's contention that "the financial industry has viewed itself as mostly just passively tied to its clients and borrowers, and not meant to meddle in their policies

---

218. Watchman, *supra* n. 164, at 36.
219. A.B. Coulson, "How Should Banks Govern the Environment? Challenging the Construction of Action Verses Veto" (2009) 18 *Business Strategy and the Environment* 149, 154.
220. Rainforest Action Network, "Wall Street Backs Away From Mountaintop Removal Coal Mining: Top 4 US Banks Curb Loans for Destructive Practice; Cut Financing for Massey Energy," *ran.org* (Aug. 11, 2010).
221. Richardson, *supra* n. 116, at 353.
222. UNEPFI, *supra* n. 10, at 12.
223. Richardson, *supra* n. 107, at 55.
224. Westpac, *Westpac Climate Change Position Statement: Financing the Transition to a Low Carbon Economy* (Westpac: Sydney, undated), 10.

and operational practices unless [they] jeopardize the industry's financial interests."[225] He conjectures that the capacity of responsible investing to generate social returns is presently rather limited.[226]

Nonetheless, some finance actors *have* decided to veto fossil fuel projects. Examples include: Deutsche Bank's refusal to finance the Abbot Point coal port; Goldman Sachs' equity divestment from Carrix; Bank of America's decision to reduce the GHG emissions rate of its energy-lending portfolio; and separate decisions by Bank of America, Citi, JPMorgan Chase, Wells Fargo, Credit Suisse and Morgan Stanley to limit their relationships with companies that practice mountain top removal coal mining.

Yet it is unclear why decisions to veto certain projects and clients are taken by private finance actors. Hence the need for qualitative empirical research in order to pinpoint and understand subjective motivations: is it driven by senior management, business case factors, and/or external pressure such as NGO campaigns? And how can those motivations be harnessed by policy-makers?

[3]   *New Market Entry and Innovative Product Design*

Many of the aforementioned challenges can be transformed into opportunities. Indeed, the finance sector is aptly placed to exploit climate-related opportunities in their roles as creators of innovative insurance products, brokers in global carbon markets, and as financiers and investors in renewable energy and clean technology markets.

Regarding insurance activities, the Geneva Association has noted that "the only way to ensure that ambiguous risks remain insurable is to promote risk mitigation today" by disseminating high-quality risk information and supporting climate-related measures through innovative product design.[227] For example, Aviva, a leading UK insurer, offers insurance products that "support commercial development of low-carbon products including wind farms and biomass energy conversion plants."[228] Similarly, innovative insurance products can assist climate adaptation through, for example, insurance-linked securities like natural catastrophe bonds, and parametric insurance that can replace lost tax revenue or fund increased insurance premiums due to unprecedented wild weather.[229] Not only do these products facilitate efforts in public and private sectors and also communities but, in so doing, they enhance the positive reputations of insurers and reinsurers.

Regarding carbon markets, industry reports have estimated that carbon trading could reach US$3trillion by 2020 and eventually total US$10trillion a year, which

---

225. Richardson, *supra* n. 107, at 45.
226. *Ibid.*, 46-100.
227. Niehörster, *supra* n. 89, at 21. See also R.E.T. Ward, C. Herweijer, N. Patmore and R. Muir-Wood, "The role Of Insurers in Promoting Adaptation to the Impacts of Climate Change" (2008) 33(1) *The Geneva Papers on Risk and Insurance – Issues and Practice*, 133, https://www.genevaassociation.org/media/246469/ga2008_gp33%281%29_ward_et_al.pdf.
228. Aviva, *Annual Report and Accounts 2010* (Aviva: London, 2011), 78.
229. Swiss Re, *Closing the Financial Gap: New partnerships between the public and private sectors to finance disaster risks* (Swiss Reinsurance Company Ltd: Zurich, 2011).

would make it the largest commodity market in the world.[230] Many large private finance actors are established in carbon markets and responsible for much of the traded volume within the EU ETS.[231] Despite the lack of international agreement on climate change, there remains an industry expectation of regionally linked and national carbon markets[232] with profitable brokering and arbitrage opportunities.

Moreover, a variety of carbon funds have been created by and for the private finance sector together with public entities. For example, the *Post-2012 Carbon Credit Fund* was an initiative developed by the KfW, European Investment Bank, Caisse des Dépôts, Instituto de Crédito Oficial and the Nordic Investment Bank with a budget of €125million to promote projects generating carbon credits beyond the Kyoto Protocol.[233] Importantly, these sorts of funds generate brokering and capital investment opportunities for private finance actors. Indeed, from 2005 to 2010 the cumulative value of origination and trade in Clean Development Mechanism primary and secondary carbon credits was US$95billion.[234]

A concurrent global market in renewables and clean technology similarly presents lucrative opportunities for private finance actors. For example, the World Bank has estimated that returns on exchange-traded funds in clean energy in 2013 reached up to 140 percent.[235] Also in 2013, the WilderHill New Energy Global Innovation Index (NEX), which tracks the performance of 102 clean energy stocks worldwide, rose almost 54 percent, which exceeded wider market indices such as the S&P500 and the FTSE100; and equity raisings by quoted clean energy companies more than doubled.[236] In January 2014 the NEX stood at its highest level since 2011.

According to Ethical Markets Media, private sector green investment for the six years from 2007 to 2013 totaled US$5.3trillion globally in the areas of Renewable Energy, Green Construction, Energy Efficiency, Corporate Green R&D, Cleantech and Water.[237] Specifically, US$2.6trillion was privately invested and committed in renewable energy (excluding biofuels), which comprises private technology development, equipment manufacturing, project finance, and M&A activity. For the same period, private investment into energy efficiency, including Smartgrid, exceeded

---

230. Carr, *supra* n. 8.
231. Carbon Disclosure Project (CDP), *supra* n. 84, at 92.
232. *Ibid.* See also Deutsche Bank, *Investing in Climate Change 2010: A Strategic Asset Allocation Perspective* (Deutsche Bank: New York, 2010), 10.
233. UNEP, *Bilateral Finance Institutions and Climate Change: A Mapping of 2009 Climate Financial Flows to Developing Countries* (United Nations Environment Programme: Nairobi, 2010), 14, http://www.unep.org/pdf/dtie/BilateralFinanceInstitutionsCC.pdf.
234. The World Bank, *State and Trends of the Carbon Market Report 2011*, http://www.ifc.org/wp s/wcm/connect/Topics_Ext_Content/IFC_External_Corporate_Site/CB_Home/Mobilizing + C limate + Finance/CarbonFinance/.
235. The World Bank, "World Bank Group President: This is the Year of Climate Action, *world-bank.org* (Jan. 23, 2014), http://www.worldbank.org/en/news/feature/2014/01/23/davos-w orld-bank-president-carbon-pricing.
236. Bloomberg New Energy Finance, "Clean Energy Investment Falls for Second Year," *bnef.com* (Jan. 15, 2014), http://about.bnef.com/press-releases/clean-energy-investment-falls-for-seco nd-year/.
237. Henderson et al., *supra* n. 9.

US$1.1trillion; green construction investment exceeded US$5billion; and clean technology investment (excluding coal carbon sequestration or CCS) exceeded $US2.5billion.[238] These are aggregate figures for private sector investment only and do not include investments or subsidies from national governments.

Some industry commentators describe these developments as an "important milestone in measuring the increasing economic viability of the CleanTech universe"[239] and "tangible evidence ... [that] momentum is building for a green economic transition."[240]

Certainly, renewable energy and clean tech markets create opportunity for "large investment banks that have the financial firepower to unleash mature [green] technologies globally."[241] This role is not new for private finance actors, which provide critical financial support where governments lack:

> As the financial heft of national governments hit their limits, we'll need the big financial machines – which funded the Internet, cellular networks, highways, bridges, water systems, and mass transit systems – to make clean technology part of the global fabric.[242]

Accordingly, there is a clear role for private finance actors to assist our transition to a low-carbon economy, and it is a role that they have played in previous transitions. Moreover, doing so presents them with new opportunities including reputation enhancement and profit and value increase through new lucrative carbon and renewable energy markets.

Early-moving finance actors are already active in these markets. Yet the finance sector as a whole faces "an immense but as yet largely untapped opportunity to enter new markets and develop more efficient and environmentally sound industries."[243] The question this book seeks to address is what is required for that to happen?

§2.04 BANKS AS CORPORATE CHANGE AGENTS?

Importantly, through their own efforts as advisors, risk assessors, lenders and investors, private finance actors can facilitate the "greening" of other corporate actors. Indeed, the influential capacity of the finance sector generally is recognized in environmental-financial soft law instruments. For example, the aim of the Climate Principles included a "common global standard of best practice...to assist the [finance] sector in supporting its clients and stakeholders in managing their own impacts."[244] That the industry is self-aware of its influence is illustrated by Westpac's statement that:

---

238. *Ibid.*
239. Per Stuart Valentine quoted in Editorial, "Private Investment in Green Business Tops $1.6T Since 2007," *GreenBiz.com* (Aug. 4, 2010).
240. *Ibid.*, per Timothy Nash.
241. J. Makower, "Financing Our Cleantech Future," *GreenBiz.com* (Jan. 18, 2010).
242. *Ibid.*
243. Cogan, *supra* n. 194, at ii.
244. The Climate Group, *supra* n. 209, at preamble.

> The role of the leader is not just to look at themselves but to also look outwardly
> at who they can influence. As a bank we are in a good position to do that because
> nearly all businesses have a banking relationship.[245]

To a large extent, changes that private finance actors make to their own behaviors will impact upon clients and other businesses, for example, through risk assessments and finance/lending conditions that account for carbon intensity and CO_2 mitigation strategies.

This is where the banking industry comes to the fore. Conley and Williams note that when banks "operat[e] with their own risk" in carrying out ESG risk assessments, then "they can exercise a moderating effect on the most destructive aspects of unconstrained finance and development."[246] Similarly, Lundgren and Catasús observe that:

> Banks have large contact nets on all levels of society . . . [and] may intentionally,
> or unintentionally, affect other actors . . . in terms of attitudes toward the
> environment and environmentally responsible management.[247]

Clearly then, private sector banks are influential even when they are not engaged as lenders or creditors. They set ESG standards for their suppliers and provide advice to corporate clients as well as governments and local communities on climate change mitigation strategies. These advisory and benchmarking activities have concentric flow-on effects to a range of actors outside the direct purview of banks. In short, due to their unique business practices and their influential role on other corporations, banks have what I term "network change potential" to facilitate climate change mitigation.

Naturally there are limitations to banks' influence in light of increasing competition in the global financial economy. Specifically, multinational corporations have a growing capacity to self-finance, which minimizes influence from bank lenders. Moreover, some non-bank financial intermediaries, such as money market funds, hedge funds, and structured investment vehicles, provide credit and other services similar to traditional commercial banks. This phenomenon is known as the shadow banking system. Nonetheless, private sector banks can and do still operate within it. For example, shadow banks can facilitate credit in financial markets by becoming part of a chain that involves mainstream banks; and investment banks may conduct much of their business in the shadow banking system, including the use of hedge funds and special purpose entities to enhance client returns on investment.

The influential capacity or "network change potential" of banks is little considered to date and worth focusing on. It is evidenced in their roles as creditors, investors, advisors and heads of supply chains.

---

245. Per Graham Paterson quoted in A. De Lore, "Leaders of the Pack Unite to Inspire Others: A
     Network of High Achievers is Showing the Way," *The Age: Special Report on Corporate
     Responsibility Index* (May 20, 2008), 2.
246. Conley and Williams, *supra* n. 167, at 571.
247. Lundgren and Catasús, *supra* n. 164, 189.

[A]   Banks as Creditors and Investors

Traditional corporate governance theory in Anglo-American jurisdictions contends that banks do not have control over borrowers' management other than at times of distress. Standard "shareholder influence" accounts of corporate governance conceive of creditors as outsiders to firm management while shareholders are active monitors of the firm.[248] Similarly in these jurisdictions, where large equity stakes are not permitted, the traditional view is that a bank's influence on investee corporate governance is relatively limited.[249]

In contradistinction, banks in Germany and Japan hold or control major stakes in the largest companies and exert direct influence over their corporate governance.[250] Under the German *Hausbank* phenomenon, corporate loans to a firm are secured by the bank holding substantial shares.[251] Similarly, German banks retain a strong presence on and over corporate boards even as German corporate governance and financing may be shifting more toward a UK or American model.[252]

Accounts vary of why differences exist in corporate governance structures between Europe and Japan on the one hand and the United States, UK and Australia on the other, but the most prominent commentator is Mark Roe who uses an economic model to highlight differing cultures, political histories, and paths of economic development.[253] Moreover, David Skeel Jr. and also Christopher Bruner seek to explain commonalities in American and UK corporate governance through shared national common law histories, a lack of controlling shareholders in large corporations, and a more shareholder-oriented governance approach.[254] For these commentators, financial institutions have long been prevented from actively controlling Anglo-American companies due to "legal prohibitions on their stock ownership and by strong norms against their exerting control."[255]

However, empirical evidence has revealed that Anglo-American banks routinely and pervasively influence firm managerial decision-making in companies at much earlier points in the creditor-debtor relationship than borrowers' distress.[256] Frederick

---

248. Eg: D.G. Baird and T. Henderson, "Other People's Money" (2008) 60 *Stanford Law Review* 1309, 1343; D.G. Baird and R.K. Rasmussen, "Private Debt and the Missing Lever of Corporate Governance" (2006) 154(5) *University of Pennsylvania Law Review* 1209, 1215.
249. Schmidheiny and Zorraquin, *supra* n. 66, at 101. See also C. Mallin, A. Mullineux, and C. Wihlborg "The Financial Sector and Corporate Governance: the UK case" (2005) 13(4) *Corporate Governance* 532, 536.
250. M.J. Roe, "Some Differences in Corporate Structure in Germany, Japan, and the United States" (1993) 102 *Yale Law Journal* 1927.
251. A. Onetti and A. Pisoni, "Ownership and Control in Germany: Do Cross-Shareholdings Reflect Bank Control on Large Companies?" (2009) 6(4) *Corporate Ownership and Control* 54.
252. *Ibid.*, 59-60, 73.
253. Roe, *supra* n. 250; M.J. Roe, *Strong Managers, Weak Owners: The Political Roots of American Corporate Finance* (Princeton University Press: Philadelphia, 1994).
254. D.A. Skeel, Jr. "Corporate Governance and Social Welfare in the Common Law World" (2014) 92 *Texas Law Review* 973; C. Bruner, *Corporate Governance in the Common-Law World* (Cambridge University Press: Cambridge MA, 2013).
255. Skeel, *supra* n. 254, at 48-49.
256. F. Tung, "Leverage in the Board Room: The Unsung Influence of Private Lenders in Corporate Governance" (2009) 57 *UCLA Law Review* 115, 118-119.

Tung terms this influence "lender governance," contending that it rivals conventional corporate governance mechanisms and is comparable to the influence of boards of directors. He demonstrates that bank creditors have significant influence over managerial decisions and monitor financing, investment, and operational decisions of investee companies. Banks not only constrain managerial decisions but also dictate them, even the hiring and firing of a firm's chief executive and board.[257] In so doing, private lender power has been described as the "missing lever" of corporate governance.[258]

Similarly, empirical evidence has been adduced regarding the influence of banks as investors. John Holland has evidenced that Anglo-American financial institutions exercise "private corporate governance influence" over investee companies and that they do so even when the company is performing well.[259] This influence affects a range of internal company matters from corporate governance in relation to board composition and top management recruitment to accountability issues regarding wealth creation.[260]

These empirical studies show that banks as creditors and investors are motivated to influence the corporate governance of their companies to improve financial performance. The implication is that banks can exercise influence regarding GHG emission mitigation and management strategies if to do so were to affect a company's bottom line in a carbon-constrained economy. Within the lifetime of loans granted today, climate change will have a profound impact on company operations and investments. The fact that climate change will affect the pecuniary interests of a bank as creditor, lender and investor as evidenced throughout this chapter could motivate heightened involvement by them in the climate-related corporate governance of other companies.

In this way, bank influence could precipitate changes to corporate carbon accounting, mitigation and management with the potential for wide-spread corporate GHG emissions reductions. The potential for widespread change is due to the pervasiveness of the banking industry: private lending is the single largest source of external financing for public corporations being larger than public debt and equity combined,[261] and private credit agreements are maintained by 80 percent of public companies.[262]

This influential capacity of banks is also seen in their roles as advisers and heads of supply chains.

[B]   Banks as Advisers and Heads of Supply Chains

Banks provide advice on issues within their expertise and of concern to their stakeholders, which include clients, local communities and governments. Climate change

---

257. *Ibid.*, 119-120.
258. Baird and Rasmussen, *supra* n. 248, at 1215.
259. J. Holland, "Financial Institutions, Intangibles and Corporate Governance" (2001) 14(4) *Accounting, Auditing & Accountability Journal* 497.
260. *Ibid.*, 514.
261. Tung, *supra* n. 256, at 121.
262. G. Nini, D.C. Smith and A. Sufi, "Creditor control rights and firm investment policy" (2009) 92 *Journal of Financial Economics* 400.

has become a relevant issue for research and dissemination by banks. Some banks acknowledge that part of their role is to help the transmutation of corporate clients to become low-carbon emitters and to succeed in a carbon-constrained economy. For example, Deutsche Bank has stated that "thought leadership requires us to strengthen awareness of climate change issues with all our clients and advise them on strategic opportunities and challenges."[263] This advisory role is discursive and involves a partnering or mentoring mentality through which banks can influence corporates in a more subtle way than by setting benchmark standards or imposing finance conditions.

For some banks, their advisory capacity extends to sharing knowledge with local communities "to further advance ... understanding of potential [climate change] impacts and issues."[264] It can also include dialogue with governments regarding policy formulation on climate change.[265] Similarly, banks may partner with business and/or NGOs to commission and disseminate research that influences policy-makers as well as other corporate actors.

As heads of supply chains, many banks acknowledge that the goods and services supplied to them by other businesses impact upon their own social and environmental performance. As such, some banks have instigated Sustainable Supply Chain Networks (SSCNs), which benchmark social and environmental standards for bank suppliers. The SSCN process is likely to involve a risk assessment, a checklist of compliance with minimum standards, a remedial action plan to remedy any gaps, and a Certificate of Compliance valid for a limited time.[266] For some banks, suppliers must be compliant with their SSCN code to be eligible for a tender or to remain or become a bank supplier.[267]

SSCN standards may include reducing carbon emissions of a car fleet or, given that IT operations contribute significantly to corporate environmental footprints, selecting highly-efficient IT hardware.[268] Minimum-standard compliance is not only significant for a bank's emissions but also those of other corporations. As stated by ANZ: "ANZ works with more than 15,000 suppliers around the world...The Code aims to improve the social and environmental performance of our supply chain."[269] Similar to their advisory role, banks with SSCN codes will often work together with suppliers to assist their improvement of SSCN performance and reporting. Importantly, some banks stipulate that their suppliers must communicate these standards onward to their own corporate suppliers and customers.[270]

---

263. Deutsche Bank, *supra* n. 141, at 4.
264. Westpac, *supra* n. 224, at 12.
265. Richardson, *supra* n. 107, at 100.
266. E.g. Westpac, *Westpac's Sustainable Supply Chain Management Code of Conduct*, August 2014 (Westpac, Sydney, 2014).
267. See e.g. ANZ, *A Global Approach to Responsible Sourcing* (ANZ: Melbourne, undated), 1, http://www.anz.com.au/resources/d/5/d5c64f0047508cdab7c2bff55bff9ae9/ANZ_Global_A pproach_Sourcing.pdf?MOD = AJPERES (accessed Aug. 30, 2014).
268. E.g. Westpac, *Pact: Sustainability and Community News from Westpac* (Issue 4, May 2008), 2.
269. ANZ, *supra* n. 267, 1.
270. See e.g. Westpac, *Embracing The Corporate Journey Into Sustainability: Stories From Westpac's Supply Chain* (Westpac: Sydney, 2008).

Accordingly, banks have influence over the climate-related practices and policies of other corporate actors in their capacity as advisers and heads of supply chains. Consequently, private sector banks are an important unit of study regarding the move to a low-carbon global economy. Through the banking industry's practices and its influence on other corporate actors, there is the possibility of more than just linear flow-on effects from banks to clients and suppliers. There is also the potential for wide-spread change through client and supplier networks as more businesses set benchmarks, not only for their own corporate governance climate strategies but also for their clients and suppliers, which then set such standards for their own clients and suppliers, and so on in an ever-widening web of corporate change.[271]

This "network change potential" of banks – and other private finance actors – is important given the uncertain governmental regulatory environment at both international and domestic levels. Nonetheless, early-moving banks are *voluntarily* adopting and outputting climate-related practices and products. This raises questions as to why they are doing this when not legally compelled to do so; and what might be the implications for other private finance actors.

---

271. Bowman, *supra* n. 143, at 464.

CHAPTER 3

Why Do Companies Go Green?

§3.01 OVERVIEW

Two points are evident from Chapters 1 and 2. First, corporate climate finance is an essential part of the solution to dangerous global warming, and secondly, early moving private finance actors are voluntarily entering climate-related markets and adopting climate-related products and practices.

The key question is *why* are they making these changes voluntarily? Through an investigation of this question we can better understand the drivers for and barriers to climate-related uptake, the potential implications of these motivations for mainstreaming "enlightened" practices in the sector more broadly, and likely outcomes for the timely move to a low-carbon economy.

The next three chapters address that question from both theoretical and empirical perspectives. Chapters 4 and 5 present the case study data and analysis on the specific issues of levers and limits to voluntary corporate greening. For necessary background prior to the empirical study, this chapter provides a review of theoretical approaches to why companies go green.

§3.02 THEORETICAL FRAMEWORK

This chapter conceptualizes theory using three lenses of focus: micro (intra-organizational), meso (inter-organizational) and macro (socio-cultural). All three levels can provide insights about the internal and external influences on corporate environmental behavior:

- First, the risks and opportunities identified in the previous chapter suggest that a meso-level business case drives climate-related action by early moving private finance actors. As such, the largest portion of this chapter, and the focus of this book, concerns the business strategy and the business and society literatures that give particular attention to business case and CSR motivations for voluntary corporate change.

- Second, at a macro level, institutional theory goes beyond business case theorizing to highlight the importance of rules and norms in shaping the behavior of corporate actors. In particular, it focuses on isomorphism and mimetic behavior as institutional processes of corporate change.
- Finally, the chapter reviews theories about micro level managerial decision-making as an internal driver of corporate environmental change.

Naturally, these theoretical lenses or levels are not neatly self-contained or mutually exclusive. Levels are often "nested" within each other and can influence each other,[272] and therefore there is merit in integrating different levels of analysis and theoretical perspectives to enrich our understanding of change processes. For example, individual preferences (micro) are often shaped by socio-cultural influences (macro) and/or firm policy (meso). Therefore, in terms of drivers, there may be multiple levels operating simultaneously. As such, multi-level interactions revealed by the case study data are highlighted where appropriate in subsequent chapters.

§3.03   MESO LEVEL: ORGANIZATIONAL FIELD LEVEL DRIVERS OF CHANGE

For decades now, business and society scholars have endeavored to explicate voluntary corporate change that responds to environmental and/or social issues. This business and society literature postulates that companies can and should engage with CSR in order to be socially or environmentally responsible beyond their legal or economic obligations. A related body of scholarship is evinced in the business strategy literature, which focuses on business case reasons – such as profit enhancement and risk mitigation – for "doing good"; in so doing it investigates whether adopting social or environmental initiatives is "win-win," "win-lose" or variable for the corporate bottom line.

Accordingly, a good place to start is with a review of the business and society literature, with particular attention to ethical versus pragmatic conceptions of CSR, before turning to the business strategy literature on the business case as a driver of corporate greening.

[A]   CSR: Short History and Lack of Conceptual Clarity

The literature reveals CSR as a moving target that defies universal definition.[273] CSR embodies a spectrum of meaning and activity, with ethically-based philanthropic "do-gooding" at one end and "sound business management" at the other. At the heart

---

272. W.R. Scott, *Institutions and Organizations: Foundations for Organisational Science* (Sage Publications, Thousand Oaks, CA, 2008), 193. See also M.A. Delmas and M.W. Toffel, *Institutional Pressures and Organizational Characteristics: Implications for Environmental Strategy* (Harvard Business School Working Paper 11-050, 2010).
273. See e.g. N. Churchill, "Toward a Theory of Social Accounting" (1974) 15(3) *Sloan Management Review* 1; A. Dahlsrud, "How Corporate Social Responsibility is Defined: An Analysis of 37 Definitions" (2008) 15 *Corporate Social Responsibility and Environmental Management* 1.

of the concept lies reference to voluntary and unenforceable action beyond that which is required by law. Yet definitional debate remains as to whether such action must be the product of ethical motivation or if "enlightened self-interest" is sufficient (such as protecting reputation, relieving social pressures, or simply cutting costs).[274]

The notion of CSR dates back to the 1950s, with Bowen's exploration of the obligation of business to make decisions and take actions that are "desirable in terms of the objectives and values of our society."[275] By the 1970s, and against a background of protests against capitalism,[276] publications coalesced around the notion that business had "some form of social responsibility over and above its responsibility to perform economically"[277] and beyond "narrow... technical and legal requirements of the firm."[278] van Oosterhout and Heugens contend that an implicit "justificatory stance" of CSR research began at this time, whereby authors sought to prove "that CSR is *desirable*, either in its own right...or because it is in the long-term economic interest of corporations" and that this normative bias has shaped CSR scholarship and debate ever since.[279]

Since the 1970s, there has been a proliferation of theories, terminology and approaches in the field of CSR scholarship. While a number of authors have made recent attempts to map the terrain and classify CSR theories into coherent groupings,[280] there has been little progress in constructing a systematic conceptual framework. As such, "[n]o consensus yet exists... about an appropriate taxonomy for CSR, let alone its main forms and ends."[281]

This lack of conceptual clarity around CSR is largely due to a lack of definitional consensus on its composition. In their article *Much Ado About Nothing*, van Oosterhout and Heugens assert that "no satisfactory intensional definition of CSR – one that

---

274. G.D. Keim, "Corporate Social Responsibility: An Assessment of the Enlightened Self-Interest Model" (1978) 3(1) *Academy of Management Review* 32; A.B. Carroll, "Corporate Social Responsibility: Evolution of a Definitional Construct" (1999) 38(3) *Business and Society* 268; G. Auld, S. Bernstein and B. Cashore, "The New Corporate Social Responsibility" (2008) 33 *Annual Review of Environment and Resources* 20.1.
275. H.R. Bowen, *Social Responsibilities of the Businessman* (Harper and Row, New York, 1953), 6.
276. D. Melé, "Corporate Social Responsibility Theories," in A Crane, A McWilliams, D Matten, J Moon and DS Siegel (eds) *The Oxford Handbook of Corporate Social Responsibility* (Oxford University Press, Oxford UK, 2008), 50.
277. J. van Oosterhout and P. Heugens, "Much Ado About Nothing: A Conceptual Critique of Corporate Social Responsibility," in A Crane, A McWilliams, D Matten, J Moon, and DS Siegel (eds) *The Oxford Handbook of Corporate Social Responsibility* (Oxford University Press, Oxford UK, 2008), 200. See WC Frederick, "The growing concern over business responsibility" (1960) 2(4) *California Management Review* 54; K. Davis and R.L. Blomstrom, *Business and its Environment* (McGraw Hill: New York, 1966); C Walton, *Corporate Social Responsibilities* (Wadsworth: Belmont CA, 1967).
278. K. Davis, "The Case for and against Business Assumption of Social Responsibilities" (1973) 16(2) *Academy of Management Journal* 312, 312.
279. van Oosterhout and Heugens, *supra* n. 277, at 200 (emphasis added).
280. See e.g. Melé, *supra* n. 276; D. Windsor, "Corporate Social Responsibility: Three Key Approaches" (2006) 43(1) *Journal of Management Studies* 93; E. Garriga and D. Melé, "Corporate Social Responsibility Theories: Mapping the Territory" (2004) 53(1-2) *Journal of Business Ethics* 51; K. Basu and G. Palazzo, "Corporate Social Responsibility: A Process Model of Sensemaking" (2008) 33(1) *Academy of Management Review* 122.
281. B. Horrigan, *Corporate Social Responsibility in the 21st Century* (Edward Elgar: Cheltenham UK, 2010), 34.

specifies with precision and clarity which conjunction of attributes makes up the concept – is available" nor, they argue, can one be expected.[282] But this is not for want of trying. An often-cited starting point is Archie B. Carroll's definition of CSR as comprising "the economic, legal, ethical and discretionary expectations that society has of organizations at any point in time."[283] Yet the conceptual difficulties do not lie with the economic and legal duties of companies. As noted by van Oosterhout and Heugens, even "a one-eyed neoclassical economist like Milton Friedman"[284] can acknowledge a company's responsibility to increase its profits "within the rules of the game."[285] The real debate is about the discretionary and ethical responsibilities of business. And this brings us to the normative dimension of CSR.

For those who situate CSR within a normative framework, CSR is intimately connected to issues of social justice and environmental ethics. From this perspective, "business...is an activity *embedded* in the larger society with an *obligation* to the common good of society."[286] This is a distinctly prescriptive, philosophical approach[287] that entreats corporations to voluntarily exhibit greater responsibility and accountability to society at large.[288] On this view, CSR can be defined as "the notion that corporations have an obligation to constituent groups in society other than [share-]holders and beyond that prescribed by law or union contract"[289] and that "business and society are interwoven rather than distinct entities."[290]

Some authors observe that this definition of CSR emphasizes obligation[291] and implies a social contract between business and society.[292] Others have described the

---

282. van Oosterhout and Heugens, *supra* n. 277, at 216. See also Dahlsrud, *supra* n. 273, 2.
283. A.B. Carroll, "A Three-Dimensional Conceptual Model of Corporate Performance" (1979) 4(4) *Academy of Management Review* 497, 500. See also A.B. Carroll and K.M. Shabana, "The Business Case for Corporate Social Responsibility: A Review of Concepts, Research and Practice" (2010) 12(1) *International Journal of Management Reviews* 85, 95.
284. van Oosterhout and Heugens, *supra* n. 277, at 202.
285. M. Friedman and R.D. Friedman, *Capitalism and Freedom* (University of Chicago Press, Chicago, 1962), 133.
286. C.R. Solomon, *Ethics and Excellence: Cooperation and Integrity in Business* (Oxford University Press, New York, 1992), cited in JM Shepard, M Betz and L O'Connell, "The Proactive Corporation: Its Nature and Causes" (1997) 16 *Journal of Business Ethics* 1001, 1006 (emphasis added).
287. L.E. Preston, "Corporation and Society: the Search for a Paradigm" (1975) 13 *Journal of Economic Literature* 434.
288. Carroll, *supra* n. 274; Davis, *supra* n. 278; T.M. Jones, "Corporate Social Responsibility revisited, redefined" (1980) 22(2) *California Management Review* 59; A.G. Steiner, *Business and Society* (Random House, New York, 1975).
289. Jones, *supra* n. 288, at 59-60.
290. D.J. Wood, "Corporate Social Performance Revisited" (1991) 164 *Academy of Management Journal* 691, 695.
291. B. O'Dwyer, "Conceptions of Corporate Social Responsibility: The Nature of Managerial Capture" (2003) 16(4) *Accounting, Auditing & Accountability Journal* 523, 526-27; B. Mitnick, "Systematics and CSR: The Theory and Processes of Normative Referencing" (1995) 34(1) *Business and Society* 5.
292. R.H. Gray, D.L. Owen, and CA Adams, "Corporate Social Reporting: Accounting and Accountability" (Prentice-Hall, London, 1988); S.L. Wartick and P.L. Cochran, "The Evolution of the corporate social performance model" (1985) 10(4) *Academy of Management Review* 758.

basis of ethical theories in CSR as "the right thing to do"[293] with some empirical studies finding that companies have acted on this basis. For example, Kesidou and Demiral found that UK companies that had adopted environmental management systems were "driven by ethical/strategic (i.e. CSR)" motivations as well as regulatory ones.[294] González-Benito and González-Benito similarly found that Spanish companies that adopted ISO14001 were driven by both ethical and competitive motivations.[295] Moreover, Bansal and Roth found that a small number of firms in their sample were motivated to "go green" on the basis of "ecological responsibility," a term which emphasized the ethical/obligation aspects of CSR rather than self-interest.[296]

In this vein, "corporate citizenship" has emerged as a prominent term in CSR-management literature to describe the "'rightful' place [of business] in society, next to other 'citizens,' with whom the corporation forms a community."[297] Whilst this literature lacks definitional clarity,[298] the pervasive image of the firm as a "good corporate citizen" is of "the 'good guy' next door who cares for you" as opposed to an impersonal or inhuman power-player.[299] Indeed, some management scholars have presented an "extended view" of corporate citizenship whereby corporations "enter the arena of citizenship at the point of government failure to protect citizenship"[300] and thereby fulfill a quasi-governmental role in solving social problems.[301]

Yet broad, normative notions of CSR have been accused of ambiguity and vagueness,[302] which confound efforts to operationalize CSR by business. Concepts of

---

293. Garriga and Melé, *supra* n. 280, at 60. See also: J.C. Borck and C. Coglianese, *Beyond Compliance: Explaining Business Participation in Voluntary Environmental Programs* (University of Pennsylvania Institute for Law & Economics Research Paper No. 12-04, University of Pennsylvania Law School Public Law Research Paper No. 12-06, Jun. 7, 2011); I. Jackson and J. Nelson, *Profits with Principles* (Currency/Doubleday, New York, 2004).
294. E. Kesidou and P. Demirel, *Motivations for Organisational Eco-Innovations: Adoption of Environmental Management Systems by UK Companies* (Nottingham University Business School Research Paper No. 2012-01, Jan. 10, 2012), 1.
295. J. González-Benito and O. González-Benito, "An Analysis of the Relationship Between Environmental Motivations and ISO14001 Certification" (2005) 16(2) *British Journal of Management* 133.
296. P. Bansal and K. *Roth*, "Why Companies Go *Green*: A Model of Ecological Responsiveness" (2000) 43(4) *Academy of Management Journal* 717, 728.
297. A. Crane, D. Matten and J. Moon, "The Emergence of Corporate Citizenship: Historical Development and Alternative Perspectives," in AG Scherer and G Palazzo (eds) *Handbook of Research on Global Corporate Citizenship* (Edward Elgar, Cheltenham UK, 2008), 28. See also: Solomon, *supra* n. 286, 184; and S. Waddell, "New Institutions for the Practice of Corporate Citizenship: Historical, Intersectoral, and Developmental Perspectives" (2000) 105(1) *Business and Society Review* 107.
298. Crane et al., *supra* n. 297, at 28. See also: Melé, *supra* n. 276, at 70, 74; and Garriga and Melé, *supra* n. 280, at 57.
299. Crane et al., *supra* n. 297, at 26-27.
300. Melé, *supra* n. 276, at 70.
301. D. Matten and W. Chapple, "Behind the Mask: Revealing the True Face of Corporate Citizenship" (2003) 45(1) *Journal of Business Ethics* 109; D. Matten and A. Crane, "Corporate Citizenship: Toward an Extended Theoretical Conceptualization" (2005) 30(1) *Academy of Management Review* 166; Crane et al., *supra* n. 297, at 30-31, 36-38.
302. See, e.g., M.B.E. Clarkson, "A Stakeholder Framework for Analysing and Evaluating Corporate Social Performance" (1995) 20(1) *Academy of Management Review* 92; M.T. Jones, "Missing the Forest for the Trees: A Critique of the Social Responsibility Concept and Discourse" (1996) 35(1) *Business and Society* 7.

ethical CSR and idealized citizenship emphasize altruism and voluntary self-restraint, which demand "undesirably broad" corporate duties and stakeholder rights.[303] The World Business Council for Sustainable Development found that a common business reaction to CSR in practice is: "Please give us something we can do differently on Monday morning to make things happen."[304]

More specifically, Henry Bosch AO, former Australian National Companies and Securities Commission chairman, has opined that "that the only common denominator [in CSR discourse] is a warm cuddly glow."[305] Holliday, Schmidheiny and Watts similarly call for a more pragmatic and less "fuzzy" approach, arguing that implementation of CSR must be business-directed: "Leading companies are making up their own versions of CSR as they go along, and this is as it should be, for companies must guarantee a good fit with their own market realities."[306]

Arguably, a pragmatic, self-interested approach to CSR transmutes its ethical nature. For example, Gustavson argues that the network of entities supporting CSR in Australia was framed within the context of ethics until the mid-1990s since which time there has been a notable "shift away from CSR as ethics to CSR as risk management and as a tool for driving profit."[307] But why might this be? Windsor asserts that a corporate preference for an "economic approach" to CSR belies a political philosophy of efficiency-oriented utilitarianism and a belief that companies contribute to society just by being profitable.[308] For some other authors the conceptual shift (or "revision")[309] in CSR discourse away from ethics and toward pragmatism and self-interest, is seen as a maturing of CSR. For example, William C. Frederick traces the transition from "CSR1" being "the philosophical-ethical concept of corporate social responsibility" as corporations' obligation to work for social betterment to "the action-oriented managerial concept of corporate social responsiveness" or the capacity of a corporation to respond to social pressure being "CSR2."[310] He contends that the latter is more proactive and pragmatic than the former. Intriguingly, Frederick went on to third and fourth iterations of CSR, namely CSR3 "corporate social rectitude" and CSR4 "cosmos, science, religion";[311] however these iterations did not gain widespread acceptance.

---

303. Windsor, *supra* n. 280, at 96.
304. World Business Council for Sustainable Development (WBCSD) *Corporate Social Responsibility: Making Good Business Sense* (WBCSD: Geneva, 2000), 5.
305. H. Bosch, "Corporate Social Responsibility: Submission in Response to the Invitation of the Corporations and Markets Advisory Committee from Henry Bosch AO" (Feb. 24, 2006), 4: http://www.camac.gov.au/camac/camac.nsf/byHeadline/PDFSubmissions_2/$file/HBosch_CSR.pdf viewed Jan. 22, 2012.
306. C.O. Holliday, S. Schmidheiny and P. Watts, *Walking the Talk: The Business Case for Sustainable Development* (Greenleaf: Sheffield, 2002), 113.
307. R. Gustavson, "Australia: Practices and Experiences," in SO Idowu and WL Filho (eds) *Global Practices of Corporate Social Responsibility* (Springer-Verlag, Berlin, 2009) 477, 472.
308. Windsor, *supra* n. 280, at 96. See also The Economist, "Survey: The good company," 374(8410) *The Economist* (Jan. 22, 2005).
309. van Oosterhout and Heugens, *supra* n. 277, at 216.
310. W.C. Frederick, "From CSR1 to CSR2: The Maturing of Business-and-Society Thought" (1994) 33 *Business & Society* 150, 150.
311. W.C. Frederick, "Toward CSR3: Why Ethical Analysis is Indispensable and Unavoidable in Corporate Affairs" (1986) 28(2) *California Management Review* 126; and WC Frederick, "Moving to CSR4: what to pack for the trip" (1998) 37(1) *Business and Society* 40.

Similarly, Carroll has argued that the lack of ethical or moral threads running through "social responsiveness" makes it less problematic than a broad (ethics based) conception of CSR.[312] Indeed, Sethi has recommended that a broad conception of CSR be supplanted by a narrower notion of managerial "corporate social responsiveness" due to its operational focus.[313]

However, protagonists of the ethical dimension contend that such a narrow view of CSR is "insufficient and myopic";[314] and that the fundamental nature of CSR is lost when an ethical dimension is replaced with the unexceptional notion of "business as usual" (BAU). For example, O'Dwyer notes that the "core concern" of "corporate social responsiveness" is "society's impact on business rather than business's impact on society" with the result that "it is business which decides on the level of its social response and economic issues take clear precedence over social [or environmental] issues."[315] As such, responsiveness is not viewed as complementary to responsibility but rather as displacing it in "an effort to treat as a management issue one which had been predominantly treated as a social and/or ethical issue."[316]

A number of scholars, particularly in the social environmental accounting field, have termed such displacement as "middle-way" framing[317] and "managerial or corporate capture"[318] of the debate over the meaning of CSR. In short, this comprises "a perceived necessity or desire to capture and control conceptions of CSR within conventional business norms,"[319] that is, within the parameters of BAU. In so doing, corporations may "subtly dismiss"[320] the tensions between environmental conservation and development[321] and/or the societal and ethical obligations implicit in a broad conception of CSR.[322] For example, in cross-industry interviews at 25 UK companies, Spence found that:

---

312. Carroll, *supra* n. 283, at 500-502.
313. E.g. S.P. Sethi, "Dimensions of Corporate Social Responsibility" (1975) 17(3) *California Management Review* 58.
314. Windsor, *supra* n. 280, at 96.
315. O'Dwyer, *supra* n. 291, at 527. See also: Frederick 1986, *supra* n 311; and S.P. Sethi, "A Conceptual Framework For Environmental Analysis of Social Issues and Evaluation of Business Response Patterns" (1979) 4(1) *Academy of Management Review* 63.
316. O'Dwyer, *supra* n. 291, at 527, citing R.W. Ackerman and R.A. Bauer, *Corporate Social Responsiveness* (Reston Publishing, Reston, VA, 1976), vii.
317. M.J. Milne, H. Tregidga and S. Walton, "Words not actions! The Ideological Role Of Sustainable Development Reporting" (2009) 22(8) *Accounting, Auditing & Accountability Journal* 1211.
318. Egs: O'Dwyer, *supra* n. 291; R. Gray, R. Kouhy and S. Lavers, "Corporate Social and Environmental Reporting: A Review of the Literature and a Longitudinal Study of UK Disclosure" (1995) 8(2) *Accounting, Auditing & Accountability Journal* 47; T. Tinker, C. Lehman and M. Neimark, "Falling Down the Hole in the Middle of the Road: Political Quietism in Corporate Social Reporting" (1991) 4(2) *Accounting, Auditing & Accountability Journal* 28; C. Spence, "Social and Environmental Reporting and Hegemonic Discourse" (2007) 20(6) *Accounting, Auditing & Accountability Journal* 855.
319. O'Dwyer, *supra* n. 291, at 532.
320. *Ibid.*, 524.
321. K. Kearins, E. Collins and H. Tregidga, "Beyond Corporate Environmental Management to a Consideration of Nature in Visionary Small Enterprise" (2010) 49(3) *Business and Society* 512.
322. O'Dwyer, *supra* n. 291, at 524.

> All socio-environmental [or non-commercial] concerns had to be harnessed to a business case in some fashion to the extent that it seemed as though the starting point for CSR is not any notion of social responsibility as such, but that anything that organisations do in the CSR field must bolster their own interests in some way.[323]

At the heart of the CSR conceptual debate is the issue of what distinguishes CSR initiatives from everyday business or BAU.[324] Clearly, this debate about the discretionary and ethical responsibilities of corporations over and above their legal and economic obligations shows that the features of CSR are still ambiguous. CSR appears to be a contested concept despite its existence in academic and industry discourse for over 60 years.[325]

Some scholars have postulated more recently that the main reason for conceptual dissonance is because "the CSR concept evolved from predominantly *normative* origins"[326] which has created a normative bias in CSR theorizing.[327] In other words, the CSR concept arose not from the question of why companies *actually* go green but from normative discussion about why they *ought* to go green.

[B]   The Business Case for "Doing Good"

Literature that endeavors to explain and theorize actual or "real-life" corporate change revolves predominantly around "the business case." Referred to by some authors as "the business case for (corporate) sustainability"[328] it is described as "a company... voluntarily doing something that improves not only its economic performance in absolute terms, but also its environmental and social performance."[329] Some commentators focus on a "value" increase, namely increase to shareholder value through dividends and share prices.[330] Others focus on "profit" increase by defining the business case for sustainability as a "strategic and profit-driven corporate response to environmental and social issues caused through the organization's primary and secondary activities."[331] Regardless, the dominant thread of business case research is that CSR is something business should do in order to get ahead. That is, business will do well by doing good.[332]

---

323. Spence, *supra* n. 318, at 869.
324. As noted in van Oosterhout and Heugens, *supra* n. 277, at 202.
325. *Ibid.*, 202, 216.
326. *Ibid.*, 201 (emphasis in original).
327. *Ibid.* See also: Melé, *supra* n. 276, at 76; and Matten and Chapple, *supra* n. 301.
328. Salzmann et al., *supra* n. 130, at 27; U. Steger (ed.), *The Business of Sustainability: Building Industry Cases for Corporate Sustainability* (Palgrave MacMillan: New York, 2004), 39.
329. Steger, *supra* n. 328.
330. E.g. *ibid.*, 38.
331. E.g. Salzmann et al., *supra* n. 130, at 27.
332. C. Laszio, *How the World's Companies Are Doing Well by Doing Good* (Greenleaf: Sheffield, 2008).

[1] The "Win-Win"/ "Win-Lose" Literature

Salzmann et al. note that the business case has been approached by business, NGOs and academics "in many different ways to prove or disprove the sound economic rationale for corporate sustainability management."[333] In particular, debate has focused on whether adopting and incorporating environmental and social initiatives into business activities has a positive or negative impact on the corporate bottom line. In other words, whether "decreasing the environmental impact and improving the social impact creates economic value, or an economic advantage over social and environmental laggards."[334] This encapsulates the notion of CSR – and its more environmental-specific equivalent "corporate environmental responsibility" (CER)[335] – as "win-win" or "win-lose" for company profitability.

The literature on this topic has a pragmatic undergirding based on the recognition that corporations in capitalist nations are required to maximize profits and thus engage only with socio-environmental activities that acquit this imperative.[336] This situation reflects Milton Friedman's famous assertion that:

> Few trends could so thoroughly undermine the very foundations of our free society as the acceptance by corporate officials of a social responsibility other than to make as much money for their stockholders as possible.[337]

In the years following Friedman's assertion it became *de rigueur* for commentators to condemn it. Yet now, as David Vogel points out:

> [M]any contemporary advocates of CSR have implicitly accepted Friedman's position...But they have added a twist: in order for companies to [create wealth for their shareholders], they must now act virtuously.[338]

Notions of enlightened self-interest (whereby responsibility and profitability are compatible) had been circulating since 1954;[339] however the "win-win/win-lose" debate did not begin in earnest until the 1990s. Eco-efficiency was one of the early frameworks for propounding the relevance of environmental considerations to business operations.[340] Eco-efficient companies were defined in 1991 by Schmidheiny as "those which create ever more useful products and services – in other words, which

---

333. Salzmann et al., *supra* n. 130, at 27.
334. *Ibid.*
335. A. Jamison, M. Raynolds, P. Holroyd, E. Veldman and K. Tremblett, *Defining Corporate Environmental Responsibility: Canadian ENGO Perspectives* (The Pembina Institute and Pollution Probe, Toronto, 2005), iv. N. Gunningham, "Shaping Corporate Environmental Performance: A Review" (2009) 19 *Environmental Policy and Governance* 215, 215.
336. C. Adams and G. Whelan, "Conceptualising Future Change in Corporate Sustainability Reporting" (2009) 22(1) *Accounting, Auditing & Accountability Journal* 118, 134.
337. Friedman and Friedman, *supra* n. 285, at 133.
338. D. Vogel, *The Market for Virtue: The Potential and Limits of Corporate Social Responsibility* (Brookings Institution Press: Washington DC, 2005), 26.
339. E.g. P. Drucker *The Practice of Management* (Harper, New York, 1954).
340. S. Waage, "Reconsidering Business from a Systems Perspective: The Shift to Sustainability-Oriented Enterprises and Financial Systems," in S Waage (ed) *Ants, Galileo and Ghandi: Designing the Future of Business Through Nature, Genius and Compassion* (Greenleaf: Sheffield, 2003) 45.

add more value – while continuously reducing their consumption of resources and their pollution."[341]

Schmidheiny and Zorraquin provided case study evidence that beyond-compliance actions such as minimizing emissions and energy use, recycling, and substituting hazardous materials, can have bottom-line benefits by cutting operating budgets and improving profit margins.[342] In relation to environmental regulation, Porter and van der Linde contended that such regulation, which had been traditionally regarded by economists as an inhibitor of innovation and profitability, could incentivize novel corporate approaches to pollution reduction and also provide savings on raw material, energy costs, and future compliance costs.[343]

Conversely, industry consultants Walley and Whitehead argued that while the win-win case is laudable it is also illusory. In their view, win-win outcomes are exceptional and "insignificant in the face of the enormous environmental expenditures that will never generate a positive financial return" and, as such, should not be the driver of a company's environmental strategy.[344] Akin to Friedman, they exhorted companies to "minimize the destruction of shareholder value" likely to be caused by such expenditure.[345] They further argued that win-win advocates offered little practical guidance to managers about how to implement win-win strategies.

Enter theories of natural capitalism and also ecological modernization, both of which have built on eco-efficiency literature by attempting to specify strategies that will "reduce environmental harm, create economic growth and increase meaningful employment."[346] The central tenets of ecological modernization are that environmental and economic imperatives can be reconciled and that reducing costs can increase profits.[347] Natural capitalism specifically posits that humans need to harness life-supporting eco-services and enhance "resource productivity," which roughly equates to eco-efficiency.[348] In relation to climate change, Hawken et al. use natural capital and eco-efficiency arguments to contend that the earth's climate can be protected not at a cost but at a profit. They argue that the climate threat "disappears if customers use

---

341. S. Schmidheiny and The Business Council for Sustainable Development, *Changing Course: A Global Business Perspective on Development and the Environment* (MIT Press: Cambridge MA, 1991).
342. Schmidheiny and Zorraquin, *supra* n. 66. See also: Waage, *supra* n. 340, at 46; Holliday et al., *supra* n. 306, at 83-95, 96-102.
343. M. Porter and C. van der Linde, "Green and Competitive: Ending the Stalemate" (1995a) (September/October) *Harvard Business Review* 120; M. Porter and C. van der Linde, "Toward a New Conception of the Environment-Competitiveness Relationship" (1995b) 9 *Journal of Economic Perspectives* 97. See also K. Green, P. Groenewegen and P.S. Hofman (eds) *Ahead of the Curve: Cases of Innovation in Environmental Management* (Kluwer Academic Publishers: New York, 2001).
344. N. Walley and B. Whitehead, "It Isn't Easy Being Green" (1994) (May/June) *Harvard Business Review* 46, 46.
345. *Ibid.*, 47, 52.
346. P. Hawken, A.B. Lovins, and L. Hunter Lovins, *Natural Capitalism: The Next Industrial Revolution* (Earthscan, London, 1999), 11.
347. See A. Gouldson and J. Murphy, *Regulatory Realities: The Implementation and Impact of Industrial Regulation* (Earthscan, London, 1998), 2-3; and J. Dryzek, *The Politics of the Earth: Environmental Discourses* (Oxford University Press, Oxford UK, 1997).
348. Hawken et al., *supra* n. 346.

energy as efficiently as is cost-effective," which creates win-win results.[349] Thus, whether a business "believes" in climate change or not is irrelevant; companies should take these steps "simply because they make money."[350]

Theories of natural capital and eco-efficiency are still popular today, with some authors proposing energy efficiency as the first step to climate change mitigation because it can radically reduce GHG emissions without radically altering BAU or financial output.[351] These theories have been complimented by the term "triple bottom line," which denotes single (financial) bottom line accounting as insufficient for sustainability and requiring the inclusion of environmental and social costs and returns.[352] Yet some ecological socialists and political economists have argued that such theories are fundamentally flawed because they sit within a conventional economic paradigm of greed and growth such that attempts to internalize "green" costs can only be win-lose.[353]

Given the polarity of the win-win/win-lose debate and the reality of capitalist imperatives, a body of theoretical and empirical literature has focused on linking corporate social and environmental performance (CSP) to corporate financial performance (CFP) with the aim of proving categorically whether environmental uptake creates a win-win or win-lose result. A prominent definition of CSP is "the configuration in the business organization of principles of social responsibility, processes of response to social requirements, and policies, programs and tangible results that reflect the company's relations with society."[354]

The theoretical literature has tended to revolve around stakeholder theory as espoused in R. Edward Freeman's seminal work.[355] Stakeholder theory promotes consultation and dialogue between a company and its stakeholder groups in order to shape socially responsible corporate behavior. Stakeholders may include government, unions, employees, consumers and clients, NGOs, and even competitors. This theory suggests that business strategies intended to add social and/or environmental value to external stakeholders will also provide value to shareholders. "Ergo, there is no necessary tension between economic and ethical considerations."[356]

---

349. *Ibid.*, 243.
350. *Ibid.*
351. See e.g. R.U. Ayres and E. Ayres, *Crossing the Energy Divide: Moving from Fossil Fuel Dependence to a Clean-Energy Future* (Prentice Hall: New York, 2010); T.R. Casten, *Pro-Profit, Pro-Planet: How Adam Smith Can Help Us Solve the Climate Crisis* (Recycled Energy Development: Westmont IL, 2010).
352. See generally, J. Elkington, *Cannibals with Forks: The Triple Bottom Line of 21st Century Business* (Capstone Publishing, Oxford UK, 1997). Also, Waage, *supra* n. 340, at 48.
353. See J. Kovel, *The Enemy of Nature* (Fernwood Publishing, Halifax, NS, 2002); D.H. Meadows, D. Meadows and J. Randers, *The Limits to Growth* (Universe, New York, 1972); F.E. Trainer, "The Limits To Growth Case Now" (1999) 19(4) *The Environmentalist* 325; G. Liodakis, "The People-Nature Relation and the Historical Significance of the Labour Theory of Value" (2001) 73 *Capital and Class* 113; and L Brown, *Eco-Economy: Building an Economy for the Earth* (WW Norton, New York, 2001).
354. Wood, *supra* n. 290, at 693.
355. R.E. Freeman, *Strategic Management: A Stakeholder Approach* (Pitman: Boston, MA, 1984).
356. J.C. Ludescher, A. McWilliams and D.S. Siegel, "The Economic View of Corporate Citizenship," in AG Scherer and G Palazzo (eds) *Handbook of Research on Global Corporate Citizenship* (Edward Elgar: Cheltenham UK, 2008), 320.

However, stakeholder theory makes a number of assumptions, including that (a) *all* stakeholder interests have intrinsic value, (b) interests of various stakeholders are commensurate and equally-valued by the company, and (c) any disparate interests can be compromised or balanced against each other.[357] None of these assumptions are practical given that a wide definition of stakeholders includes "any group or individual who can affect or is affected by the corporation."[358] Moreover, in capitalist markets, it must be acknowledged that a company responds to the most powerful and influential (financial) stakeholders. Indeed, Adams and Whelan opine that stakeholder theory simply offers a different way of actioning shareholder wealth maximization.[359] Orts and Strudler go further to argue that stakeholder theory is so expansive as to be unworkable.[360]

Some CSR scholars have started to focus on stakeholder theory and "value creation." In so doing, they differentiate between "profit" and "value," which are terms that tend to be conflated in the business strategy literature.[361] Conflation is unsurprising given the conventional financial wisdom that a rational investor desires only an increase in shareholder return. In contrast, some authors argue that the concept of "value" is broader than exclusively economic returns; managers need to focus on the relationship between a company and its stakeholders to ensure "total long-run value of the firm."[362] Arguably, the very existence of green investment funds and SRI demonstrate that some individual investors are willing to take lower financial returns if they receive personal value from investing ethically. But the fact that lower returns can be derived from socially responsible investments, even if investors are comfortable with that result, only undermines a CSP-CFP positive correlation.

In terms of empirical literature, Margolis and Walsh identified 127 studies from 1972 to 2002 that focused on linkages between corporate environmental uptake and

---

357. T. Donaldson and L.E. Preston, "The Stakeholder Theory of the Corporation: Concepts, Evidence, and Implications" (1995) 20(1) *Academy of Management Review* 65, 81; W.M. Evan and R.E. Freeman, "A Stakeholder Theory of the Modern Corporation: Kantian Capitalism," in T Beauchamp and N Bowie (eds) *Ethical Theory and Business* (Prentice Hall: Englewood Cliffs, NJ, 1988), 151. See generally A.M. Marcoux, "A Fiduciary Argument against Stakeholder Theory" (2003) 13(1) *Business Ethics Quarterly* 1; D.A. Gioia, "Response: Practicability, Paradigms, and Problems in Stakeholder Theorizing" (1999) 24(2) *Academy of Management Review* 228.
358. Melé, *supra* n. 276, at 64.
359. Adams and Whelan, *supra* n. 336, at 134.
360. E.W. Orts and A. Strudler, "Putting a Stake in Stakeholder Theory" (2009) 88 *Journal of Business Ethics* 605, 607.
361. See e.g. Ernst & Young, *supra* n. 132, at 6, 27.
362. See e.g. E. Kurucz, B. Colbert and D. Wheeler, "The Business Case For Corporate Social Responsibility," in A Crane, A McWilliams, D Matten, J Moon and DS Siegel (eds), *The Oxford Handbook of Corporate Social Responsibility* (Oxford University Press: Oxford UK, 2008); A. Argandoña, "Stakeholder Theory and Value Creation" (IESE Business School Working Paper No. 922, 2011); and M.C. Jensen, "Non-Rational Behavior, Value Conflicts, Stakeholder Theory, and Firm Behavior" (2008) 18(2) *Business Ethics Quarterly* 167; C. Wheeler, B. Colbert and R.E. Freeman, "Focusing on Value: Reconciling Corporate Social Responsibility, Sustainability and A Stakeholder Approach in a Network World" (2003) 28(3) *Journal of General Management* 1, 20.

financial outcomes.[363] They noted that while the studies revealed reasonable evidence of a positive correlation between CSP and CFP, there were numerous methodological problems particularly in establishing causality. A number of other authors have noted methodological shortcomings in extant empirical research, including: not empirically testing definitions and concepts; inadequate sampling techniques; limited data availability; and the variety of CFP measures used.[364] They conclude that the CSP and CFP causal sequence remains unresolved given that studies have not identified a simple positive or negative association between CSP and CFP, or if causality does exist, in which direction it actually flows. Some authors reason that definitional uncertainty of CSR necessarily ensures that CSP-CFP studies will remain inconclusive.[365]

Yet these limitations in CSP-CFP research reveal insights about the appropriateness of a one-size-fits-all approach to business case theorizing. Salzmann opines that instrumental studies show that the CSP and CFP relationship is "complex and contingent on situational, company- and plant-specific factors."[366] Similarly,in exploring both the potential and limits of CSR, Vogel uses the ambiguity of the CSP-CFP correlation as evidence that there is no clear-cut business case for virtuous corporate behavior.[367] He claims that uncertainty is unsurprising:

> [T]he risks associated with CSR are no different than those associated with any other business strategy [such as advertising or marketing expenditures]; sometimes investments in CSR make business sense and sometimes they do not. Why should we expect investments in CSR to consistently create shareholder value when virtually no other business investments or strategies do so?[368]

As such, it is unrealistic to expect that behaving more responsibly is in the self-interest of *all* firms or that engaging in CSR will *always* make business sense.[369] Specifically, Vogel contends that the market or consumer demand for virtue is not sufficiently important to incentivize *all* companies to voluntarily behave more responsibly.[370]

[2]   The "It Depends" Literature

The theses of scholars such as Salzmann and Vogel – that the business case for going beyond compliance will only yield benefits to specific businesses in certain

---

363. J.D. Margolis and J.P. Walsh, "Misery Loves Companies: Rethinking Social Initiatives by Business" (2003) 48 *Administrative Science Quarterly* 268. See also De Bakker, P. Groenewegen, F. Den Hond, "A Bibliometrical Analysis of 30 Years of Research and Theory on Corporate Social Responsibility and Corporate Social Performance" (2005) 44(3) *Business & Society* 283.
364. Margolis and Walsh, *supra* n. 363, 277-278. Salzmann et al., *supra* n. 130, at 30; van Oosterhout and Heugens, *supra* n. 277, 205-210; J. Griffin and J. Mahon, "The Corporate Social Performance and Corporate Financial Performance Debate: Twenty-Five Years of Incomparable Research" (1997) 36(1) *Business & Society* 5.
365. See e.g. van Oosterhout and Heugens, *supra* n. 277, 204.
366. Salzmann et al., *supra* n. 130, at 30.
367. Vogel, *supra* n. 338, at 17, 29-34.
368. *Ibid.*, 33.
369. *Ibid.*, 34.
370. *Ibid.*, 17, 46. See also D. Vogel, "Is There A Market For Virtue? The Business Case for Corporate Social Responsibility" (2005) 47 *California Management Review* 19, 20.

circumstances – is central to a second body of CSR/CER literature that moves beyond the win-win/win-lose polemic. This body of literature focuses on more nuanced notions of what drives corporate uptake of environmental and social strategies. I term this body of scholarship the "it depends" literature, which is an unsophisticated but accurate label to acknowledge that whether or not a business will benefit by adopting an environmental strategy depends on a variety of internal and contextual factors.

The "it depends" literature emerged at the beginning of the millennium when a number of authors attempted to explain the circumstances in which CSR/CER might be best used to gain competitive advantage. Forest Reinhardt sought to specify the circumstances in which CER uptake would provide strategic advantage and concluded that whether or not beyond compliance environmental actions could deliver corporate bottom-line benefits depended on "the structure of the industry in which the business operates, its position within that structure and its organizational capabilities."[371] In order to turn environmental strategy into competitive advantage, Reinhardt entreated companies to identify the fundamental business logic, in terms of risk management and value enhancement, peculiar to their company and context.

Writing specifically on the drivers of voluntary corporate climate-related measures, Andrew Hoffman asserts that the question "does it pay to be green?" is the wrong one. He writes that the correct question is "whether there exists an economic opportunity for your company to be green vis-à-vis your competitors and then... how and when that opportunity can best be achieved."[372] In particular, Hoffman notes that despite the absence of mandatory GHG emission reductions legislation in the United States, businesses have been engaging in voluntary climate change-related initiatives and that "many of these companies are agnostic about the science of climate change or the social responsibility of protecting the global climate."[373] His conclusion is that companies go beyond compliance for "decidedly strategic" reasons[374] that vary with company-specific cost/benefit analysis and business logic. In particular, companies wish to be prepared in the long term (in case a carbon price becomes mandatory) while simultaneously reaping short term economic and strategic benefits (if such regulation is delayed or unrealized).[375] Hoffman contends that corporate change will not occur without a business case to drive it. In short, companies "must have a bottom line rationale or such efforts will be financially unsustainable."[376]

It is apparent that a company must be clear about its core business strategy in order to benefit from adopting green practices. A joint cross-industry study edited by Ulrich Steger in 2004 investigated the business case for corporate sustainability by

---

371. F. Reinhardt, "Market Failure and the Environmental Policies of Firms: Economic Rationales For "Beyond Compliance" Behavior" (1999) 13 *Journal of Industrial Ecology* 9, 18. Also F. Reinhardt, *Down to Earth: Applying Business Principles to Environmental Management* (Harvard University Press, Cambridge MA, 2000).
372. A.J. Hoffman, "Climate Change Strategy: The Business Logic Behind Voluntary Greenhouse Gas Reductions" (2005) 47(3) *California Management Review* 21, 38.
373. *Ibid.*, 22. See also Ernst & Young, *supra* n. 132, at 19.
374. Hoffman, *supra* n. 372, at 22.
375. See A. Kolk and J. Pinkse, "Market Strategies for Climate Change" (2004) 22(2) *European Management Journal* 304.
376. Hoffman, *supra* n. 372, at 23.

focusing on "value drivers" being factors that significantly contribute to shareholder value.[377] Value drivers included license to operate, reputation and brand value, attraction and retention of talented employees, risk management and cost reductions. Similar to Reinhardt's work, this study concluded that there is not a universal blueprint for every company in every sector around the world. It concluded that "all business cases require the support of a core business strategy," which will be peculiar to a specific company.[378] Similarly, Porter and Kramer[379] and Esty and Winston[380] also focus on the link between core business and competitive advantage created by CSR/CER. For example, Porter and Kramer state that companies need to:

> analyze their prospects for social responsibility using the same frameworks that guide their core business choices... [so as to harness CSR as] a source of opportunity, innovation and competitive advantage.[381]

The importance of ensuring CSR is part of a company's core agenda or philosophy is also made clear in the change management literature. Defining seven common pitfalls on the road to corporate change, Bob Doppelt writes how "deep-rooted cultural transformation is necessary to overcome the resistance inherent to the profound changes necessary to achieve true sustainability."[382] To this end:

> the ultimate success of a change initiative occurs when sustainability-based thinking, perspectives, and behaviors are embedded in everyday operating procedures, policies, and culture.[383]

Hans-Joerg Hess's empirical study of why financial actors adopt green strategies similarly found that "real changes should take place in the core business" if the business case as driver is to generate a positive sustainability impact.[384] He notes that the existence of a sustainability unit is no guarantee for effective changes at the core.[385]

---

377. Steger, *supra* n. 328. See also U. Steger, A. Ionescu-Somers and O. Salzmann, "The Economic Foundations of Corporate Sustainability" (2007) 7(2) *Corporate Governance: The International Journal of Business in Society* 162.
378. Steger et al., *supra* n. 377, at 62.
379. M.E. Porter and M.R. Kramer, "Strategy and Society: The Link Between Competitive Advantage and Corporate Social Responsibility" (2006) (December) *Harvard Business Review* 78.
380. D. Esty and A. Winston, *Green to Gold: How Smart Companies Use Environmental Strategy to Innovate, Create Value and Build Competitive Advantage* (Yale University Press, New Haven CT, 2006).
381. Porter and Kramer, *supra* n. 379, at 78.
382. B. Doppelt, "Overcoming the Seven Sustainability Blunders" (2003) 14(5) *The Systems Thinker* 2, 3. See also D. Smith, "Engaging In Change Management: Transformation Through Sustainability Strategy at Norm Thompson Outfitters," in S. Waage (ed) *Ants, Galileo & Ghandi: Designing the Future of Business through Nature, Genius, and Compassion* (Greenleaf: Sheffield, 2003); R.J. Welford, *Environmental Strategy and Sustainable Development: the Corporate Challenge for the 21st Century* (Routledge: London, 1995), 114.
383. Doppelt, *supra* n. 382. See also B. Doppelt, *Leading Change Toward Sustainability* (Greenleaf Publishing: Sheffield UK, 2010).
384. H.J. Hess, *CSM/WWF Research Project: The Business Case for Sustainability/ Financial Services Sector Report* (IMD Working Paper Series, IMD 2003-8, 2003), 16.
385. *Ibid.*, 18. See also: C. Lins, D. Wajnberg, U. Steger and A. Ionescu-Somers, *Corporate Sustainability in The Brazilian Banking Sector* (IMD International, IMD 2008-07, Lausanne, 2008), 2; and Doppelt, *supra* n. 383, at 50.

In other words, sustainability must become built in to corporate life and not just bolted on to it. Certainly, Hess's study stressed that altering the values and culture of a company was difficult and that changes have limited effects if firm values are incompatible with short-term business goals.[386]

Similarly, Doppelt argues that a company's vision must regard sustainability as equally or more important to the goals of profitability or shareholder value.[387] This is a challenging proposition for companies in capitalist economies. It is hardly surprising that Doppelt found that none of the companies he studied were truly sustainable because each was "plagued by inconsistencies between their vision and current practices."[388]

Thus, the business strategy literature posits that competitive advantage is the key motivator for beyond-compliance corporate action. However, there are some studies that suggest more complex motivations for corporate environmental behavior, such as the studies authored by Bansal and Roth,[389] and Gunningham, Kagan and Thornton.[390]

Bansal and Roth's cross-industry, cross-national study of the motivations and contextual factors that drive corporate ecological responsiveness also features "competitiveness" (being a meta-label for business case motivations) as one of three categorized drivers of corporate change. Yet, this study concluded that firms were primarily motivated by concerns for "legitimacy," not competitiveness; and that a small number of companies were motivated by "ecological responsibility" or a company's concern for social good. Legitimacy is a company's desire "to improve the appropriateness of its actions within an established set of regulations, norms, values, or beliefs."[391] Legitimation is a common thread throughout macro theories and will be elaborated upon in the next section. At the meso-level, Bansal and Roth's classification of "legitimation" translated into managing downside risk, particularly risk to a company's reputation or social license, whereas "competitiveness" focused on fiscal benefits to the company.

Indeed, while theorists such as Reinhardt regard risk mitigation as one motivator of corporate greening, there are several studies that, consistent with Bansal and Roth's conclusions, prioritize risk mitigation as the central driver. Bryan W. Husted writes that CSR as a real option is oriented toward downside prevention of risk, that is, the containment of possible losses, rather than capturing opportunities.[392] He argues that a company's decision whether to invest in CSR projects should be based on the ability of that project to generate indirect benefits or resources for the firm, being goodwill and/or trust, as opposed to direct pecuniary benefits. He advocates that such decisions

---

386. Hess, *supra* n. 384, at 16.
387. Doppelt, *supra* n. 382, at 6.
388. *Ibid*. See also S. Waage, "Conclusion: A Shift Towards Sustainability Within Companies and the Financial Services Sector," in S Waage (ed) *Ants, Galileo and Ghandi: Designing the Future of Business Through Nature, Genius and Compassion* (Greenleaf: Sheffield, 2003), 243.
389. Bansal and Roth, *supra* n. 296.
390. N. Gunningham, R. Kagan and D. Thornton, *Shades of Green: Business, Regulation and Environment* (Greenleaf: Sheffield, 2003).
391. Bansal and Roth, *supra* n. 296, at 726.
392. B.W. Husted, "Risk Management, Real Options, and Corporate Social Responsibility" (2005) 60 *Journal of Business Ethics* 175, 179-181.

be based on managerial perceptions of future risk (as opposed to historical records) in terms of potential reputation loss and financial cost to the firm. In this way, as stated by Beck, risk assessments "promise the impossible: events that have not yet occurred become the object of current action."[393]

These findings on risk mitigation are echoed in some previous studies of financial actors' adoption of green strategies. For example, in 2003 Hess conducted an empirical study of the exploitation of the business case for sustainability in the finance sector.[394] Hess's report focused on the credit business within corporate and investment banks and activities of insurance companies. Importantly, interviews were conducted with sustainability and environmental units only (and not transactional units). He found that financial companies mostly view sustainability issues as risks with the most important value driver for a business case being risk management, particularly risk to reputation (as leveraged by NGOs). He also found that new products and business opportunities were not seen as key value drivers but rather niche markets of the future.[395] Similarly, empirical studies of why banks adopt the Equator Principles (for example, by Watchman in 2005[396] and Conley and Williams in 2011[397]) have concluded that a key reason for uptake relates to reputation risk management. Moreover, in 2008 Lins et al. found that the main motivation for climate change action by Brazilian banks was mitigating "risks against reputation" due to inertia on the issue.[398] Conversely, a 2009 empirical study by Neuhoff et al. found that large risks were a deterrent to investors engaging in climate finance, particularly regulatory risk and financial risks associated with unproven clients and uncertain technology.[399] Accordingly, previous studies have yielded mixed results on the role of risk as a driver or deterrent for actors to adopt green strategies, particularly in the finance sector.

Importantly, these studies have tended to highlight corporate reputation as a leverage point for change; it is regarded as a kind of sweet or "soft" spot of corporations. Some have found that there are key drivers, referred to as "soft issues" or "intangibles," which include license to operate, leadership, brand value, and reputation.[400] Specifically, Steger's empirical study showed that brand value and reputation are key concerns for industries that have "specialized products and brands that are in direct contact with the consumer," which include financial services industries.[401] A 2012 Ernst & Young report found that, in particular, "[r]eputation issues arise when independent organizations rate or rank companies on climate emissions and goals."[402]

393. U. Beck, "From Industrial Society to the Risk Society: Questions of Survival, Social Structure and Ecological Enlightenment" (1992) 9 *Theory, Culture and Society* 97, 100.
394. Hess, *supra* n. 384. See also Steger, *supra* n. 328.
395. Hess, *supra* n. 384, at 13.
396. Watchman, *supra* n. 164, at 50.
397. Conley and Williams, *supra* n. 167.
398. Lins et al., *supra* n. 385, at 30.
399. K. Neuhoff, S. Fankhauser, E. Guerin, J.C. Hourcade, H. Jackson, R. Rajan and J. Ward, *Structuring International Financial Support to Support Domestic Climate Change Mitigation in Developing Countries* (Climate Strategies: Cambridge UK, 2009), 10.
400. Steger, *supra* n. 328, at 39; Salzmann et al., *supra* n. 130, at 33; Porter and Kramer, *supra* n. 379.
401. Steger, *supra* n. 328, at 43.
402. Ernst & Young, *supra* n. 132, at 18.

Yet most authors agree that "soft issues" cannot be quantified easily if at all. Hess states that "risk and reputation management ... only contribute indirectly to a company's business success" and there is generally no direct feedback mechanism associated with them.[403] Similarly, Spence found that the business case for producing social environmental reports lay more in soft issues "such as managing relationships and responding to social pressures," which "were presumed to yield an advantage in the long-term" but not necessarily amenable to cost/benefit analysis.[404]

Accordingly, despite assertions that a company's reputation and the value of its brand are "among its most valuable assets,"[405] it is still not exactly clear how such soft issues fit into the business case. Several business strategy scholars, including Steger and Salzmann et al., have noted that the salience of soft drivers to a business case is under-researched in the literature.[406] In particular, a company's reputation is intuitively important but difficult to quantify, with Steger noting that: "the value of [reputation] is probably only known once it is lost."[407] As such, the precise nature of "reputation" is not clear from the business strategy literature. Mostly, it is assumed that a company's reputation is synonymous with its social reputation, which fits with legitimacy thinking. Yet this assumption may not be correct. Important questions remain about the role and nature of "corporate reputation" as a green driver.

In keeping with a more complex approach that goes beyond "competitive advantage" as the central driver of corporate greening, Gunningham, Kagan and Thornton's empirical cross-national study of the pulp and paper industry also evidenced that economic motivations are not necessarily separate from or privileged over social or regulatory ones. These authors proposed a conceptual model of corporate environmental behavior that accounts for variation by addressing the complex interaction between a firm's social license, tightening regulations, economic constraints and also internal managerial decision-making processes. They termed this model a socially constructed three-pronged "license to operate" comprised of economic, social and regulatory licenses, which reflect the expectations of various company stakeholders that monitor, enforce and leverage the licenses. Further, the authors found that the different ways that managers interpret, operationalize and negotiate external license pressures creates variable environmental performance by companies in the same industry and subject to similar external pressures.[408]

So it appears from the business strategy literature that the business case as driver is important to green uptake but that it may not be the only or even the determining driver, with notions of legitimacy and social license also taking center stage. There is also some uncertainty around the importance of risk mitigation in driving corporate greening. In particular, the need to mitigate risk to a company's reputation needs elucidation. From the literature, it appears that reputational risk is synonymous with

---

403. Hess, *supra* n. 384, 21.
404. Spence, *supra* n. 318, at 866.
405. T. Smith, "Institutional and Social Investors Find Common Ground" (2005) 14 *Journal of Investing* 57, 60.
406. Salzmann et al., *supra* n. 130, at 33. Steger, *supra* n. 328, at 41, 62.
407. Steger, *supra* n. 328, at 62.
408. Gunningham et al., *supra* n. 390, at 17, 132-33.

risk to a company's social license. Yet concepts such as "reputation" and "branding" need further deconstruction if we are to understand their strength and limitations as drivers. Specifically, the role of soft factors in making a business case – such as "reputation," "leadership" and "branding" – remains unclear in extant literature.

§3.04 MACRO LEVEL: SOCIO-CULTURAL DRIVERS OR "EXTERNAL" FACTORS

It is clear from the business strategy literature outlined above that CSR is a socially constructed and contested concept. Institutional theory has some relevance here as it investigates how external macro pressures shape organizational behavior by messaging what is "appropriate" and "responsible." As such, it has a role in helping to explicate why companies, including private finance actors, might go green.

Specifically, institutional theory posits that what constitutes responsible or irresponsible corporate behavior, and what are deemed legitimate activities or goals around which to build a business case, depends very much on the wider socio-cultural setting. The idea is that macro socio-cultural and political factors influence organizational behavior by defining and regulating "appropriate" corporate action. For this reason, several authors have asserted that academic focus on the content of CSR activities has led to an intellectual neglect of institutional factors that shape the construction of such activities in the first place.[409] Moreover, corporations may not always or only make decisions based on "atomistic, financial self-interest" pursuant to a business case imperative.[410] Indeed, Dillard et al. write that institutional theory is increasingly appearing in organization theory literature due to its wide range of applicability.[411] This is largely due to the work of W. Richard Scott, who showed that all organizations are "institutionalized" to varying degrees: that is, all organizations are socially constituted and subject to regulative processes which define "how they may operate legitimately."[412]

Institutional theory focuses on the process of norm formation by which certain actions become acceptable or normal. "Institutions" are macro influences on organizational behavior that shape "how important issues are perceived and how appropriate actions are developed."[413] "Institutionalization" is the process by which practices, as responses to these influences or pressures, are developed and adopted within a social setting.[414] Importantly, organizations are situated (or institutionally embedded) within

---

409. M.T. Jones, "The Institutional Determinants of Social Responsibility" (1999) 20 *Journal of Business Ethics* 163; J.L. Campbell, "Institutional analysis and the paradox of corporate social responsibility" (2006) 49 *American Behavioral Scientist* 925; Margolis and Walsh, *supra* n. 363.
410. C. Parker and J. Braithwaite, "Regulation," in P Cane and M Tushnet (eds) *The Oxford Handbook of Legal Studies* (Oxford University Press: Oxford UK, 2003), 131.
411. J.F. Dillard, J.T. Rigsby and C. Goodman, "The Making and Remaking of Organization Context: Duality and the Institutionalization Process" (2004) 17(4) *Accounting, Auditing & Accountability Journal* 506, 508.
412. Scott, *supra* n. 272, at 136. See also Dillard et al., *supra* n. 411, at 508.
413. A.J. Hoffman, "Institutional Evolution and Change: Environmentalism and the US Chemical Industry" (1999) 42(4) *Academy of Management Journal* 351, 353.
414. Dillard et al., *supra* n. 411, at 508-9.

their "organizational field" or inter-organizational context. This means that individual organizations must conform to the rules and requirements of their field in order to receive support and legitimacy.[415] Meyer and Rowan state that when an organization adopts institutionalized elements then it protects itself from having its conduct questioned because it can give "prudent, rational and legitimate accounts" of its activities.[416]

So what form do institutions take, and how might we recognize their influence on a company's willingness to go green? Scott describes how a social framework comprises three institutional elements, being regulative, normative and cognitive "pillars."[417] Hoffman further states that the pillars form a continuum moving "from the conscious to the unconscious, from the legally enforced to the taken for granted,"[418] as illustrated below.

Regulative aspects of institutions are often regulations that "guide organizational action and perspectives by coercion or threat of legal sanctions."[419] Hoffman gives the example of corporations adopting new pollution control technologies in order to conform to environmental regulations.[420] Institutional theorists are less interested in law as a fact, and more concerned with law as a social construct and how it shapes corporate behavior and managerial decision-making. The idea is that prevailing socio-economic norms and values are sometimes codified in laws and regulations.[421] For institutionalists (as opposed to lawyers), the focus is not on fines or inspections as a deterrent threat *per se* but on how fines and inspections bring management attention to larger social expectations that may affect the legitimacy and operation of their company.[422] As such, regulative institutions can shape CSR activities and corporate attitudes. With this in mind, a detailed analysis of the impacts of corporate law and banking rules on bank corporate governance is provided in Chapter 5.

Normative aspects of institutions guide organizational action and belief via social obligation or professionalization. Scott states that norms "specify how things should be done; they define legitimate means to pursue valued ends."[423] Normative or value systems define goals and objectives (for example, making a profit) but also designate appropriate ways to pursue them (for example, via ethical business practices).[424] Pattberg and Stripple posit that new "norm-setting devices" have deepened the

---

415. W.R. Scott and J.W. Meyer, "The Organization of Societal Sectors," in JW Meyer and WR Scott (eds) *Organizational Environments: Ritual and Rationality* (Sage, Beverly Hills, 1983), 149.
416. J. Meyer and B. Rowan, "Institutionalized Organizations: Formal Structure as Myth and Ceremony," in WW Powell and PJ DiMaggio (eds) *The New Institutionalism in Organizational Analysis* (University of Chicago Press: Chicago, 1991), 50.
417. Scott, *supra* n. 272, at 52-59.
418. A.J. Hoffman, *From Heresy to Dogma: An institutional history of corporate environmentalism* (Stanford University Press: California, 2001), 36.
419. Hoffman, *supra* n. 413, at 353.
420. *Ibid.*
421. Dillard et al., *supra* n. 411, at 527.
422. Parker and Braithwaite, *supra* n. 410, at 131.
423. Scott, *supra* n. 272, at 54-55.
424. *Ibid.*, 55.

institutionalization of beyond-the-state approaches to climate change.[425] They cite civil society driven campaigns, the rise of non-state market- and information- based mechanisms such as the Carbon Disclosure Project and the Investor Network on Climate Risk, and the rising proliferation of voluntary codes such as the Equator Principles and Carbon Principles, as evidence of the institutionalization of "new norms at the transnational level."[426] These norms affect how corporations respond to climate change and legitimacy threats by, for example, reducing and/or reporting their emissions, and signing onto voluntary "green" codes.

Finally, the cognitive pillar embodies behavior that is "taken-for-granted" and attention to this third pillar is a key distinguishing feature of neo-institutionalism.[427] Cognitive beliefs are the most entrenched as they are essentially unconscious "ways of doing things"[428] such that it is "almost unthinkable to do anything else."[429] For example, in market economies, it is regarded as natural that corporations care about profits and pursue economic goals.[430] "Institutional isomorphism" arises out of this cognitive element and is defined as "a constraining process that forces one unit in a population to resemble other units" that face the same set of external conditions.[431] Institutional scholars use isomorphism to explain the phenomenon of organizational homogeneity amongst competing firms.

So how does institutional theory help to explain organizational change given its apparent preoccupation with reproduction and inertia? DiMaggio and Powell use the concept of isomorphism to explain organizational and even field-level change. They give three reasons or mechanisms for why and how organizations copy and adopt institutional practices.[432] First, "coercive isomorphism" results when (political) pressure is exerted on an organization by other organizations upon which it is dependent and by expectations of communities within which it operates. For example, organizational change may be a direct response to government mandate; indeed DiMaggio and Powell note that "a common legal environment" creates increasingly homogenized organizations within specific jurisdictions.[433] Second, "normative isomorphism" takes place through professionalization, whereby members of an occupation collectively struggle to define the modes and conditions of work.[434] Finally, "mimetic isomorphism" occurs when an organization imitates a more successful organization in their field, which often occurs when there is a lack of certainty and guidance in its own environment. DiMaggio and Powell describe uncertainty as "a powerful force" that

---

425. P. Pattberg and J. Stripple, *Remapping Global Climate Governance: Fragmentation Beyond the Public/Private Divide* (Global Governance Working Paper No 32, November 2007), 17, 30.
426. *Ibid.*, 12.
427. Scott, *supra* n. 272, at 57-58.
428. Hoffman, *supra* n. 413, at 365; L. Zucker, "Organizations as Institutions," in S Bacharach (ed) *Research in the Sociology of Organizations* (JAI Press: Greenwich CT, 1983).
429. Parker and Braithwaite, *supra* n. 410, at 132.
430. Hoffman, *supra* n. 413, at 353.
431. P.J. DiMaggio and W.W. Powell, "The Iron Cage Revisited: Institutional Isomorphism and Collective Rationality in Organization Fields," in WW Powell and PJ DiMaggio (eds) *The New Institutionalism in Organizational Analysis* (University of Chicago Press, Chicago, 1991), 66.
432. *Ibid.*, 67-74.
433. *Ibid.*, 67.
434. Dillard et al., *supra* n. 411, at 509.

encourages imitation or "modeling" of competitors that are perceived as more legitimate or successful.[435]

This third mechanism of mimetic isomorphism highlights Aldrich's assertion that "the major factors that organizations must take into account are other organizations."[436] In other words, business organizations often act on the basis of peer pressure in order to be seen as "normal" and "appropriate" so as to survive.[437] As part of this mimetic process, businesses adopt the practices, rhetoric and symbols of "being part of the club."[438] In the institutional theory and diffusion literatures, authors label early moving organizations as "innovators," being the organizations that develop new organizational practices within the boundaries of organizational field practices, and label the mimicking organizations as "late adopters."[439] Dillard et al. explain the process of corporate change and resulting "normalizing" in this way: late adopters uptake and implement the new practices of innovators, and their effect of establishing what is legitimate and rational is further reinforced at the field organizational level, which in turn encourages more late adopters.[440]

As such, corporations can be influenced by the enlightened or green actions of other firms. In this way, corporations model or copy each other in order to enhance their social reputation and their economic and political legitimacy.[441] In particular, Hoffman's longitudinal study of environmental change in the chemical industry showed how cognitive beliefs changed over a 30 year period "as particular rules or norms became less contested through enduring persistence"; and new practices can thus become "accepted as the legitimate form of organizational action and ingrained in the cognitive institutional realm."[442] In other words, *over time*, new practices can become the industry norm. Other authors have similarly found that if successful competitors adopt widely publicized greener technologies or environmental management plans then pressure is created for all firms in the same industry to follow suit.[443] Lundgren and Catasús further note that embedding the "green value" at the corporate

---

435. DiMaggio and Powell, *supra* n. 431, at 69-70.
436. H.E. Aldrich, *Organizations and Environments* (Prentice Hall, Englewood Cliffs NJ, 1979), 265.
437. See Meyer and Rowan, *supra* n. 416; L. Zucker, "Institutional Theories of Organization" (1987) 13 *Annual Review of Sociology* 443; M.C. Suchman and L.B. Edelman, "Legal Rational Myths: the New Institutionalism and the Law and Society Tradition" (1996) 21 *Law and Social Inquiry* 903.
438. R. Friedland and R.R. Alford, "Bringing Society Back In: Symbols, Practices and Institutional Contradictions," in WW Powell and PJ DiMaggio (eds) *The New Institutionalism in Organizational Analysis* (University of Chicago Press, Chicago, 1991), 232. See also Meyer and Rowan, *supra* n. 416, at 50.
439. E.g. E.M. Rogers, *Diffusion of Innovations* (Free Press: New York, 1995).
440. Dillard et al., *supra* n. 411, at 515.
441. Suchman and Edelman, *supra* n. 437.
442. Hoffman, *supra* n. 413, at 365.
443. Eg: R.S. Marshall, M. Cordana and M. Silverman, "Exploring Individual and Institutional Drivers of Proactive Environmentalism in the US Wine Industry" (2005) 14 *Business Strategy and the Environment* 92; B. Cashore and I. Vertinsky, "Policy Networks and Firm Behaviours: Governance Systems and Firm Responses to External Demands for Sustainable Forest Management" (2000) 33 *Policy Sciences* 1.

level through mimetic isomorphism may, in turn, create positive changes for the natural environment.[444]

However, it is not clear from the literature whether there is some magic number or critical mass of early-moving firms required to create such a domino effect. Arguably the tipping point may vary within different organizational fields or depend upon the "stickiness" of current norms. It is also not certain whether early-moving firms acting alone or in small numbers that move too far ahead of the curve are at risk of exclusion rather than celebration for their leadership. Finally, even when organizations adopt institutionally acceptable practices in order to be perceived as legitimate, it is not certain that they will integrate those practices into their "essential machinery" so as to create substantial change within the firm, the field or the natural environment.[445]

In summary, the concepts of isomorphism and mimetic behavior help to explain the process of corporate change as a response to external socio-cultural and field-level pressures. However, some scholars have critiqued institutional research as tending to under-emphasize the political nature of organizations, which includes notions of power and the role of special interest groups, and to over-emphasize maintenance of the status quo due to the constraining nature of institutional practices and beliefs.[446] As such, institutional theory, on its own, may have only limited explanatory power regarding heterogeneity within and between firms and internal drivers of change.

§3.05   MICRO LEVEL: INTRA-ORGANIZATIONAL DRIVERS OR "INTERNAL" FACTORS

Much of the meso and macro level literatures encompass external drivers of corporate environmental change. However, internal drivers may also be important and are a particular focus of micro level theories, which identify multiple variables for corporate greening that are internal to a firm. These include coalition-building within an organization,[447] the availability of intra-organizational resources (human and financial) to invest in green strategies,[448] an organization's capacity to learn, and also how it acquires, interprets and internally distributes knowledge.[449] Yet the one consistent theme in the micro literature is that a central internal driver of corporate environmental change is managerial decision-making. Adams and Whelan write that "human action

---

444. Lundgren and Catasús, *supra* n. 164, at 189.
445. See B.E. Ashforth and B.W. Gibbs, "The Double-Edge of Organizational Legitimation" (1990) 1(2) *Organizational Science* 177; O'Sullivan and O'Dwyer, *supra* n. 166; Dillard et al., *supra* n. 411; Meyer and Rowan, *supra* n. 416.
446. E.g.: Scott, *supra* n. 272; Dillard et al., *supra* n. 411; Hoffman, *supra* n. 413; P. Hirsch and M. Lounsbury, "Ending the Family Quarrel: Toward a Reconciliation of 'Old' and 'New' Institutionalisms" (1997) 40 *American Behavioral Scientist* 406.
447. See e.g. A. Prakash, *Greening the Firm: The Politics of Corporate Environmentalism* (Cambridge University Press: Cambridge UK, 2000).
448. See e.g. R. Hartman, M. Hug and D. Wheeler, *Why Paper Mills Clean Up: Determinants of Pollution Abatement in Four Asian Countries* (World Bank Policy Research Paper 1710, 1997).
449. A. Schaefer and B. Harvey, "Stage Models of Corporate 'Greening': a Critical Evaluation" (1998) 7 *Business Strategies and the Environment* 109-123.

is of obvious importance to understanding organizational action [as]...it is always individuals, or a given number of individuals, that *decide* to act in a certain way."[450]

Previous studies on corporate greening have demonstrated that the attitude and actions of senior management are crucial. For example, Hoffman found that "senior-level support and engagement are the most critical components of any successful climate strategy"and that, more specifically, "CEO leadership was identified as a key driver at all stages of [climate] program development and implementation."[451] Similarly, Thomas Gladwin argued that "charismatic green leadership" is required for proactive corporate environmental greening; [452] and in their study of Swedish banks, Lundgren and Catasús found that it was important for the CEO of a bank to "use the word environment...in a way that signals it is a sincere belief rather than a mere intellectual exercise."[453] In other words, senior management sets the tone for firm policy from which flows actual practices, employee mind-sets and also public perceptions.

Some corporate environmental management literature asserts that a crucial variable in uptake or avoidance of environmental strategies is whether senior managers perceive environmental protection issues as an opportunity or a threat.[454] This literature interacts with the business strategy literature. For example, Carroll's conception of "social responsiveness" embodies a continuum of managerial business strategies or philosophies "from no response (do nothing) to a proactive response (do much)" in the face of social pressures.[455] Moreover, perception of risk or opportunity varies markedly between individual managers. For example, Aseem Prakash found that risks and opportunities were interpreted differently by different managers even within the same firm.[456] Gunningham et al. also evidenced that different managers can interpret similar external demands differently and that "skilful corporate officials" can reshape some demands by, for example, negotiating with regulators and activists or engaging in outreach with local communities.[457] Yet this literature does not explain *why* a manager would act in one or more of these ways, or whether and how a manager might change their own response over time or in the face of new influences.

To this end, some scholars have identified specific organizational factors, such as identity and culture, and organizational structure, which help explain how managers perceive external pressures and develop their motivations for action.[458] However,

---

450. Adams and Whelan, *supra* n. 336, at 120 (emphasis in original).
451. A.J. Hoffman, *Getting Ahead of the Curve: Corporate Strategies that Address Climate Change* (Pew Center on Global Climate Change, Arlington VA, 2006), 37.
452. T.N. Gladwin, "The Meaning of Greening: A Plea for Organizational Theory," in K Fischer and J Schot (eds) *Environmental Strategies for Industry: International Perspectives on Research Needs and Policy Implications* (Island Press, Washington DC, 1993), 52-54.
453. Lundgren and Catasús, *supra* n. 164, 192.
454. See e.g.Walley and Whitehead, *supra* n. 344 *cf.* Porter and van der Linde 1995a, *supra* n. 343.
455. Carroll, *supra* n. 283, at 501. See also T.W. McAdam, "How To Put Corporate Responsibility Into Practice" (1973) 6 *Business and Society Review/Innovation* 8.
456. Prakash, *supra* n. 447.
457. Gunningham et al., *supra* n. 390, at 38.
458. E.g. J. Howard-Grenville, J. Nash and C. Coglianese, "Constructing the License to Operate: Internal Factors and Their Influence on Corporate Environmental Decisions" (2008) 30 *Law &*

Borck and Coglianese note that empirical testing of these frameworks remains incomplete.[459]

As part of this burgeoning scholarship, researchers have attempted to explain how managers make decisions. Traditionally, managerial decision-making has been conceptualized as a rational and linear process.[460] Yet recent theories espouse that a manager's past experience, risk expertise, personal values, and even gender bear upon their decision-making so that decisions follow an inner logic rather than an objective one. Klein's Recognition-Primed Decision Model indicates that experienced decision-makers are often influenced by what they judge as familiar rather than rationally weighing options.[461] On the specific topic of low-carbon investment, Neuhoff states that "investment decisions are influenced to a larger extent by traditional approaches and intuition rather than by fundamental analysis."[462] As such, there is a concern that senior finance managers might have a natural tendency to support conventional technologies over and above cleaner and greener ones on the basis of familiarity and competitive advantage.

Moreover, some scholars contend that managers' personal morals or ethics may influence their decision-making, particularly as it pertains to socio-environmental deliberations and corporate uptake. Beach's Image Theory contends that personal values or principles are "the foundation of one's decisions" such that "potential goals and actions must not contradict them or those goals and actions will be deemed unacceptable."[463] Wood goes one step further in proffering the "Principle of Managerial Discretion," which posits that "managers are moral actors, [so] they are obliged to exercise [managerial] discretion, within the very domain of CSR, as is available to them, towards socially responsible outcomes."[464] In other words, managers cannot avoid exercising their individual discretion to "do the right thing" even when faced with contrary corporate rules, procedures or policies.

In this context, the concept of cognitive dissonance is relevant to human and organizational change. Festinger explains that humans experience a "cognitive dissonance" and psychological discomfort when ideas/values (internals) do not match with an actual state of affairs (externals) and we experience "wrongness."[465] Festinger contends that when a person experiences dissonance they will act to remove the discomfort in order to return to a level of harmoniousness. Drawing on Festinger's theory and Lewin's three-step model of change, Adams and Whelan contend that

Policy 73. See also J. Howard-Grenville, *Corporate Culture and Environmental Practice: Making Change at a High-Tech Manufacturer* (Edward Elgar, Northampton, 2007).

459. Borck and Coglianese, *supra* n. 293, at 6.
460. I.L. Janis and L. Mann, *Decision Making: A psychological Analysis of Conflict, Choice, and Commitment* (Free Press, New York, 1977), 172.
461. G. Klein, *Sources of Power: How People Make Decisions* (MIT Press, Cambridge MA, 1998), 30. See also P. Slovic, "Perception of Risk" (1987) 236(4799) *Science* 280.
462. K. Neuhoff, "Investment Decisions Under Climate Policy Uncertainty: Based On Workshop Discussions And Interviews With Sector Participants" (University Of Cambridge Working Paper, Electricity Policy Research Group, Jul. 6, 2007), 13.
463. L.R. Beach (ed), *Image Theory: Theoretical and Empirical Foundations* (Lawrence Erlbaum Associates, Mahwah NJ, 1998) at 9.
464. Melé, *supra* n. 276, 53; Wood, *supra* n. 290, at 699.
465. L. Festinger, *A Theory of Cognitive Dissonance* (Stanford University Press, Stanford CT, 1957), 3.

managers can resolve dissonance by unfreezing corporate stasis, finding new role-models and adopting a new cognitive framework, and "refreezing" a new set of cognitions once they are expressed in everyday behavior.[466]

Empirically, previous studies have revealed evidence of cognitive dissonance between respondents and their organizations on the topic of CSR. For example, in 1996 Bebbington and Thomson found that managerial conceptions were more radical than organizational conceptions of sustainable development; and O'Dwyer's 2003 examination of managerial conceptions of CSR in Irish corporations evidenced "pockets of resistance" by individual managers to a narrow corporate conception of CSR that privileged shareholder wealth maximization.[467] Yet it is not clear from these studies whether managerial dissonance then led to radical organizational change; so the potential for agency to bring about transformative corporate change is not demonstrated satisfactorily. Moreover, Spence writes that cognitive dissonance was not evidenced in his cross-industry study of UK companies in 2007, reasoning that companies had become more proficient at expounding a business case for sustainability that eradicates conflict.[468]

In summary, theories about managerial decision-making acknowledge the "human" element in corporate activity and help to explain variation in corporate green uptake. However, this literature often raises more questions than it answers and, as such, does not provide a comprehensive theory of corporate environmental behavior. Specifically, it does not assist with predicting organizational change because the ethical dimensions of managerial decision-making are intrinsic to each individual and therefore defy consistency. Moreover, it is still not clear how individual perceptions of CSR are shaped or, more importantly, whether leadership alone is sufficient to induce far-reaching corporate change.

Furthermore, while previous studies highlight the combined influence of external pressures and internal variables on a company's commitment to beyond-compliance measures,[469] it is not obvious how they interact and influence each other. The complex and sometimes confounding interplay between internal and external drivers was acknowledged by Gunningham et al. who stated that individual managers do not operate in a vacuum: their decisions and personal values are reflective of bigger picture influences. They assert that attempting to "disentangle environmental management style from the external pressures that a firm faces" is overly simplistic.[470] Indeed, it appears from the literature that the relationship between internal and external drivers can be difficult to define or predict systematically.

---

466. Adams and Whelan, *supra* n. 336, 125-126; K. Lewin, "Frontiers in Group Dynamics: Concept, Method, and Reality in Social Science" (1947) 1 *Human Relations* 5.
467. J. Bebbington and I. Thomson, *Business Conceptions of Sustainability and the Implications for Accountancy* (ACCA Research Report No. 48, ACCA, London, 1996). O'Dwyer, *supra* n. 291, at 548.
468. Spence, *supra* n. 318, at 875.
469. Gunningham et al., *supra* n. 390; N. Roome, "Developing Environmental Management Strategies" (1992) 1 *Business Strategy and the Environment* 11-24; A. Ghobadian, H. Viney, J. Lui and P. James, "Extending Linear Approaches to Mapping Corporate Environmental Behavior" (1998) 7 *Business Strategy and the Environment* 13.
470. Gunningham et al., *supra* n. 390, at 133.

§3.06   SUMMATION: KNOWLEDGE GAPS AND EMPIRICAL NEXT STEPS

This chapter has provided three different theoretical perspectives from meso, macro and micro levels on why companies voluntarily adopt environmental and social initiatives. Yet there are several gaps in the extant literature, particularly at the meso level, in relation to the research questions posed in this book.

In summary, at the meso level, it appears from the business strategy literature that the business case as driver is important to green uptake. However, it may not be the only or even the determining driver, with notions of legitimacy and social license also taking center stage. There is also some uncertainty around the importance of risk mitigation in driving corporate greening, with "reputational" risk requiring specific elucidation. From the literature, it appears that reputational risk is synonymous with risk to a company's social license. Yet concepts such as "reputation" and "branding" need further deconstruction if we are to understand their strength and limitations as drivers. Specifically, the role of soft factors in making a business case – reputation, leadership and branding – remains unclear.

At the macro level, it appears that the concepts of isomorphism and mimetic behavior help to explain the process of corporate change as a response to external socio-cultural and field-level pressures. Yet institutional theory, on its own, has limited ability to explain any heterogeneity within and between firms. While that gap can be remedied in part by theories about managerial decision-making, the micro level theory does not systemically explain how managerial perceptions of CSR are shaped or whether leadership alone is sufficient to induce far-reaching corporate change. Finally, neither institutional theory at the macro level nor decision-making theories at the micro level adequately account for how external pressures and internal variables meet and interact to shape a company's commitment to beyond-compliance measures.

Importantly, for the purposes of this book, our understanding of what motivates private finance actors to adopt climate-related strategies remains poor despite the plethora of theory about why companies do and should "go green." As such, we need a better understanding of what drives the climate-related behavior of these actors given their importance to one of the most significant economic transitions for modern human society.

Many of the studies that investigate why companies go green have either grouped finance sector actors together with other sectors, such as the food and beverage industry and primary producers, or not included finance sector actors in their data sample at all. As such, it has not been clear whether the finance sector is driven to "go green" by the same forces as other corporate sectors. Although some studies have attempted to remedy this gap, they have tended to imply or assume that the finance sector is homogenous and to elucidate *what* actors are doing and/or why they *ought* to become greener rather than investigating how and why they are actually doing so.

In summary, there is very little empirical data on why private finance actors are *actually* adopting climate-related strategies. Previous studies have created some uncertainty around risk mitigation as the main driver of change, the role and meaning of "reputation" as a soft factor in the business case, and whether and which macro influences shape corporate behavior.

The remainder of this book is predicated on the thesis that without an empirically grounded understanding of the actual motivations of regulatees, "it is impossible to disentangle wishful thinking and ideological exhortation from the kind of realistic expectations on which governmental and social policy can sensibly be based."[471]

---

471. *Ibid.*, 135.

CHAPTER 4

The Levers of Corporate Change: Case Study Evidence

§4.01 OVERVIEW

Despite regulatory uncertainty on climate change at national and international levels, early moving private sector banks around the world are making voluntary climate-related changes to their business practices. Although Chapter 3 outlined a plethora of theory regarding why companies go green, it is still not clear why they are doing so when not legally compelled. In other words, what drives early moving banks to adopt climate-related practices? And what are the implications for other private finance actors?

At the outset, it must be acknowledged that the word "driver" is an imperfect descriptor for what shapes (and is shaped by) bank behavior because it implies that a motivator is both unidirectional and independent. Yet there is complex interplay between actors in the field (banks, corporate clients, NGOs, regulators) and reasons for behavioral change. Nonetheless, viewing the data in terms of "drivers" provides a useful lens through which to identify and disaggregate influences on leading banks in order to prompt new theoretical and practical insights.

In summary, the case study data reveal that climate-related changes to bank business are driven primarily from three directions: bottom-up via clients seeking solutions; top-down via senior management initiatives or support; and middle-out via mimicking of competitors' behavior due to peer pressure. As such, all three levels of theoretical lens – meso, micro and macro – from Chapter 3 are evidenced and engaged by the data. When taken together however, reported drivers of bank action yield a single strong message: the overarching driver of climate-related initiatives by early moving banks is business case logic. The business case comprises two facets. First, profits are enhanced: directly via fee generation and indirectly via competitive edge. Second, risks (credit, investment, litigation, reputation, regulatory) are minimized:

what is known as "downside prevention." Behind each direction of change lies a common, ulterior motivation for banks: make money and do not lose money.

Arguably, these findings are unsurprising in a capitalist paradigm where shareholder return is paramount. Yet this case study informs and challenges parts of the business strategy and CSR literatures by revealing a deeper and more complex understanding of "corporate reputation" as a powerful motivator of bank behavior. Moreover, it elucidates the role of external drivers such as NGO pressure and the importance of legal and policy contexts that shape bank decision-making. This yields jurisdictionally-specific findings that have important implications for practitioners and policy-makers.Finally, I deconstruct the role and meaning of CSR as an actual driver of climate uptake by banks, which reveals valuable insights about managerial conceptions of CSR and the limited ability of CSR to mobilize corporate change in actuality.

§4.02   QUALITATIVE METHODOLOGY

Answering the question of what drives early moving banks to adopt climate-related practices required empirically-based "real life" investigation. I conducted interviews with senior departmental managers regarding organizational structures and managerial decision-making processes in early moving banks that had been leading on the issue of climate change. The logic was to understand why, in their own words, these banks had adopted climate-related practices and to draw implications for other private finance actors and also policy-makers. This type of investigation has not been done before in the field of corporate climate finance.

[A]   Sourcing the Banks and Respondents

In a qualitative case study design the sampling logic is purposive not random.[472] I targeted early moving banks due to their climate-related leadership and network change potential. My sample comprised seven banks that were early movers on the issue of climate change, being a mix of investment banks and brokerages, diversified commercial banks, private banks, and retail banks.

I conducted 32 semi-structured interviews, comprising 19 interviews with internal (bank) respondents for their insider knowledge and 13 interviews with external (third-party) respondents for additional perspectives on the role of banks in climate change mitigation and the practices of these early moving or "leading" banks specifically. Interviews were conducted in the United States, Australia and Europe during May-June 2010 and September 2011.

---

472. R. Yin, *Case Study Research: Design and Methods* 31-33 (2d ed., Sage: Thousand Oaks CA, 2003); K.M. Eisenhardt, *Building Theories from Case Study Research*, (1989) 14 Academy of Management Review 532, 536-537. See also D. Silverman, *Interpreting Qualitative Data: Methods for Analyzing Talk, Text and Interaction* 250-252 (2d ed., Sage:Thousand Oaks CA, 2001); M.Q. Patton, *Qualitative Evaluation and Research Methods* 230-240 (3rd ed., Sage: Thousand Oaks CA, 1990).

I pinpointed the early moving or leading banks by cross-referencing the seven highest ranked banks in two contemporary reports on climate change endeavors in the banking industry.[473] I then contacted the authors of those reports to confirm the rankings and clarify the leadership criteria used, which included: climate-related corporate governance mechanisms (such as an environment charter, board oversight, and specific focus on climate change as opposed to the more general "environment"); risk management and project financing/lending practices that account for carbon intensity of clients and projects; climate-oriented corporate investment products (such as clean tech equities); high activity in carbon markets and funds; and high activity in advising/knowledge-sharing on climate change with clients, community and government.

These banks were further confirmed as leaders in the space by multiple other sources, namely bank respondents, NGOs, consultants, and mass media. These banks are transnational with offices throughout the world, but their head offices are located in the United States, Europe/UK, and Australia.

Bank respondents had intimate knowledge of the climate-related initiatives in each leading bank. Two to four people from different units within each leading bank were interviewed so as to ensure a variety of perspectives and responses for each case. The majority of respondents were managers (titled "managing director," "executive director," "director," "vice president" or "senior manager") who headed up units for renewable and clean technologies, equity and investment research, carbon trading, project and structured finance, and energy and power. I term these respondents collectively in this book as "transactional bankers." I targeted this group of managers for their hands-on knowledge and implementation of climate-related practices, as well as their link to board management and central policy within the bank. A minority of respondents headed up CSR and corporate sustainability units. I label these respondents collectively as "CSR managers."

In this way I had access mostly to upper management transactional bankers and analysts, not public relations officers. This was an important and distinctive aspect of the study design given that qualitative studies on corporate sustainability often focus on employees within PR and CSR departments, leaving the perspectives of employees "at the coal face" relatively unexplored in the literature.

I also interviewed external respondents who could give additional perspectives on the role of banks in climate change mitigation generally and the leading banks' practices specifically. In qualitative research, validity refers to whether the findings of a study are true and certain as opposed to replicable: "'true' in the sense that research findings accurately reflect the situation, and 'certain' in the sense that research findings are supported by the evidence."[474] Data triangulation is a method used by qualitative researchers to check and establish validity in their studies. I adopted this approach by

---

473. Cogan, *supra* n. 110; B. Furrer, V. Hoffmann and M. Swoboda, *Banking and Climate Change: Stumbling into Momentum? An Analysis of Climate Strategies in More Than 100 Banks Worldwide* (SAM, ETH, ZHAW, 2009).
474. L.A. Guion, D.C. Diehl and D.McDonald, *Triangulation: Establishing the Validity of Qualitative Studies* (University of Florida, IFAS Extension, FCS6014, 2011), 1.

using different sources of information to enable analysis of the research questions from multiple perspectives and thus to increase the validity of the study. To this end, I purposively selected external respondents on the basis of their independence and expertise as evidenced by their public contribution to and critiques about corporate climate finance. External respondents comprised: expert consultants on climate change to the finance sector via the UNEPFI; academics and consultants who had authored reports on the topic; relevant staff at several transnational NGOs and environmental activist groups; staff at a niche climate change bank; and the editor of a leading industry journal.

Respondents and their organizations will not be identified in this book. It became clear during data analysis that naming individuals or even individual banks would likely distract from my findings and their import for good practice and effective policy-making. This is because the literature, particularly from NGO and media sources, tends to single out and judge individual banks; and the data confirmed high competition within the banking industry. In order to ensure confidentiality, interview quotes are cited only with a bracketed letter-number code in-text, for example (C1) or (NGO-A1).

[B]   Data Analysis

My interview question guide reflected the core research themes for the study regarding "drivers," "conceptions of CSR," "regulatory context," and "composition of the business case." I sifted the interview data into key comparative categories which included: risk averse versus opportunity-driven; business as usual versus entrepreneurial; economic versus moral/ethical; regulatory versus social pressures. I reduced the data by selecting out all mentions of these terms, as well as inferences related to each.[475] Included in the reduced data were statements of contextual relevance relating to green achievements and constraints, conceptualizations of "sustainability" and "CSR," and divergence or adherence to industry standards. I prepared detailed matrices to summarize the categories identified in each interview, which then displayed the core coding categories.[476]

This process of categorization, reduction and coding facilitated cross-case analysis by looking for intra-case similarities coupled with inter-case differences.[477] I did this by comparing: interviews within the same case; cases within the same jurisdiction; and all cases with each other as an aggregate. In order to provide some protection against "the presentation of 'unreliable' or 'invalid' evidence,"[478] I placed emphasis on locating interviews and cases that would tend to conflict with the primary coding categories that had developed from earlier stages of analysis.

---

475. Kearins et al., *supra* n. 321, 526 (2010).
476. M.B. Miles and A.M. Huberman, *Qualitative Data Analysis* (2d ed. Sage:Thousand Oaks CA, 2001).
477. Eisenhardt, *supra* n. 472, at 540.
478. M.B. Miles, "Qualitative Data as Attractive Nuisance: The Problem of Analysis" (1979) 24 *Administrative Science Quarterly* 590, 590. See also: Silverman, *supra* n. 472, at 250-252; Patton, *supra* n. 472, at 230-240.

This highly iterative process of sifting, cutting, coding and comparison, enabled the identification of key themes from the data, which formed the basis for the analysis captured in initial drafts of the findings. These drafts were then revisited and revised leading to more refined descriptions of the findings, which underpin the empirical exploration and analysis.

§4.03   DRIVERS: BOTTOM-UP, TOP-DOWN AND MIDDLE-OUT

Client interest in business aspects of climate change was one of the prime motivators for a bank to create new products and enter new markets. Large corporate and/or wealthy individual clients approached banks to seek solutions: not only for mitigating regulatory risk associated with extant or threatened carbon pricing, but even more so to capitalize on opportunities created by new markets. For example, "Climate change is a big theme for our clients so it makes sense for [the bank] to focus on it" (B1). Similarly:

> It's incredibly client driven. We wouldn't do it unless clients wanted it. Clients see an economic opportunity for them to do it and their clients and customers want to do it. So it's end-user driven. (B1).

One bank had built up a specialist climate-related service because client interest in climate change had "motivated [the bank] to build successful products" (A1). These climate-related products included new investment indices, a research center dedicated to climate knowledge and policy, and development of "investable ideas" in the renewables/clean tech spaces. Specifically, the new investment indices re-classify industrial sectors into four climate-related themes: low carbon energy production, energy efficiency, climate finance, and water/waste/pollution control. In so doing, the indices help to evaluate and funnel investment into companies that are focused on addressing, combating or developing solutions to offset and overcome the effects of climate change. An analyst explained their creation:

> My work started end-2006 with the request of a very large pension fund that came to my team saying that "we've been working with you for a long time and [the bank] is climate friendly, can you help us with a problem?" The Trustees had set a mandate to identify parts of the [fund] portfolio affected by climate change and the client didn't know where to start. Neither did we! But we ended up creating the Indices which map the world for different investors to identify different [climate-related] investment opportunities. (A2)

At the same time, new initiatives required senior management support. In order to adopt or develop new products and enter new markets, additional human and financial resources were needed. Yet bank unit budgets were fixed; if unit managers wished to go beyond the budget for new initiatives then they needed to make a business case to senior management. As stated by one banker: "[The Board has] a clearly defined strategy: if it makes good business sense then we will do it" (G1). And senior management had become more informed about when and how climate-related initiatives make good business sense. An investment banker explained his Board's rationale for getting into the renewables space:

> What's different now compared to the '90s is that the senior management team is more focused on the economics of it all. Distribution strategies, differentiation strategies from our competitors, levelized cost of energy or "LCE"; that is, creating a business model that makes sense. In the '90s we were dealing with ideas of what might happen in the [renewables] space in the future, it wasn't concrete yet. Now it is concrete and investors like that. (C3)

Similarly, a carbon research analyst described why his bank had become active in carbon trading:

> I wrote a report in 2005 when no-one else was in carbon emissions trading and I spelt out the (positive) impacts of it on profitability...I've written a lot of reports but this one had the most impact internally within the bank management and also externally with investors and clients. (E2)

In this way, climate change can be a strategic pillar of a bank's business:

> Climate change was recently placed under very senior management leadership... The first thing they will do is assess climate change markets and opportunities in each business unit and decide how to change or move business to capture that opportunity... [The bank] is saying this is not an issue that will go away anytime soon. (A1)

So we begin to see an interplay between Board decision-making, "good business sense" and client service. For banks, providing responsive and innovative client service creates two important benefits. The first consequence is direct: it generates fees. These fees are significant, coming from large corporate and/or affluent individual clients and they create profit for the bank and increase value for shareholders. In other words, fee-generation is a hard or measurable item that comprises an unambiguous business case.

The second consequence is indirect and harder to quantify, yet it ties into the first. Excellent and innovative client service creates a reputation for the bank as the "go to" bank in the space. This is crucial because the banking industry is "a very competitive industry because we are all competing for a fixed universe of clients. There is a fixed size of the purse" (A2). The effect is that "You do or die on customer perceptions of you" (F2). So reputation as the "go to" bank creates competitive advantage. Each bank works very hard to win clients from other banks whilst retaining their own.

Clearly then, retaining and gaining clients is good for profits/value increase; thus "reputation" inevitably, albeit indirectly, generates fees. In terms of the business case, it is a virtuous hard-soft-hard circle.

The previous chapter demonstrated that leadership and reputation are regarded in the business strategy literature as "soft issues," which are intangible, often unmeasurable, and therefore under-researched as factors relevant to a business case.

Yet the data from this case study suggest that (soft) "reputation" is as crucial as (hard) fee generation in driving bank behavior and innovation. Specifically, "corporate reputation" is an important albeit unquantifiable part of the business case for green uptake. First, it helps a bank to make money via being – or being seen to be – "green" and/or delivering excellent client service that puts a bank above its competitors (that is, reputation enhancement). Second, it helps a bank to not lose money via negative social or corporate perceptions (that is, prevention of reputation loss).

Naturally these statements require detailed explanation. To do that, I need to unpack a central concept in practice and theory: the meaning of "corporate reputa-tion."

[A]   Meaning and Role of Corporate Reputation as Driver

Notions of legitimacy often focus on "social license" and "social reputation" as the key to a company's right to exist. Social license is borne of compliance with regulatory standards, institutional norms and/or societal expectations and, as such, confers legitimacy upon a corporation. The idea is that:

> society gives license to business to operate and, consequently, business must serve society not only by creating wealth, but also by contributing to social needs and satisfying social expectations towards business.[479]

Complying with these elements shows that a company's values match those of the social system within which it is embedded; when there is a mismatch then an organization's legitimacy, and hence survival, is threatened.[480] Social license also comes from "the power-responsibility equation" whereby business has power which brings with it responsibility beyond economic and legal considerations.[481] This dove-tails with Davis's Iron Law of Responsibility, which states that "those who do not take responsibility for their power, ultimately shall lose it."[482] In other words, business needs social acceptance to survive.

A necessary part of social license is social reputation. Social reputation empha-sizes both the risk to which a company is exposed if its actions are contrary to societal expectations and the benefit for a company if it is accepted by the community within which it operates.[483] Social reputation also dovetails with the concept of "reputation

---

479. Melé, *supra* n. 276, at 49.
480. S.P. Sethi, "A Conceptual Framework For Environmental Analysis of Social Issues and Evaluation of Business Response Patterns" (1979) 4(1) *Academy of Management Review* 63; J. Dowling and J. Pfeffer, "Organizational Legitimacy: Social Values and Organizational Behav-ior" (1975) 18 *Pacific Sociological Review* 122.
481. K. Davis, "Understanding the Social Responsibility Puzzle" (1967) 10 *Business Horizons* 45. See also K. Davis, "Can Business Afford to Ignore Social Responsibilities?" (1960) 2(3) *California Management Review* 70; Walton, *supra* n. 277.
482. Davis and Blomstrom, *supra* n. 277, at 50.
483. S. Lewis, "Reputation and Corporate Responsibility" (2003) 7(4) *Journal of Communication Management* 356.

capital," which "represents a communications bridge that predisposes NGOs, communities, and other groups to enter into open discussion rather than hostile opposition" with a corporation and therefore engenders its credibility.[484]

By building reputation capital, the up-front risk and costs associated with gaining social acceptability are reduced and access to profits is enhanced through local community tolerance, having the trust and ear of regulators, and not being targeted by NGOs.[485] In this way, firms can regard reputation capital as a good economic investment.

Some CSR scholars have found that a company's social license and social reputation can be leveraged by NGOs and civil society to pressure individual firms into becoming more socially and environmentally responsible. For example, in their cross-national study of pulp and paper mills, Gunningham et al. found that "the social license was the primary source of [corporate] beyond-compliance measures of the good citizenship variety" due to pressure from societal stakeholders.[486]

Similarly, a number of previous bank studies have highlighted that bank reputations are at risk from public naming and shaming campaigns by NGOs that target financiers of environmentally 'dirty' borrowers. Such campaigns have potential to undermine the trust and reputation deemed so crucial to banks' operations. Arguably, the fear of that outcome motivates attention to CSR in downside prevention. As such, protection of social reputation would seem to be a logical driver of socially and environmentally responsible bank behavior.

Yet, in reality, social reputation is not the only type of reputation to motivate bank behavior, and trust has a number of facets. The terms "social reputation" and "corporate reputation" are often used interchangeably in the literature,[487] which indicates that a company's reputation has been conflated with its social standing. However, this case study reveals that the concept of "reputation" from the perspective of banks and bankers is multi-dimensional.

In summary, "corporate reputation" has more than one meaning for banks: it encapsulates "social reputation" in a way that accords with the literature; but it also encapsulates reputation for good business sense, which is termed in this study as "client service reputation." Both types of reputation can co-exist in the same firm; they are not mutually exclusive. Yet their context and effect differ. This nuanced conception of corporate reputation as a disaggregated phenomenon is captured in Figure 4.1. In summary, "client service reputation" is a powerful motivator of bank behavior, as acknowledged by all respondents regardless of their type of bank or jurisdiction.

---

484. S.A Joyce and I. Thomson, "Earning a Social License to Operate: Social Acceptability and Resource Development in Latin America" (1999) (June) *Mining Journal* 441.

485. Gunningham et al., *supra* n. 390, at 53.

486. *Ibid.*, 51.

487. See e.g.: Melé, *supra* n. 276, at 49; C. Wright and A. Rwabizambuga, "Institutional Pressures, Corporate Reputation, and Voluntary Codes of Conduct: An Examination of the Equator Principles" (2006) 111 *Business and Society* 89 (2006); Lewis, *supra* n. 483; J.F. Mahon, "Corporate Reputation: Research Agenda Using Strategy and Stakeholder Literature" (2002) 41 *Business and Society* 415; D. Windsor, "Corporate Citizenship: Evolution and Interpretation," in *Perspectives on Corporate Citizenship* (J. Andriof & M.McIntosh eds., Greenleaf: Sheffield UK 2001), 39.

Conversely, regard for "social reputation" – and the concomitant role of NGO pressure – as a driver of bank behavior was variable. It *is* important to banks, but its salience as a driver depends on a number of factors.

Figure 4.1   A New Conception of "Corporate Reputation"

[1]   *Client Service Reputation*

Since client service excellence is top priority for banks, and given that bank corporate clients are also motivated by the business case, it is at this level that banks pitch their services. Often, the approach is one of "strategic partnership" or that a bank can be a "trusted advisor" to the corporation. The notion of trust here pertains to the bank being an excellent service provider that will help the client's business to flourish. The inner logic is that "we need our clients to do well in order for us to do well. Helping them helps us" (G3). So client service reputation has two key components, namely "trust" in the context of "good business sense."

A number of banks have moved to a "relationship model" with corporate clients that goes beyond doing deals to understanding client needs and designing solutions for them. In the words of one CSR manager: "We are a strategic partner as well as a financial partner with our clients. All clients need to reduce their carbon footprint so we help to solve clients' hurdles" (D2). A relationship of trust augments corporate client satisfaction; and satisfied clients keep coming back. Thus, client service reputation is crucial because it builds bank business. "Leadership and trustworthiness are very important to [the bank]. Clients come to me because of our reputation. Reputation creates business and can sustain business" (A2).

Client service reputation is crucial in a highly competitive market. It helps a bank to differentiate itself from its competitors; it creates hard profits from soft factors:

> Australia is an emerging market regarding carbon so we are carving out our leadership now... Our perceived expertise, understanding the energy market, guiding clients in the right products etc. is critical. So reputation from a customer service perspective is vital. (F2)

And, as described by one banker, the clear upshot to having a reputation for good business sense is "the bottom-line" because it "creates client business: sector expertise puts us above our competitors, which means that clients keep coming back and they bring others with them" (G2).

99

This is very important for an industry in which banks are motivated by profit imperatives and jostling each other for a bigger slice of the same pie.

Accordingly, appropriate branding strategies and image management are vital to differentiation; and the hook for marketing to corporate clients is good business sense. A number of bankers and analysts explained how they were blending climate-related strategy with good business sense to enhance client service reputation and competitive edge for their bank. One example is the climate investment indices discussed above. A project financier at another bank described how he worked closely with the commodities group to use hedges in order to facilitate lucrative returns for the bank and clients; this had made his bank a leader in energy/renewables projects in the United States.

Another bank coordinated its environmental activities through a senior management committee. It was explained that this "improves [the bank's] ability to sell itself to clients as an integrated climate change aware entity" (E2), which comes back to good business sense and appropriate branding. Bankers in different departments were coordinated in their dealings with the same client in order to provide better "one stop shop" service:

> We give integrated advice and client service – it is important that [the bank] is seen as integrated... It is also part of our branding. We make clients aware [of climate-related opportunities] because it's good for business and also we can be seen as a socially responsible corporate entity. (E2)

This notion of being *seen* to be green came up most clearly when respondents discussed branding and differentiation strategies at other banks. Many respondents had worked at other leading banks and, armed with inside information and now having changed alliances, were quite willing to reveal competitors' strategies. At these times, respondents tended to be quite frank about un-green motivations for going green, specifically as they related to image:

> [One bank] is at the start of the cycle so it's very busy raising its profile regarding climate change initiatives. We pulled back from sponsoring climate-related leadership conferences; [the other bank] is doing it now. Why are they doing it? Because it makes money... Their message will be "we are doing this to be a good corporate citizen." But I argue that at the heart of it there is a strong economic rationale. (B1)

Sometimes it is not in a bank's interest to be seen as a global warming advocate, yet it can still push climate-related products and services. An analyst recalled a client meeting where this became apparent:

> We were having a meeting with a large US pension fund and one of their executives said "we don't believe in the science" and got very angry about it... [My bank's] response is that we don't make any assumptions regarding the science of climate change. It is purely quantitative research. For example, in 2004 there were 154 companies delivering goods and services in low-carbon. In March 2010 there were 385 such companies. So obviously companies are moving on the climate change issue. (A2)

The implication is that banks and their clients need to get on board with climate change in order to not get left behind.

Thus, for a climate-related initiative to make economic sense and/or enhance the brand, banks can be agnostic about climate change. This supports the postulations of Hoffman and also Hawken et al. that climate-related corporate changes can be viewed as a strategic issue rather than an ethical response to a scientific fact. There were some individuals within the banks who did profess to care about climate change on a moral level, as elucidated below. However, the overarching characterization of climate change by respondents was in terms of "risks/benefits," which embodies business case logic. A bank can get ahead or be left behind depending on its approach to climate-related strategy. As such, "going green" is therefore viewed as appropriate and even necessary for bank and client businesses. Put simply: "whether or not you want to be interested in the climate issue, you've got to be. This is the way the world is moving" (E2).

Figure 4.2   Constitution of Social Reputation versus Client Service Reputation

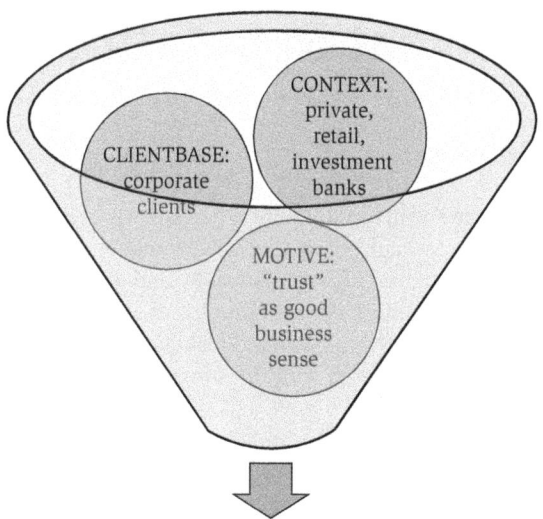

Client Service Reputation

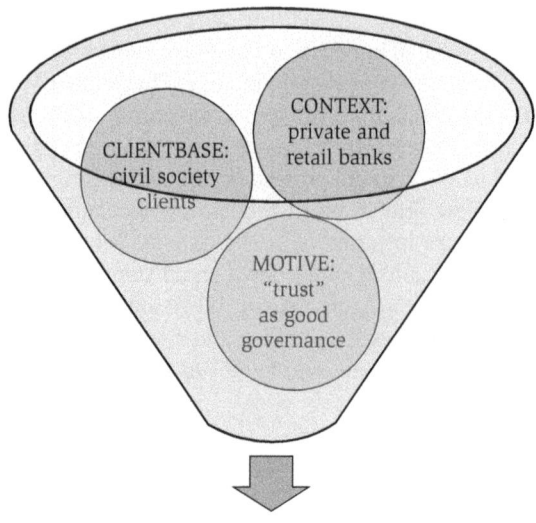

Social Reputation

[2]   *Social Reputation*

Social license – and hence social reputation – is particularly salient for companies that produce environmentally damaging products or are highly visible polluters and are therefore targeted by environmental activists locally and abroad.[488] Banks are not big polluters nor are they high-emitting entities; indeed, they have been peripheral actors until relatively recently. However, bank relationships with GHG-intense clients and projects have come under increasing scrutiny from NGOs that have started to look behind polluters to their financiers. Banks, and particularly their CSR managers, are quite aware of this change: "The public and also advocacy groups are more sophisti-cated about following the money … . Banks are no longer invisible" (C2).

    Thus, two factors combine to provide the "social reputation" of banks. First, banks are in a highly competitive service industry and reliant on customers' goodwill for survival. Second, they are now visible and their financing activities can be scrutinized publicly. Yet there is subtlety and variation associated with social reputa-tion beyond that described in the literature. The data reveal that the salience of social reputation and the concomitant power of NGOs as a driver of bank behavior vary with three factors: the type of bank, the type of employee, and the regulatory context. Appreciating this variation means that social reputation for banks can be understood more accurately and, therefore, leveraged more effectively.

---

488. E.g. Gunningham et al., *supra* n. 390, at 31-32, 52.

[a]   *Type of Bank*

Social reputation has greater salience for private banks and retail banks than for investment banks without consumer branches. This is because private banks deal with wealthy individuals and are therefore very careful to check a potential client's background for risks that may cause credit or reputational damage to the bank. For banks with retail branches, their relationship is with the general public because people invest their savings with them. As such, both private and retail banks have a relationship with civil society.

Trust is crucial to a bank's relationship with civil society account-holders. Yet, unlike client service reputation, this type of trust relates less to a bank's business acumen and more to it being a good corporate citizen:

> We want to be a force for good but there is probably a pragmatic component to this. We have a large retail base in Australia. We have lots of customers and shareholders who are a bit like voters – we can make an analogy to government here – and we want to be well-perceived by them because they hold accounts with us and own our shares…We don't want our retail base to think we are a complete pack of bastards, well, any more than they already might. We want them to know we are not entirely evil all the time. (G4)

Bank respondents emphasized their concern that if retail clients distrust the bank or dislike its behavior, then they can switch across to another bank taking their accounts with them. Thus, capturing the hearts and minds of the general public is a crucial component of consumer relationships and social reputation for retail banks.

Accordingly, branding and image management are also very important to the social dimension of reputation. Yet, unlike client service reputation which requires good business sense branding, social reputation requires good conscience branding:

> With our bank logo we want an emotional response. It's about trust – put your money with our logo, or borrow from us for "the biggest purchase you'll ever make" as the marketers say. (G1)

Notions of sustainability, good corporate citizenship and CSR become important branding tools here. One banker evidenced this by using the marketing strategy of a competitor bank where he had previously been employed. He stated that the competitor bank was the biggest home lender with the biggest branch network in that jurisdiction and therefore it had a commercial rationale underlying its eco-friendly marketing. That rationale was to keep retail customers happy (so they kept their mortgages and accounts with the bank) and NGOs happy (so they did not campaign against the bank and turn customers away). I asked an institutional banker at said competitor bank about this assessment and he agreed with it, adding:

> Our ad campaign aimed at consumers regarding climate change was highly successful. We are in the market as a good corporate citizen and we are doing our bit for a sustainable future. It resonated well with the moms and dads… many people came to us because of it. (F1)

So it is at the retail/consumer level – where banks have a relationship with civil society as opposed to the corporate sector – that social reputation manifests for banks.

Concomitantly, it is at this level that NGOs have the most influence as a driver of bank behavioral change. A risk manager related how a large environmental NGO (ENGO) had launched a "nasty campaign" against his bank for its funding and support of fossil fuel and extractive industries. The campaign involved customer boycotts, during which activists visited college campuses and advocated for university students to cut up their credit cards. Similarly, another large ENGO had been campaigning against another leading bank for its financing of coal. As part of that campaign activists had protested outside local branches and the ENGO had engaged media publicity to distort the bank's marketing slogan. Such campaigns had hit at the heart of these banks' social reputations.

The campaigns had also clearly rattled staff and senior management. Bankers commented that: "our retail arm is important enough – especially when ENGOs are hassling branch staff – for us to want to mitigate protests" (G1); "We have customers and we want to keep them" (G2); and "we don't want customers to go through that – it's not sustainable for a business" (G3). The bank that was targeted by the campaign, and indeed all major retail banks in the jurisdiction, had initiated discussions with that ENGO to explain their approach to coal financing. Moreover, senior management started ensuring that new renewable energy deals were announced publicly in order to reinvigorate the bank's image. The aim of this "green" marketing approach was to keep customers with the bank, and hostile ENGOs away from it.

In contrast, banks without retail arms, such as investment banks, are not so easily moved by public protests. One project financier at an investment bank gave me his perspective of the 2007 TXU campaign during which RAN had targeted financiers of the TXU proposal to build 11 coal-fired plants in Texas.[489] Activists had described the outcome as a success because TXU ended up being bought out by Kohlberg Kravis Roberts (KKR) and Texas Pacific Group with an agreement to build only three of the coal-fired power plants and to invest in "green" energy technology. However, this banker gave me a different perspective:

> TXU was sold via consortium to KKR. KKR then lobbied government and said "we'll only build three plants" and got everyone happy and comfortable with it. But TXU was never going to build 11 plants, it was just negotiating for three. So [the bank] did the deal and was a winner anyway. (C1)

Indeed, the TXU deal at US$45billion was described as the largest-ever acquisition by private equity firms. In short, NGO influence on a bank without consumer branches is negligible. "There were protestors outside [the bank's] office over TXU. But that won't stop us from financing as long as we are doing the work necessary to be done" (C1).

So, NGOs can drive greener corporate consciousness and behavior via campaigns directed at private banks and banks with retail arms. They do not have such leverage with investment banks. Moreover, the data also show that, after a certain point, even retail banks demonstrate resistance to hostile campaigns.Bankers acknowledged the

---

489. See Chapter 2 *supra*.

importance of discussions with NGOs but emphasized that: they would "not be held to ransom" by them; NGOs would "not dictate" bank policy; and capitulating entirely to NGO demands would be analogous to "negotiating with terrorists." One CSR manager communicated resistance in these terms:

> [The NGO] wanted the bank to get out of the forestry sector. We said "no, that doesn't make sense." We did a forestry policy instead. Then they said "get out of all coal-fired deals." Our response was "no, but we will ask better questions and understand the risk." European NGOs have recently said "get out of all nuclear." It won't happen. (D1)

In other words, banks would not comply with NGO demands if they contradicted the bank's conception of good business sense:

> [The NGO] wanted us to put in writing that we would not fund new coal, and I refused. It is ludicrous to say that I will not do coal today when there might be less polluting technology tomorrow... There is so much coal in this country that it is not very smart to ignore it. (F1)

Similarly, "Will we fund new coal? Maybe not. But we decide each case on its merits in accordance with regulation and good business sense" (G1).

Indeed, good business sense may dictate that, even if new coal financing slows down domestically, it can continue and even escalate overseas where there is not NGO pressure or media interest. For example, respondents at one bank noted their large presence in Asia where there is a growing market for coal-generated energy and minimal NGO resistance.

Banks had a clear preference for partnerships and dialogic engagement with well-known NGOs. Partnerships can reduce hostility, protect social reputation and, interestingly, enhance client service reputation. On this latter point, a number of CSR managers noted that partnering with a bona fide ENGO facilitated research collaboration and helped the bank to manage projects and clients better, which in turn gave it a competitive advantage.

This is a "sweet spot" of banks – both retail and investment – that has not had much attention to date. Mostly, discussion of NGO leverage has focused on exploiting social reputation. Yet NGOs can also leverage banks' client service reputation in order to facilitate social goals. For example:

> Large NGOs want to come in the door and have conversations with financial institutions. We can partner with them which can work to our [corporate] clients' advantage too. (G4)

Overall, NGOs help banks to "focus on the cost/benefits" (C2) of bank activities and can thereby exploit business case logic, and not just social reputation, for socio-environmental gains.

[b]   Type of Employee

Regard for social license and NGOs also depended on the type of bank employee. A marked difference existed between the attitudes of transactional bankers/analysts and CSR managers.

CSR departments are a bank's contact point with NGOs and there can be frequent interaction between CSR managers and environmental activists. The descriptor "CSR departments" is an all-encompassing term for the bank units responsible for engaging with external stakeholders and making policy on issues of corporate responsibility, CSR and sustainability. CSR managers tended to view constructive relationships with NGOs as preferable to fighting public campaigns. Indeed, coalition building with externals, including NGOs, was a key performance indicator (KPI) for many of them.

The approach toward NGOs by CSR managers followed a stakeholder theory model.They balanced the interests of NGOs and civil society with those of clients and shareholders; they engaged in consultation and dialogue; and they developed corporate strategies, such as green policies and enhanced due diligence, to meet the approval of a broader conception of key stakeholder groups:

> There is a process of original demand by NGOs, then discussion with the bank, engagement and disagreement, but we see if a robust policy can come out of it which satisfies the NGOs. (D1)

These managers understood the importance of engagement. They realized that shareholder and customer satisfaction was partly informed by their bank's social reputation, and that social reputation needs to be managed. As such, they appreciated the value of reputation capital. For example: "We came out of the NGO's [coal] campaign and report much better than [other banks] partly because we fund less coal, partly because we engage actively with NGOs" (F1). Similarly:

> We try to proactively engage with NGOs – we want long term relationships with them – you can build up credit with them in order to move forward in the lean times… it reduces conflict and increases productive dialogue. (F2).

Thus, a dialogic approach is central to the job of the CSR manager. They are a bank's front of house to NGOs: "[the bank] has a long history of engagement with NGOs and clients to create better understanding of where they are coming from. We will talk when someone knocks on the door" (D1)

In contradistinction, transactional bankers and analysts were not motivated by social reputation or NGO pressure. They viewed it as a "policy thing" (B1) and outside their remit. These managers did not interact directly with NGOs. They were motivated by profit imperatives, client/external acknowledgment and by "finding and winning every good renewable deal" (C2). An investment banker in renewable energy described his motivation clearly:

> You eat what you kill in the banking business. In order to keep your job and get paid bonuses you need to keep bringing home the bacon. You need to make revenue, transaction-oriented revenue, which is then reflected in your compensation and the fact that you still have a job at the end of the day. (E1)

Some transactional bankers did not consider NGOs or even civil society to be bank stakeholders. They were concerned with balancing the needs and interests of four main constituents: shareholders, regulators, clients and employees.

For bankers, legitimacy comes not from social acceptance by giving back to society but from market acceptance by creating wealth. This is done via fee generation and excellent client service. Contrary to previous studies, they do well by doing their job properly, not by doing good for society:

> The remit from our shareholders is to make money... [The bank] does some socially responsible activities, for example microfinance, but for renewable energy the core remit is "what is it that's good for the bank?" (B1)

So transactional bankers can acknowledge social license, but it does not drive them. Indeed, some view it as inappropriate that it should: "Are banks there to support society? No they are not. They are economic, commercial experts" (B1). In contrast to their CSR counterparts, these bankers – wittingly or not – subscribe to Levitt's 1958 exhortation on the dangers of CSR that "... if something does not make economic sense, sentiment or idealism ought not to let it in the door."[490]

Were there hybrid positions within each group of managers? Were they giving me the expected answers for their role?

I asked CSR managers about the relevance of making a business case for extra funding for their unit and/or new policy initiatives. Business case logic was still relevant for them even when addressing CSR issues.

Similarly, I asked transactional bankers and analysts about the relevance of their bank's social standing to their own job satisfaction. The extent to which a bank's social reputation mattered to these managers varied with each individual. In the words of one banker: "Risk is everything, it's just a matter of managing and pricing it" (G1). Some bankers commented that the wealth of a client was irrelevant if that client would cause serious reputational damage to the firm; however, such decisions were made by a risk reputation committee, not the bankers, so they did not have a strong view about that process. Other bankers were able to explain the nature of reputational risk in the ESG risk assessment process: invariably, if a client had good governance structures in place, even if it engaged in socially unacceptable activities, the bank could work with them. Importantly, environmental risk was not synonymous with social reputational risk:

> A company can be involved in quite controversial activities but if those risks are well-managed then they do not pose a reputational risk for the bank; the risks might be environmental or social, but not necessarily reputational. (F1)

Finally, social reputation was not the overarching driver of green activity for transactional bankers, although it was sometimes a background factor or reason to feel good about their work:

> From a business perspective it gives me the opportunity to feel good regarding deals and the environment, so when I talk to friends and relatives about what I do tangibly it is very satisfying. (F2)

---

490. T. Levitt, The Dangers of Social Responsibility (1958) 36(5) *Harvard Business Review* 41, 42.

The disparate approaches and motivators between transactional bankers/analysts and CSR managers created occasional friction within the same bank. One CSR department had "head-butted with the mining bankers for months" (D1) before agreeing on and implementing an enhanced due diligence policy around MTR coal mining. The reason for struggle was simple: the bankers feared a negative reaction by their client base and a slow down on deals. To the bankers, it did not matter that financing such deals compromised the bank's social reputation due to NGO campaigns let alone that such practices irrevocably damaged local environments and communities. CSR did not factor for the bankers or, if it did, they conceptualized it very differently to their counterparts. This notion of contest or debate was reiterated by a CSR manager who described how difficult it could be to "convince" colleagues, especially bankers, of the importance of sustainability. Similarly, another CSR manager noted that he had "better conversations with other people like me at other banks than with bankers at [my bank]" (G2). Hence, CSR is a contested concept within firms, which is an issue to which I return toward the end of this chapter.

[c]   Regulatory Context

NGOs can better and successfully leverage the social license of banks when there is a lack of government regulation or industry standards and/or the regulator is weak. A risk manager explained it this way: "We need to make our own decisions when [regulation] doesn't exist – then civil actors and common standards become quite important" (D1).

The changing attitude of American leading banks toward the practice of MTR coal mining provides a good example of this dynamic. NGO pressure had motivated one bank to initiate an enhanced due diligence process for clients and projects involved in this form of coal mining:

> We realized that these were scary practices to try and get coal. But the issue is that MTR is legal in the US, companies can get permits under certain conditions. And under the Bush Administration the regulations were watered down. Then NGOs started focusing on financiers, and we started getting [adverse] attention. (D1)

This threat to social license motivated the involvement of the bank's Chairman and precipitated changes to bank practice:

> In the past we would simply get representations from a client that they were complying with their licenses etc. Now we want to see every permit, every asset, sometimes even an independent review. (D1)

Several leading banks based in the United States had voluntarily instigated internal MTR policies ranging from enhanced due diligence to outright veto.

Moreover, respondents at NGOs and banks alike told similar stories about the creation of voluntary industry standards. The Carbon Principles were described as "a direct outcome of the RAN-TXU campaign" (NGO-B2). A risk manager described how that campaign had "received major attention" in the *Wall Street Journal* and *New York Times*:

> The United States had no carbon mitigation standards and banks were in a tough situation because NGOs and civil society were asking "what are you using to judge investments in the coal sector?" We got together with [other leading banks in the financing syndicate] and developed the Carbon Principles for coal fired plants in the United States, which is an enhanced due diligence process. (D1)

Similarly, the Equator Principles had been drafted by four banks that were the target of NGO campaigns. Previous studies suggest that the Equator Principles were part of a strategy by "exposed" banks to "deflect NGO criticisms" and create a level playing field with competitors who were less vulnerable to campaigns.[491] So out of NGO pressure on the social license of vulnerable banks arises evidence of the role and effect of bank peer pressure. That is, the first stage – *creation* of voluntary Principles – had been driven by NGOs targeting retail and private banks that had a civil society client base and therefore a vulnerable social license to leverage. The second stage – *adoption* of the Principles within the broader industry – occurred when primary banks peer-pressured competitors in their syndicates and consortia, particularly investment banks that, without a civil society base, were immune to NGO naming and shaming campaigns. In legitimacy theory parlance, the creation of the principles by primary banks mended their social reputational damage; the adoption of the principles by secondary banks lent legitimacy to their operations and decision-making. As described by one project financier:

> With the Carbon Principles we were driven by what our competitors do. I went to a whole bunch of those meetings and we were all looking at each other and wondering why we were there. Its peer pressure, we don't wanna be singled out. (B2)

Peer pressure is important in bank life. This is due to the competitive nature of the banking industry and the fact that there is no regulatory certainty across jurisdictions on climate change let alone climate finance. A CSR manager stated: "finance is fungible so banks cannot be an island. For example, if one bank won't finance coal then the client will go to other banks. Money is money" (C2).

So regard for client service reputation, not just social reputation, helps to forge and embed voluntary Principles. These soft codes reduce competition by setting a common industry standard so that "all banks are asking for the same thing when pitching a deal" (D1).

Moreover, yielding to peer pressure (not just NGO pressure) can help to protect and enhance a bank's social reputation. For example, signing onto voluntary standards can be a strategic part of image management and risk mitigation; it can be "a nice PR thing" (B2). Thus, using language from O'Sullivan and O'Dwyer, creating voluntary standards can "repair perceived reputational damage caused to [banks] by NGO campaigns" and adopting them can help competitor banks "gain legitimacy for their financing activities."[492]

---

491. F. Amalric, *The Equator Principles: A Step Towards Sustainability?* 6 (Centre for Corporate Responsibility and Sustainability Working Paper Series No 1/5, University of Zurich 2005); O'Sullivan and O'Dwyer, *supra* n. 166, at 565.
492. O'Sullivan and O'Dwyer, *supra* n. 166, at 565.

In this way, the creation and then adoption of voluntary industry standards demonstrates the combined power of social reputation and client service reputation in shaping bank behavior.

Figure 4.3   Modality and Effects: Social Reputation versus Client Service Reputation

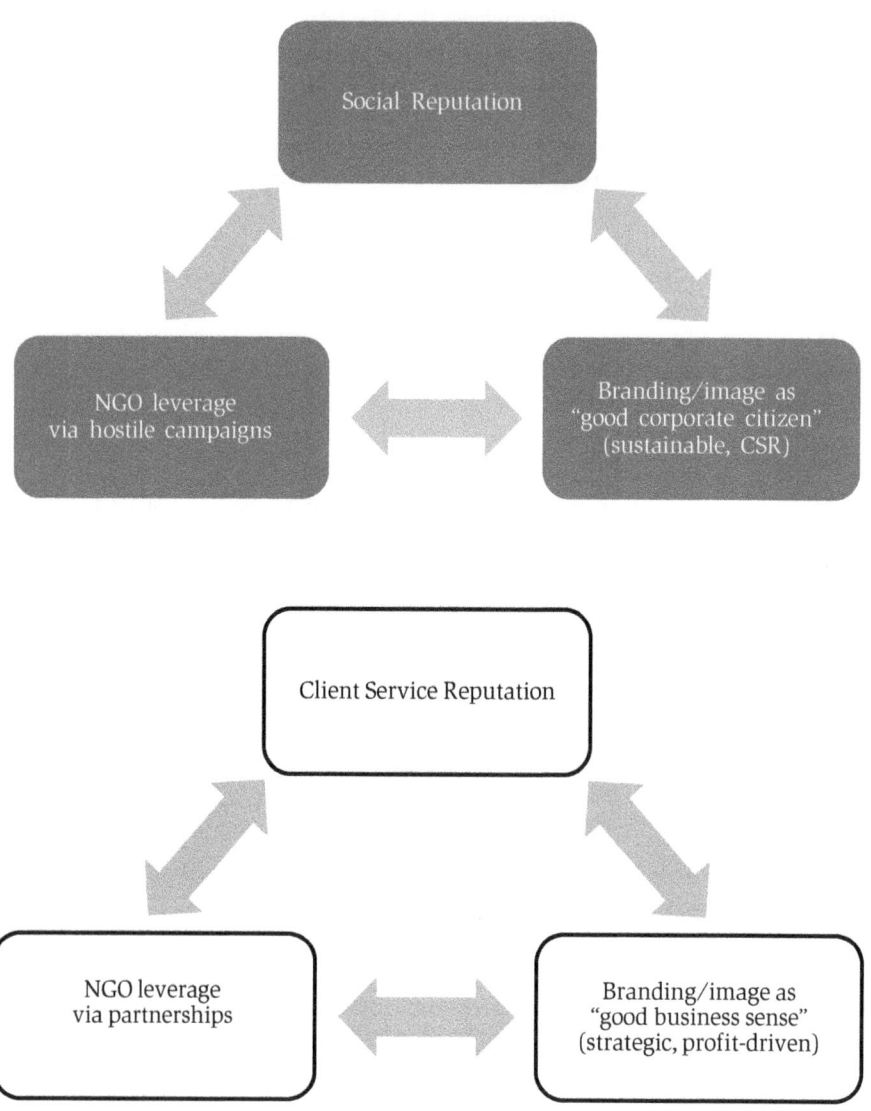

[B]   Risk Mitigation as Driver

The business case as driver comprises not only upside enhancement but also downside prevention. Contrary to previous studies where finance actors apparently viewed sustainability issues as threats, the data in this case study revealed that risk mitigation was not the main driver for all leading banks. For many, capturing climate-related opportunities was the prime motivation.

The banks could be placed along a spectrum – with risk mitigation as the main driver of climate-related behavior at one end, and opportunity (for profit, reputation enhancement, and competitive edge) at the other end. While individual banks were at slightly different points on this spectrum, data analysis revealed that overwhelmingly the placement of banks on the spectrum correlated to jurisdiction.

In summary: for Australian leading banks, risk mitigation was the prime driver of climate-related behavior with opportunities just beginning to emerge; for leading banks in the United States, risk mitigation was a driver of climate-related behavior but the opportunities created by climate change were a stronger driving force for uptake; and for leading banks in Europe, risk mitigation did not drive climate-related behavior because they were primarily focused on capturing opportunities created by adopting and innovating green strategies.

This variability of risk mitigation as driver and the differing placement of where leading banks sat on the risk-opportunity spectrum related more specifically to the regulatory environment in each jurisdiction. Specifically, regulation that (a) set a carbon price, (b) incentivized investment in renewables and clean tech, and/or (c) directly coerced bank behavior in a socially-focused way, informed bank perceptions toward risk and opportunity. Moreover, the weighting that a bank placed on risk or opportunity was strongly informed by the type of risk (regulatory, credit/financial, reputational).

[1]   Types of Regulation and Risk

An actual or even threatened carbon price whether in the form of a tax or ETS has the effect of, at least in part, discouraging bank activity away from carbon-intense industries. This stems from banks' traditional concern to foresee, mitigate and manage risk. The main risk for banks from a carbon price is associated with their clients, especially credit and investment risks (from corporate clients that are exposed to regulatory pressure and changes in market demand in an increasingly carbon-constrained economy) and reputation risk (to both social license and client service reputation). Banks are also concerned here with regulatory risk and uncertainty particularly when a price is only threatened or newly implemented. However, carbon trading needs governments to decree a carbon price and set up the ETS architecture. Thus, carbon markets are "extremely political and dependent on politicians making very big decisions and technocrats implementing them" (E2). At the time of interviews, Europe and Australia had priced carbon; the United States had not priced it at the federal level.

Similarly, financial incentive regulation (such as tax credits, FITs and grants) motivates private funding and business activity in alternative energy sectors. Akin to pricing carbon, this type of regulation is indirect: it does not apply to banks specifically nor is it coercive. Instead, this type of regulation encourages bank activity in specific sectors by providing financial incentives. In particular, incentives are crucial in the renewables and clean tech spaces where start-up costs of projects can be prohibitive; and new technologies can be unpredictable in their early phases such that returns are uncertain. Stable financial incentive regulation addresses market risk, which encourages funding. Financial incentive regulation is also important for encouraging institutional investment because pension funds require price signals to know where to invest in the long term. At the time of interviews, both the United States at the federal level and Europe had this type of incentivizing regulation in place; Australia at the federal level did not.

In relation to socially-focused direct regulation, the United States is unique in having the *Community Reinvestment Act of 1977* (Community Reinvestment Act).[493] This federal statute imposes obligations on defined "banks," including commercial and savings banks and financial holding companies, but not traditional investment banks (section 803) to serve the credit needs of low- and moderate-income communities within which they operate (section 802). Banks can do this by, for example, providing affordable mortgage programs, small business loan products, community development financing, funding for non-profit housing, and economic development programs. The Community Reinvestment Act was introduced by President Carter to serve a social objective through financial means. It was enacted to help revitalize inner cities by redressing inadequate and discriminatory lending to low socio-economic neighborhoods known as "redlining" by which depository institutions reputedly "drew a red line around certain neighborhoods on the basis of the racial composition, age of housing stock, or other factors regardless of the creditworthiness of individual loan applicants, and declined to make loans in those neighborhoods."[494] National banks that are subject to the Act are examined every three years and smaller banks are evaluated every four to five years by federal financial agencies in order to determine levels of compliance. Examination culminates in a rating and a written report that becomes part of the supervisory record for that bank (section 803). Sanctions, resulting from an unsatisfactory rating due to inadequate or non-compliance, consist of government denial or delay of bank mergers, acquisitions and new branch openings (section 804).

American banks that are subject to this legislation were extending their activities to include climate-related initiatives. Europe and Australia did not have equivalent legislation.

---

493. *Community Reinvestment Act of 1977*, Pub. L. No. 95-128, 91 Stat. 1147, 12 USC. § 2901, implemented by Regulations 12 C.F.R. parts 25, 228, 345, and 195.
494. K. Ardalan, "Community Reinvestment Act: Review of Empirical Evidence" (2006) *Academy of Banking Studies Journal* 1, 1.

[2]   Leading Bank Approaches by Jurisdiction

[a]   Europe / UK

Risk mitigation was not a prime driver of climate-related uptake by leading European and UK banks. Rather, leading banks in this region were focused squarely on capturing business opportunities created by climate change.

Europe is a mature jurisdiction with regard to climate change regulation. Its carbon pricing mechanism is well-established; the EU ETS moved into Phase III in 2013 with a more stringent GHG reduction target and the UK ETS now dovetails with it. Concomitantly, several European countries have financial incentives in place for renewables, mostly in the form of FITs to guarantee the price for a specified duration to generators of renewable electricity.

The most relevant risk for leading European banks was social reputation risk. This is perhaps unsurprising given that all leading banks were either private or had retail arms. All banks had risk reputation committees sitting above operational units to vet clients and projects prior to engagement. As described by a banker: "the focus is protecting reputation, not enhancing it" (B1). Due diligence is an entrenched process guided by internal bank standards or Equator Principles benchmarks to identify and assess a range of risks including ESG risks. If risks can be managed or mitigated then the client/project is approved.

Yet bankers emphasized the opportunities and benefits arising from the adoption of climate-related strategies instead of the risks to be avoided in doing so. This approach was driven by the desire to capture both direct and indirect financial opportunities. Direct financial opportunities included carbon trading which presented "attractive returns for an early mover" (E2). Indirect financial opportunities came from enhanced client service reputation, as derived from having expertise and offering innovative investment products in the carbon and renewables spaces.

Financial incentive regulation motivated the development of innovative products and renewable energy uptake:

> I'm an arch capitalist; let the markets decide. But we still need government intervention...It is very difficult to drive it solely by purchasing-power. We need government to kick-start the process, for example FITs for solar and wind. Then it can become viable on its own merits. (B1)

An investment analyst further explained:

> There is correlation between sector performance and government subsidies. We watch very carefully for regional and national statements so as to ascertain investment targets. (A2)

The attractiveness of regulatory incentives was brought into sharp relief when they were taken away. At the time of data collection, Spain had threatened to retroactively remove FITs for renewable projects, which had created investment uncertainty:

> At a time when long term finance capital is in short supply due to the GFC these [renewable] projects are struggling...The bullet proof solution is good tariffs to

ensure high returns but now there is pressure on tariffs which is leading to decreased returns. And if the return is not strong then you need strong regulation. If there's not strong regulation then things slow down. (A1)

Thus, incentivizing regulation needed to be stable and long-term for financial actors to obviate market risks and capitalize on opportunities associated with renewables. Germany's EEG provided an excellent example.

In summary, the well-established regulatory environment in Europe, which combined a regional carbon price with an ambitious reduction target and financial incentives for renewables, provided an amenable context for leading banks to focus on opportunities as the main driver of climate-related uptake. Although banks conduct enhanced due diligence to ascertain potential risks, it was a process that they undertook for all new clients and projects. Climate-related practices of European leading banks were motivated by opportunities to make profits, to enhance client service reputation, and to strengthen social license.

[b]   United States

Risk mitigation was relevant to climate-related bank behavior in the United States but not a prime motivator of it. Leading banks were primarily driven by the opportunities and benefits derived from adopting climate-related strategies, particularly in relation to enhanced client service reputation and profits.

Although a carbon price had been touted by the Obama Administration with an original pledge to implement an ETS by 2016, there was no federal ETS or carbon tax at the time of interviews. Despite the lack of an actual price, the credible threat of one was sufficient to mobilize risk mitigation strategies in leading American banks. Banks anticipated a carbon price in the intermediate future, and credit rating agencies had started factoring in that likelihood. Bank business in traditional fossil fuel projects had diminished since 2009 in the United States. Interestingly, this coincided with the passing through the Lower House of the *American Clean Energy and Security Act* (HR 2454) or "Waxman-Markey Bill" to price carbon (amongst other things); although it was defeated in the Senate the following year. Contracts for coal-fired plants are long term (15-20 years) and equity investors, financiers and hedge providers would not risk locking in revenue streams at today's prices in case of a carbon price tomorrow.

In addition, leading American banks had implemented enhanced due diligence processes that included climate risk after commencement of the Carbon Principles in 2008. As a risk manager stated: "Now we're asking questions like 'what's your company's carbon/GHG mitigation strategies?' No bank was asking that question previously" (D1). Nonetheless, all bankers pointed out that they would continue to finance coal-fired power plants that have (to use their language) "clean coal" characteristics. So risk mitigation drove cautious not radical green behavior.

All leading American banks had entered the renewables space due to client demand for opportunities. For example: "It was a fresh area for return; the entire area was built on government subsidies so it was more about seeing opportunities than avoiding regulatory risk" (B1); and "investment clients were attracted to the growth element... The clean tech sector resembles the Internet in this way" (C3). Those opportunities had arisen predominantly from government financial incentives in the form of tax credits, being the energy investment tax credit (ITC) and renewable energy production tax credit (PTC), pursuant to section 1102 of the *American Recovery and Reinvestment Act (2009)*.

These tax credits reduced the federal income taxes of owners of renewable energy projects, either via capital investment in renewable energy projects (in the case of ITCs) or the electrical output of renewable energy facilities (in the case of PTCs). ITCs had provided a maximum 30 percent tax credit for investments in renewable energy projects, thereby reducing federal income taxes for owners of renewable energy projects based on their capital or upfront investments in those projects (such as solar photovoltaic and fuel cells). PTCs provided a tax credit for electricity produced from renewable sources as based on the electrical output of those facilities (such as wind, biomass and geothermal).

At the time of data collection, the United States also had a federal renewable energy grant scheme in lieu of tax credits for investment in renewable energy facilities, pursuant to sections 1104 and 1603 of the *American Recovery and Reinvestment Act*. However, the grant scheme expired on December 31, 2011 as a result of section 707 of the *Tax Relief, Unemployment Insurance Reauthorization, and Job Creation Act of 2010* (HR 4853). That grant scheme had been an effective commercial incentive:

> You can build a wind farm and the government will write a check for 30% [of that expenditure]. There's no tax credits but you get 30% off plus accelerated depreciation. That government regulatory incentive is very attractive. (C1)

This suite of financial incentive regulation had encouraged bank activity in renewable and clean tech spaces: "Renewable sources are actually in [our] portfolio because tax credits make it very attractive" (C2); and "We look at deals with renewables because incentives are there so banks and business are shifting attention to it" (D1). One banker in project finance described the government grant scheme as a "30% off sale" on renewable power with the result that "everyone is rushing to do it" (C1). Capitalizing on opportunities satisfied business case imperatives for enhanced profits and client service reputation: "We can only do something if it makes sense economically. If we get government incentives then we are happy to put 95% of our financing towards renewables" (C1). And, by implication, the reverse is also true. If there are no incentives then there is little bank interest in renewables due to their inherently risky nature. This dynamic had important ramifications for the renewable sector: "Regulation helps the sector to grow; it provides a kiss to the financing" (C1).

Finally, leading banks in the United States were highly conscious of compliance with the Community Reinvestment Act because ratings impacted on business and reputation, both in terms of social license and client service. How had the Community Reinvestment Bank driven climate uptake? Proactive banks extended their legislative social-financial obligations to include environmental facets, such as micro financing low-carbon initiatives and putting tax equity credits into the renewables space. They believed that doing so satisfied regulatory obligations while creating strategic advantage in a highly competitive industry. For example: "Our attitude is that we could just give money away and regard the CRA as a tax, or we could regard it as an opportunity to do well-structured and risk-managed deals" (C2).

This behavior fits with the theses of Porter and van der Linde and also Reinhardt that there are economic rationales for beyond compliance behavior and that regulation can incentivize novel corporate approaches to mitigate environmental issues. However, the data provide an interesting twist to the theory: a driver of climate action for leading American banks was going beyond compliance with socio-financial legislation aimed directly at the banking industry, not environmental legislation aimed at the corporate sector more generally.

In summary, the culmination of a threatened carbon price, financial incentives for renewables, and direct socio-financial legislation shaped bank perceptions of risk and opportunity as drivers of climate-related behavior in the United States. This regulatory combination pushed profit-making and client service reputation opportunities into the foreground while simultaneously moving social reputation and credit risks to the background.

[c]   Australia

Risk mitigation was the main driver of climate-related behavior for leading banks in Australia. Banks in this jurisdiction were motivated by downside prevention regarding regulatory risk, credit risk, social reputation risk (from associating with coal clients) and potential loss of client service reputation (from ostracizing fossil fuel clients). Nonetheless, banks and their clients were starting to turn attention toward opportunities created by climate-related strategies, such as enhanced profits from carbon market activity and garnering social and client service reputations from increased association with renewable energy clients.

The Australian regulatory context was the least certain and least incentivizing of all three jurisdictions. Debate and discord had surrounded a carbon price since its proposal by the Rudd Labor government in 2007. Three versions of carbon legislation were drafted and debated within four years; changes in Labor Party leadership saw government support for a carbon price wax and wane. Finally, in November 2011 the Australian Senate had passed the Clean Energy Legislative Package comprising a number of initiatives including a carbon tax. Under the *Clean Energy Act 2011* (Cth) a carbon tax commenced on July 1, 2012 at a fixed price which was scheduled to move to a market-based ETS in 2015. At the time of data collection, the legislation had been passed but was yet to commence. There was still uncertainty around: the exact price; transitional arrangements from a tax to an ETS; and whether or not the scheme would be revoked by a Coalition Government (a concern that paid true when the carbon price was repealed only a few years later in 2014).

The threat and eventual implementation of a carbon price created broader risk exposure for Australian banks through their infrastructure and energy portfolios. There is an abundance of coal in Australia and thus a large number of lucrative coal clients in leading bank portfolios. As such, banks had become keenly aware of the need to rebalance their portfolios to reduce their own risk:

Ten of my clients are responsible for half of the taxable emissions in Australia. This has driven our climate change strategy in response to the Government's carbon legislative package. We are in the frontline and we need our risk management processes to be up to speed with the language of carbon and climate change. (G1)

As with the United States, there had been a decline in finance for new coal in Australia due to a mix of tougher EPA approvals, increased NGO pressure, and the advent of the carbon pricing framework. Yet, as with American and European banks, Australian leading banks were willing to continue to work with coal clients and finance new coal if technologies became available to make it "cleaner." Banks remained cautious and committed to client service: "if we can mitigate risks then we will work with the client rather than turning them away" (G3).

However, a slow movement from a risk model to one of profiteering was evidenced by the data. "Mostly we are focused on risk but we will need to focus on the opportunities, that is, increasing expertise in carbon markets and also renewables and getting new business" (G4). Profiteering included accessing carbon trading opportunities and also capturing opportunities in renewable energy:

> We need to look mid to long term on renewable energy. There will be $20billion investment in renewable energy in Australia to meet the RET. This amount of money is easily manageable. 70% of it will be funded by banks, so it is a $17billion opportunity for banks. That also includes foreign exchange opportunities and hedging opportunities (of interest rate risk and electricity risk). So renewable energy is a big opportunity for banks. (F1)

To this extent, Australian banks had started working more with renewable energy clients in order to rebalance their portfolios and to garner client service reputation in that sector: "we were the first to finance wind and solar here because we want to be well thought of as easy to work with by renewable energy clients for the future" (G1).

This interest in renewables was nascent, however, due to the lucrative nature of coal clients and a lack of financial incentives to switch support to renewables. In Australia there have been no federal-level FITs or tax credits, only solar grants and the RET.

The RET was introduced by the Howard Coalition Government in 2001 at which time only 4 percent of new electricity generation was required to come from renewable sources. In 2009 this target was expanded by the Rudd Labor Government to 20 percent by 2020. In 2011, the RET was split into two parts: the Small-scale Renewable Energy Scheme which applied to households, small business and community groups, and the Large-scale Renewable Energy Target which was intended to deliver the majority of the 2020 target by covering large-scale renewable energy projects such as wind farms, commercial solar and geothermal. Key to the RET were tradable Renewable Energy Certificates (RECs) for large-scale projects and Solar Credits for small-scale projects.

However, bankers explicitly noted that a RET was insufficient to mobilize finance at the required scale:

> The venture capital set up in Australia is bad. In the United States it is booming because venture capital providers get tax breaks. There aren't any in Australia. We need government support to get it happening. There are examples of innovative

Australian clean tech companies doing business in Israel, California and Spain instead of here. There's no support for them here...That absence of government incentives is very frustrating for banks and bankers here. (F1)

Federal-level solar grants to incentivize solar photovoltaic (PV) investment manifested as the "Solar Flagship Program." This program was described by several interviewees as a "disaster" because too many solar RECs had been issued under the scheme with the result that "even though solar has a lot of potential in this country we're still not there yet" (F1). But the focus on solar meant that little attention had been paid to wind or other forms of renewable energy generation during that period. The upshot was that neither solar nor any other form of renewable energy had taken off in Australia.

As such, some Australian superannuation funds were directing investment to large-scale foreign renewable projects instead of locally:

There aren't the right opportunities and returns in Australia...It comes back to uncertainty. We must know what the returns will be. We are focused on short-term returns but power assets are long-term investments of 20 to 30 years. (G2)

In summary, the uncertain regulatory environment surrounding a carbon price and the lack of stable financial incentives for renewables in Australia largely shaped bank preferences for downside prevention as a driver of behavior. This regulatory environment had pushed the full gamut of risks – credit, regulatory, social and (coal) client service reputation – into the foreground. As summarized by one banker: "Government support and certainty mean that banks get on board. We have not seen it in Australia" (G1). The consequence is that corporate behavior has been reactive and banks remain cautious: "there's no need to make radical changes in a time of uncertainty" (F1).

§4.04   NON-DRIVER: THE "CARE" FACTOR

The case study makes clear what motivates early-moving banks to adopt climate-related initiatives. Specifically, bank behavior regarding climate change is motivated by the business case, being: (a) the profit motive, both directly via fee generation and indirectly via reputation enhancement and competitive edge; and (b) downside prevention or risk mitigation. The preceding analysis yields a deeper and more complex understanding of reputation, one that separates "corporate reputation" from "social reputation," and introduces the notion of "client service reputation" as the most powerful motivator of bank behavior. Moreover, cross-comparisons between jurisdictions have demonstrated the role of environmental activist groups and the importance of law and policy to shaping bank decision-making. This regulatory context colors bank mindsets toward climate change as a risk or an opportunity.

Yet it is equally important to reflect on what factors are *not* powerful motivators for banks to "go green." What is revealed by silences in the interviews? Which topics were dealt with ambiguously or ambivalently by respondents? When questioned directly about the relevance or meaning of CSR and related concepts, how did they respond? Attention to these questions enabled a second layer of findings about what does *not* drive banks to adopt climate-related initiatives.

In summary, CSR considerations were subordinate (at best) and in some cases negligible. While attention to CSR was present in most banks, in no cases did CSR appear as a stand-alone driver, and in all cases it was subordinate to business case imperatives.

[A] Manifestation of CSR

Nearly one half of all bank interviewees did not mention CSR or any variation of it when describing their bank, their Board, themselves, their work, or the drivers of adopting climate-related business practices. Bank respondents that did use CSR terminology used it to describe their banks as: "having an interest in CSR and being a responsible citizen" (A1); being "a good corporate citizen and leading by example" (E1); engaging in "socially responsible activities" such as microfinance and philanthropy (B1); contributing to "a sustainable future" (F1); and "pushing a strong sustainability agenda" (F2).

The clearest manifestation of CSR was in individual action whether through a CEO or manager's actions. For two banks, a passionate senior officer had set the tone at the top, instigating and supporting climate-related initiatives. For example, the then-Chairperson at one bank was described as "very green and sustainability-minded" and "a very important driver" (A3). Similarly, a transactional banker had initiated change because he was pro-environment and had located a green gap in the market:

> I had a personal interest in the environment and environmental issues. I see myself as a responsible citizen. I pitched the clean tech policy to management because of this personal interest and because no other bank was doing it at that stage so it added value to our clients. Management was amenable and then the clean tech area in [the bank] was created. (A1)

This evidence of "charismatic green leadership" supports theory from Chapter 3 that senior-level engagement is a key driver of corporate climate-related uptake and that a manager's personal morals may influence their decision-making. It is at this micro level that climate-related practices and products were initiated as "the right thing to do" (A3). As such, actions by individuals provided the strongest examples of a genuine "care factor" in business.

However, the case study data also revealed that ethical discretion by individual managers was constrained by narrow conceptions of CSR and business case parameters, and that leadership alone was insufficient for far-reaching corporate change.

[B] Conceptualization of CSR

Bank respondents used certain CSR-related terms, namely "good corporate citizen," "CSR," "socially responsible," and "sustainability." Analyzing conceptions of these terms assists our understanding of their relevance to bank business strategy and their role in driving (or not) green uptake.

"Sustainability" was the preferred term for Australian and American banks; whereas "CSR," "socially responsible" and "good corporate citizen" were terms used almost solely by European and UK managers. Importantly, "CSR" was not used proactively by any American or Australian bank respondents.

Half of the managers who used the term "good corporate citizen" mostly used it to describe their organization. They referred to a variety of examples, from non-green philanthropy work on the one hand to reducing carbon emissions or proposing climate initiatives on the other. The other half conceptualized good corporate citizenship at the individual level, usually to describe themselves or their Chair. Overall, there was a lack of clarity and consistency about what a "good corporate citizen" actually meant: "It's an esoteric concept. It's in the hands of each individual" (A2).

In contrast, scholars tend to agree that corporate citizenship applies to corporations, not individuals. For example, Logsdon and Wood state that notions of business and corporate citizenship are separate from and secondary to individual citizenship, and Solomon claims that "the corporation *itself* is a citizen, a member of the larger community and inconceivable without it."[495] As such, there is a misalignment between practice and theory. Practitioner conceptions of good corporate citizenship as an individual preference do not accord with scholarly conceptions of it as an organizational obligation.

Similarly, the terms "CSR" and "socially responsible" had different meanings for practitioners. One banker described CSR as a "drumbeat" to which his bank was moving (A1); whereas another discarded the term as "simply PR," "dead and buried" and "too heavily oriented toward philanthropy" (F2).

Most bankers in project finance and energy and utilities departments believed that voluntarily taking radical steps toward a low-carbon economy, such as redlining fossil fuel clients and funding only clean tech was tantamount to corporate social *irresponsibility*. These bankers were in the frontline doing deals with lucrative fossil fuel clients and motivated by client service reputation. Here are some examples:

- "CSR is a big issue in the bank. We need to look at the social responsibility of pulling out of coal immediately, what negative consequences that will have for the community. Power security is a main part of that. It's not as simple as dropping coal and going straight to renewable energy." (G3)
- "We recognize that the world needs energy; we want to deliver products in a socially responsible way. If the bank turned off the tap altogether then it would bring a huge amount of trouble on the global economy. It would cause more damage to the green movement than benefit." (E1)

In other words, social responsibility meant keeping the lights on to keep the world turning. These bankers viewed their continued association with fossil fuel clients as

---

495. R.C. Solomon, *Ethics and Excellence: Cooperation and Integrity in Business* (Oxford University Press: Oxford, 1992), 184 (emphasis added). J.M Logsdon and D.J Wood, "Global Corporate Citizenship: From Domestic to Global Level of Analysis" (2002) 12(2) *Business Ethics Quarterly* 155, 167.

consistent with what society wants and needs, which made that association socially responsible, socially responsive and, most importantly, legitimate:

- "What constitutes 'fossil fuel' [that we should stop associating with]? Clients that own fossil fuels? Clients reliant on them? Even NGOs rely on fossil fuels – they're not sitting in the dark or using solar 100% of the time." (B2)
- "It's not that [the bank] is hanging on to coal; the community and society is. This is because, first, there are huge amounts of coal reserves here in Australia for hundreds of years to come and, second, it's very cheap. So we have a safe, reliable, secure energy source, which means that coal is very good. The fact that it pollutes the environment is bad, but that's a separate issue." (F1)
- "In Europe – nuclear is acceptable but not coal. In Australia it's the opposite. In Australia coal is not the dirty dirty thing it has become in Europe – it is an essential part of our economy and it's not going away tomorrow." (G4)

In contrast to their conceptions of "CSR" and "good corporate citizen," bankers were much clearer in their conceptions of "sustainability." There were three groupings of definition.

First, and overwhelmingly, "sustainability" was defined to mean bank longevity and viability. The conflation of sustainability with economic viability partly flowed from succession planning and good business sense:

Sustainability means longevity... something that can survive a long time, that is bank sustainability, that we'll still be here in 50 years beyond the lives of the current management board. (G3)

The conflation was also partly a reaction to the reputational carnage of the GFC such that "it's important [now] to show a bank as long term sustainable, that it won't disappear overnight" (B1). Bank longevity is assured by proactive risk management, high returns, regulatory compliance, and appropriate branding and image management.

Secondly, "sustainability" was conceived in socio-financial terms by some CSR managers. Particularly in America, they defined it in the context of going beyond compliance with the Community Reinvestment Act "to prove that it is sustainable – socially and financially – for the finance sector to do business with, as opposed to just granting money to, poor communities" (C2).

Finally, respondents at only one bank in Australia included longevity of the environment in their definitions of sustainability.

Clearly, corporate conceptions of "sustainability" do not accord with formulations in the Brundtland Report or the UNEP Finance Statement on Sustainability, both of which explicitly acknowledge that economic imperatives must be balanced with human welfare and a healthy environment. Specifically, according to the UNEP Finance Statement on Sustainability, signatories agree that:[496]

- "economic development needs to be compatible with human welfare and a healthy environment";

---

496. See Chapter 1 *supra*.

121

- they are "committed to working cooperatively [with government, business, individuals] toward common sustainability goals"; and
- "the sustainable development agenda is becoming increasingly inter-linked with humanitarian and social issues as the global environment agenda broadens and as climate change brings greater developmental and security challenges."

This disconnect between practice and rhetoric exists despite the fact that all leading banks bar one are signatories to the UNEP Finance Statement on Sustainability.

Importantly, in all conceptions of CSR-related terms, mention of an ethical or moral dimension was negligible. Only two respondents implied that adopting climate-related initiatives was an ethical obligation or the "right thing to do"; however it became clear after probing that doing the right thing had to make money or it would not be done. Two others claimed a moral obligation to assist poor communities; however their bank was obligated by legislation to help poor communities, so ascribing a moral motivation to that activity seemed superfluous at best. Some bankers even downplayed moral aspects of CSR in order to elevate the business case. An investment banker summarized this well when describing his bank's message to clients:

> Don't read too much into having "a higher agenda." The focus should be on "good business sense." Don't just feel good by reducing your emissions; you can make a shitload of money if you do it right. (E1)

Indeed, a number of respondents stated explicitly that a corporate-ethical agenda did *not* drive their bank's involvement in climate- or even social-related spaces. For example: "One of the core drivers in renewable energy is energy security. It is not about saving polar bears"; "We are not trading carbon to save the planet, we are doing it because of fee generation"; and "...we can engage in pseudo pro bono work, not because we are gold hearted but because it's good for [the bank] and builds its reputation" (B1). Similarly, financial engagement with disadvantaged communities can be self-serving: "we get access to more customers [and we can] make strong communities in places where we operate which creates a strong healthy economy which creates strong places to do business." (D2).

Another bank was in the renewables space not for ethical reasons but because tax credits had made it so financially and reputationally attractive: "if there were tax credits for doing couches, the bank would be doing couches" (C4).

In short, all respondents conceptualized CSR and related phrases in terms of (enlightened) self-interest and business case logic.

[C]   Subordination of CSR

Regarding CSR, two findings are clear from the data: first, in all cases CSR was conceptualized in a business case context with little regard for ethical dimensions; and second, in all cases CSR was subordinate to the bottom line. The driving imperative for leading banks is to make money and not lose it. The data show that being socio-environmentally responsible, a good corporate citizen, or "sustainable" are also on bank agendas only if it flows from or contributes to this profit imperative. In the words of one investment banker: "banks need to be more socially responsible but they can't increase costs in doing so" (B1). A clean tech specialist elaborated:

> It is important that financial actors do look at these areas [such as climate change]... As a good corporate citizen I need to help that happen but if I can't convince the board and clients of the economic viability of new strategies then it's no use. (A1)

Some Boards were willing to back new and risky initiatives, but those initiatives needed to become financially viable on their own in order to persist. "The success of this [climate research unit] is gauged by external and internal feedback and how much money it makes. We have passed the honeymoon period so it must be financially sustainable now or it wouldn't exist" (A3).

For some banks, financial success was a necessary prerequisite to engaging with CSR. That is, client service reputation and fee generation were first-tier priorities, attention to which then enabled second-tier activities such as pro bono work and green uptake.

In contrast, CSR had become more prominent at other banks that were *not* doing well due to the GFC. That is, attention to CSR was important to help "rebuild credibility and trust at a time when the industry is viewed as not credible and not trustworthy" (D2). However, one banker qualified the motivation: "part of this is genuine, part of it is image management and a marketing tool" (B1). Indeed, another banker explained that, although his bank was genuinely engaging with renewables projects, it was actively publicizing those green activities to appease NGO campaigners and to rebalance the bank's image due to its concurrent engagement with coal.

Regardless of these sentiments, the advent of the GFC provided further illustration of the secondary nature of CSR. Several banks had minimized climate-related initiatives and staff as a result of fiscal tightening. In particular, the GFC "pulled the rug from underneath the equity market" (E1), which meant less ready capital for renewable energy and clean tech projects. An investment banker explained that "Now I spend more time on IT in Europe and less time in the renewables space" (E1). Indeed, the GFC had practical effects on several of the banks' green activities. For example:

- "[The bank] was one of the first to start in clean tech and renewables in 2000. But then the GFC came and the bank focused its approach and kept only the fee earners and let go the people without a real role who were running around doing nice things." (B1)
- "Three years ago we had an idea for a clean energy fund which focused on investment in clean energy products and projects...But then the GFC happened and we did not proceed because it did not stack up from a financial perspective." (F1)

In stark contrast to the change management literature, the case study evidences that CSR is subordinate even when it is purported to be core to a bank's business philosophy. For example, "sustainability is always at the forefront of how we want to be seen in the market. We want to be seen as a good corporate citizen and that is part of our core values." (F1). Yet how do sustainability and the bottom line get balanced in practice?

> The bottom line is very important. I will not support a transaction that underper-
> forms on return on equity. If it is below the "return on equity" threshold, even if
> it's good for climate change, we won't do it. So we will not sacrifice return on
> equity just to support climate change mitigation innovations. [The CEO] would
> support and agree with that. (F1)

Furthermore, in contrast to the literature on cognitive dissonance as a modality of organizational change, the case study did not reveal pockets of micro-level resistance to or deviation from an imposed corporate perspective even if it clashed with personal perspectives. Some respondents expressed regret at the limitation of single bottom line thinking and the hypocrisy of contributing to global climate change by exporting coal or tradable emissions to other jurisdictions whilst redlining carbon pollution domesti- cally. One investment banker even acknowledged that inflexible obeisance to following the money could sometimes perpetuate societal harm. He used the example of slave trading, which, at that time in history, was the "done" thing and there were strong economic arguments to keep the status quo even though it was morally wrong and proved *not* to be necessary for economic prosperity. It was a visceral analogy for current thinking on the economic imperatives of continued fossil fuel use; yet it received only passing mention.

In short, there was no evidence of personal resistance to the status quo. Indeed, most respondents expressed satisfaction with their bank's level of engagement with climate, environmental or social considerations. Some even expressed professional and personal satisfaction that their pursuit of commercial success could have the corollary effect of "saving the planet" (C2). For example: "You can use core commer- cial skills and be a banker, but you are also on the cutting edge of a low carbon economy" (C2); and "When I sit and have drinks with my friends in the energy sector I say to them 'I may be an evil banker but at least I am facilitating some kind of global economic shift'" (B1).

Indeed, all respondents appeared to accept that doing good is secondary to doing well because it is a bank's job to make money for clients and shareholders. As such, bankers deem the status quo as reasonable and appropriate. Any request or expectation that a bank voluntarily do much more than the status quo, especially if it might cause financial loss, is seen as irrational and inappropriate.

Accordingly, conceptions of CSR are informed by a bank's self-perception of its role and purpose in society. Bankers had not apprehended their network change potential to facilitate GHG emissions reductions through enlightened practices that influence corporate networks in an ever-widening web. Only one respondent, a CSR manager, adamantly felt that banks could be "a greater force for good in the long term" by creating "uniquely value-added ways to find solutions for problems":

> [My] group asks "what are the opportunities of the future? Water, housing,
> conservation: areas that are not quite commercial yet but could be. For example,
> we need to save the rainforests, we all know that. But so far only activists and
> philanthropists are doing it – but that's not enough and government is not

fast-moving enough. There has to be a significant role for private capital to bring in funding to monetize the preservation of forests, to make it more economical than deforestation… What today seems just interesting "castles in the sky" stuff can become the clean tech banking group 10 years from now if we work on it. (C2)

Yet nearly all other respondents, being transactional bankers/analysts and CSR managers alike across all jurisdictions, viewed the role of a bank in narrow fiscal terms. Indicative of this mind-set were statements such as:

- "A bank's traditional role is to oil the cogs of industry" (B1);
- "Banks are the center of commerce: they've got to do what's best for shareholders" (C3); and
- "We are massive creators of shareholder wealth and massive contributors to the economy…That is a key role for banks: stewardship of the economy" (G1).

Contrary to Carroll's seminal definition of CSR as comprising "the economic, legal, ethical and discretionary expectations that society has of organizations at a given point in time,"[497] banks were not conceived as fitting within or benefitting society beyond their legal and economic obligations.

§4.05 IMPLICATIONS

Some may argue it is unsurprising that CSR does not drive voluntary environmental uptake, that it is subordinate to business case considerations, and that this status quo is acceptable to bank employees. What else should be expected in capitalist societies? Yet these findings *are* important for understanding and harnessing motivations of private finance actors in a way that is realistic and, therefore, accurate and fruitful. Moreover, multiple implications flow from these findings that are relevant to deepening extant business strategy literature.

First, there are clear structural restraints on when CSR is activated and how far it can go in driving corporate change. Due to the GFC, CSR had become more important as an image tool for some banks, particularly in Europe and the United States. Yet the very advent of the GFC resulted in less capital available for CSR-type endeavors, including financial support for clean tech and renewable energy projects. Indeed, when lean times had come, some leading banks shrank their CSR departments and/or curtailed their climate-related initiatives. As such, there is evidence of a "perceived impact of structural restrictions on management when addressing CSR" creating "a perception [by managers] of limited power at the corporate level."[498] This is in stark contrast to Wood's theoretical Principle of Managerial Discretion which posits that, on moral grounds, managers cannot avoid exercising their individual discretion to "do the right thing" even in the face of contrary corporate rules or policies. In actuality, the perception by managers of constrained decision-making power curtails not only individual discretion but also broad, ethics-based conceptions of CSR. Structural

---

497. Carroll, *supra* n. 283, at 500.
498. O'Dwyer, *supra* n. 291, at 538-539.

restraints – actual and perceived – ensure that CSR is conceived (if at all) in a narrow fashion consistent with corporate goals of shareholder wealth maximization.

Secondly, "doing well by doing good" is not a pervasive philosophy within the banking industry. The driving philosophy is that "we can do good *if* we are doing well." And a bank does well by doing what it does best: making money and not losing it. Attention to "green" or "sustainability" issues can buttress this goal by delivering business value, enhancing reputation, and creating business opportunities, which is why banks have adopted climate-related strategies. Yet this focus on win-win by banks arguably demonstrates two points. First, consistent with literature on corporate capture, any tension between corporate and social-centered goals are downplayed, which allows corporate actors to construct their own narrow meaning of CSR and obviate broader ethical obligations. Secondly, notwithstanding exhortations in the change management and sustainability literature, green considerations are added on and/or subordinated to fiscal considerations. The bottom line is still very much a singular and financial one. Evidence of a triple bottom line in action – where social, environmental and financial imperatives are balanced *equally* in reality and beyond rhetoric – was not apparent from the data even despite a "sustainability" mantra at some banks and my impression that those bankers genuinely believed it. Yet the supremacy of a single bottom line in practice is clearly inconsistent with the standard formulation of sustainable development, the UNEP Finance Statement on Sustainability, or triple bottom line theory. This situation only makes sense when one realizes that "sustainability" had been conceived by firms in such a way as to mean business longevity, economic viability and/or economic stewardship. In such a case there is no conflict between a bank's social responsibility and its pursuit of profit.

Clearly then, corporate and managerial conceptions of "sustainability," "CSR" and "good corporate citizenship" determine their relevance to business strategy. Definitions are conceived within the context of how a firm and its employees perceive their role in society. In the case of banks, this role is predominantly utilitarian; ethical or moral preferences are included, if at all, at a personal-individual level. And exercising ethical preferences at the individual level is "limited by macro power structures that place business case parameters around CSR."[499]

This case study disconfirms the CSR literature that posits "doing the right thing" can and does drive corporate greening. In particular, it does not support the contention from previous studies that ethical motivations drive social or environmental corporate changes. Thus, it provides an empirically-based reality check. In short, CSR is a limited explanatory tool for what drives actual corporate environmental-social uptake.

Business strategy scholars appear to agree on only one point regarding CSR: it has multiple, contested meanings. The case study confirms that point and goes one step further to evidence that CSR has contested legitimacy and that such contest occurs even within the one firm. Specifically, there are "competing priorities and different cultures" between departments (F2) and the meaning and importance of CSR (and related terms) is conceived heterogeneously – when it is thought of at all.

---

499. Spence, *supra* n. 318, at 874.

Moreover, the case study finds no support for the literature that maintains that CSR must be core to a corporation's values and culture. Such literature assumes that firms in capitalist economies have a core agenda beyond making money by legal means. In the case of banks, that assumption is largely inaccurate. More specifically, the notion of one core agenda to which CSR ought be central assumes that within one firm all departments share the same goals and are driven by the same incentives. This is not the case. In summary, how CSR manifests for *one firm* and the extent to which it drives voluntary change is varied. By implication, how CSR might, on its own, drive a *whole industry* is simply unfathomable.

Proponents of CSR have tended to focus on the "ought" and not the "is" of corporate greening. This study demonstrates that the question of why firms *actually* go green can be separated from a normative discussion about why they *ought* to go green. It is crucial to appreciate these real life aspects in order to theorize how corporate change might best occur in actuality. The profit imperative and competitive nature of the banking industry coupled with a multi-dimensional understanding of corporate reputation provides a theme for practitioners and policy-makers and anchors the regulatory recommendations in later chapters.

But first, we need to consider the implications of these findings for mainstreaming of climate-related practices in the banking industry, and for private finance actors more generally.

CHAPTER 5

The Limits of Corporate Change: Case Study Evidence

§5.01 OVERVIEW

The previous chapter evidenced that business case logic is the overarching driver of voluntary climate-related changes by early moving banks. Specifically, the business case comprises two facets: first, increasing profits directly via fee generation and indirectly via competitive edge and reputation enhancement; and secondly, minimizing financial, regulatory, and reputational risks. Moreover, a new finding has been revealed by the data: "corporate reputation" comprises not only "social reputation" but also "client service reputation," which has important and subtle implications for how and why banks are motivated to change their behavior.

This chapter considers the implications of those findings for mainstreaming of climate-related practices in the banking industry. Is it likely that mainstreaming will occur given certain limitations to business case logic? If so, how might mainstreaming occur organically? And can mainstreaming occur fast enough and help take us far enough toward a low-carbon economy?

As described in Chapter 2, the network change potential of banks is a key reason why the industry is an important and unique player in climate change mitigation efforts. In short, enlightened climate-related banking practices that are mainstreamed in the industry as a whole – and not just the praxis of early movers – have potential to facilitate widespread GHG emissions reductions through corporate networks in an ever-widening web. Therefore, we need to consider how best to mobilize the industry as a whole, which will also provide lessons for other private finance actors.

Using the empirical data, this chapter investigates three main areas in relation to mainstreaming. First, it analyzes the limitations of the business case as a driver of *en masse* green uptake, which includes materiality of the green business case and the existence of a competing non-green business case. Secondly, it theorizes how mainstreaming might occur organically via isomorphism and mimetic behavior. Thirdly, it

129

revisits the urgency of global warming and asks whether mainstreaming in the context of pure voluntarism can be sufficiently far-reaching and fast-acting to facilitate required global change. Contemplating implications of the business-case-as-driver for industry-wide change in this chapter paves the way for regulatory recommendations in the next chapter.

§5.02   LIMITATIONS OF THE BUSINESS CASE AS DRIVER

The data reveal two key limitations to the business case as a driver for mainstreaming climate-related finance practices. The first is that, by definition, if an initiative is not profitable or the costs outweigh the benefits – that is, if there is *not* a business case to do it – then banks will not make the change. The second important limitation is that, at present, there exists a competing business case to support GHG-intensive clients and projects. Each of these limitations is discussed in turn below.

[A]   The Requirement of a Business Case for Green Uptake

The first limitation – the prerequisite of a business case for green uptake – is best exemplified by looking at the bank in the case study that demonstrated the most conscientiousness toward addressing climate change. This bank had a passionate Chair who provided leadership on green issues and supported climate-related business initiatives; some of its managers and advisers had green credentials and were brought onboard to advance the bank's low-carbon agenda; other managers held strong personal beliefs about environmental stewardship and were proactive in green initiatives. This bank demonstrated innovation around climate-related services and products and it had elevated climate change to the corporate governance level under senior management oversight. Finally, this bank was the only one in the case study to have signed up to all three climate-related soft law Principles. Despite all of this, data were clear that if a green initiative did not make good business sense then it would not get traction with the Board.

For example, respondents at this bank stated that there had to be "revenue streams associated with climate change" and that green initiatives needed "a sound business case and to create value for shareholders." (A1). They stated that a climate-related unit needed to be commercially viable in order to get started otherwise "we simply couldn't do it, we wouldn't get the management funding and support." (A2). And whilst the Board had shown willingness to back new and risky climate-related initiatives, those initiatives needed to become self-sustaining in order to persist. For example: "The success of this [climate research unit] is gauged by external and internal feedback and how much money it makes. We have passed the honeymoon period so it must be financially sustainable now or it wouldn't exist." (A3).

Again we see how business case imperatives override green considerations and that this, at least in part, is a product of the prevailing bank view of its role in society. A respondent from a different bank summarized it thus: "We are massive creators of shareholder wealth and massive contributors to the economy" (G1) and therefore, as stated by yet another banker, "No-one is expecting banks to be charities...[or] to finance a bunch of unprofitable projects." (E1). Put simply: "we recognize we are a bank and we need to make money." (E1).

The problem with building a business case for climate-related initiatives is that profit increases may take time to show up on the balance books, and value increases can be difficult to measure. For example, profit momentum from gaining new large clients may take time to materialize; and some initiatives will have only indirect consequences on the bottom line, such as enhancing a company's social reputation or its capacity to attract and retain talented staff. Thus, the "materiality" of the business case can be an impediment to corporate environmental uptake.

This limitation is acknowledged in corporate sustainability literature, which highlights how the business case can be difficult to detect and/or marginal in practice given that "the economic value of more sustainable business strategies...only materializes in the long term."[500] The problem is that, as noted by Dermine, we live in "a world in which financial markets reward short-term reported profits."[501] As such, the focus on short-term profit can trump long-term value and be a significant barrier to corporate environmental change.[502] Generally, if public companies and especially financial firms, "cannot demonstrate tangible economic success in the here and now [then] there may be no long term to look forward to."[503]

[B]   The Countervailing Business Case for Non-Green Initiatives

The second key limitation of the business case as the driver for mainstreaming climate-friendly bank practices is that, at present, there is a countervailing business case to support GHG-polluting clients and projects. The data reveal that banks will continue to work with these clients, which often have relationships with a number of different departments within the same bank and are therefore a lucrative overall source of fee generation. Retaining such large clients is inextricably entwined with the highly competitive nature of the banking industry. Ostracizing a client is equivalent to handing that client on a silver platter to a competitor and missing out on a part of the purse. It also ties back to client service reputation via client satisfaction: "For any bank a huge percentage of business is mining companies. We can't afford to turn business away and we wouldn't be looking after our clients if we did."(E1)

American and Australian banks provided rich evidence of this point. Respondents in these jurisdictions described how, in recent years, they had become focused on client "relationships" and how being trusted as a good business adviser was central to that dynamic. Coal clients comprise a major part of banks' energy/power portfolios.

---

500. Salzmann et al., *supra*, n. 130, at 33.
501. Dermine, *supra* n. 129, at 268.
502. Richardson, *supra* n. 107; Hess, *supra* n. 384; N. Gunningham, "Beyond Compliance: Management of Environmental Risk," in B Boer, R Fowler and N Gunningham (eds) *Environmental Outlook: Law and Policy* (Federation Press: Sydney, 1994).
503. Gunningham, *supra* n. 335, at 218.

Making any radical voluntary changes against these clients is tantamount to compromising client service reputation and therefore compromising financial returns. Banks need to balance competing client interests with their own financial and reputational interests. Coal is a short- to mid-term reality; low carbon energy sources are a mid- to long-term proposition. Banks are willing to hedge their bets by taking on renewable energy clients, but they are not willing to drop lucrative coal clients of their own volition. For example:

> We did two [renewable] transactions that created positive feedback from the market which led to more renewable energy business that in turn lifted our reputation in that space. That's important for the future. But reputation with the coal-fired sector is also important to us, especially regarding refinancing, which is happening now. (F2).

It is here that we get an even deeper understanding of the interplay between client service reputation and the business case.As one bank respondent stated:

> We are a well-regarded energy bank, so energy clients – whether coal, gas or renewable – come to us for business...We are not going to pick sides in an emotive turf war. We need to be a trusted adviser. (G4).

It becomes clear that client service reputation is a double-edged sword in terms of voluntary corporate action. On the one hand it mobilizes banks to innovate new products and to become active in green spaces due to the desire for competitive advantage and enhanced fee generation. Yet, concomitantly and ironically, client service reputation impedes far-reaching entrepreneurialism. Even leading banks are not willing to voluntarily make radical business changes lest they compromise their competitive edge and financial returns. They will not irrationally sacrifice lucrative clients on the green altar. So, while client service reputation is a strong motivator for banks to voluntarily drive action in a greener direction, it is a simultaneous brake on driving too far down that path.

It becomes clear that business case logic is tantalizing but, due to its intrinsic limitations, also unsatisfying as an assured modality of green corporate change.

§5.03 BANKING REGULATION, CORPORATE LAW AND CORPORATE GOVERNANCE NORMS

It is useful at this juncture to consider the relevance and roles of bank regulation, corporate law and corporate governance norms regarding bank decision-making, particularly in the context of short-termism and business case imperatives. Indeed, understanding the legal duties of bank directors and the framework of bank corporate governance helps to illuminate constraints, both perceived and actual, on climate-friendly bank action.

The banking industry is highly regulated in market economies. It must comply with general corporate law as well as specific banking rules, and its activities and compliance are monitored by government authorities. The net effect of this regulation and supervision is that banks cannot do whatever they like – for better or worse. With this in mind, the next two sections set out: first, a summary of the regulatory envelope within which banks must function in the UK, Australia and the United States; and secondly, a conjectural analysis of how corporate law and corporate governance norms may impact upon bank-directed greening in these jurisdictions.

[A]   Bank Regulation and Supervision

Private finance actors are highly-regulated entities. Banks, particularly commercial banks, face the strictest regulation. Central to bank regulation and supervision are the concepts of risk-taking and prudence: that bank activities will be undertaken with reasonable care.[504] Banks must be highly regulated because their way of doing business makes them susceptible to failure; that is, they function on a small asset reserve, hold a large proportion of illiquid assets, and work within an inter-bank market that comprises a network of large, unsecured debtor and creditor relationships such that failure of one bank can lead to the collapse of others.[505] Accordingly, banking rules and supervision exist to regulate and monitor day-to-day banking activities such as bank-customer relationships, credit lending, mortgages and other securities transactions, and also financial crime.

[1]   United Kingdom

Cornerstone entities in the UK financial system are the Bank of England, which monitors systemic risk including matters of interest rate policy, and Treasury, which has overall responsibility for financial markets. In the 1980s the UK adopted a consolidated system of regulation to govern deregulated markets pursuant to: the *Banking Act 1987*, which formalized prudential supervision; and the *Financial Services Act 1986*, which introduced self-regulation into a statutory framework intended to oversee investment activities.

By the early 2000s, a third entity, the Financial Services Authority (FSA), had been given responsibility for banking supervision pursuant to the *Bank of England Act 1998* and the *Financial Services and Markets Act 2000*.[506] The FSA worked together with the Bank of England and Treasury to monitor banks and other financial institutions; and it made rules regarding authorized and prohibited activities, enforcement and appeals.[507] The objectives of FSA regulation included maintaining market (depositor and investor) confidence, protecting consumers, and reducing financial crime.[508]

---

504. D. Singh, *Banking Regulation of UK and US Financial Markets* (Ashgate Publishing Ltd: Burlington VT, 2007).
505. *Ibid.*, 8. See also: C.A.E. Goodhart and G. Illing (eds), *Financial Crises, Contagion, and the Lender of Last Resort: A Reader* (Oxford University Press: Oxford UK, 2002).
506. Bank of England, HM Treasury and FSA, *Memorandum of Understanding between HM Treasury, The Bank of England and the FSA* available at http://www.bankofengland.co.uk/financialstability/Documents/mou.pdf viewed Feb. 11, 2013.
507. Preamble, *Financial Services and Markets Act* 2000 (UK) (hereafter "Financial Services and Markets Act").
508. Financial Services and Markets Act, s. 2(2).

Thus, banks were required to comply with FSA rules in their daily operations; for example, by having appropriate record keeping, customer identification and internal reporting procedures to prevent money laundering.[509]

Ironically however, the FSA's predominant focus on prudential regulation undermined the capacity of the UK regulatory structure to deal with crises of the scale and ilk of the GFC and the LIBOR benchmark manipulation scandals. Regulatory lacunae were highlighted starkly when banks engaged in miscreant behavior that was not technically illegal.

Thus, in a significant regulatory change, the FSA was abolished by the *Financial Services Act 2012* and replaced with the Prudential Regulation Authority (PRA) and the Financial Conduct Authority (FCA), which began operating on April 1, 2013. The functions of the defunct FSA, together with additional powers, were separated out across this "Twin Peaks" system.[510] The PRA supervises deposit-taking financial firms and insurance companies and promotes the stability of the UK financial system whereas the FCA protects consumers from any malpractice while ensuring healthy competition between financial service institutions. In addition, an independent Financial Policy Committee was created within the Bank of England to maintain financial stability by monitoring and responding to systemic risks that threaten the financial system.[511]

Firms and individuals can only conduct financial service activities in the UK if they are authorized by the FCA.[512] Once authorized, they must report to the FCA and be monitored accordingly.[513] Key objectives of the FCA are to protect consumers, protect and enhance the integrity of the UK financial system, and promote effective competition.[514] To this end, the FCA inherited many of the FSA's responsibilities, including: market regulatory functions; enforcement of rules and imposing sanctions for breaches; regulatory oversight of client assets and countering financial crime; acting as the UK Listing Authority; and responsibility for e-money firms, payment service providers and mutual societies.

New powers that were not available under the previous FSA structure include:

- publishing details of disciplinary actions related to rule breaches or compliance failings (held jointly by the PRA and FCA);
- imposing a requirement upon certain unregulated parent undertakings that exert influence over authorized persons (held jointly by the PRA and FCA);

---

509. *Money Laundering Regulations 1993* (UK), rr. 5-14.
510. J. Haydn-Williams, "United Kingdom: Reorganising the Regulation of Financial Services in the UK: The Financial Services Act 2012," *Mondaq* (Jul. 3, 2013), http://www.mondaq.com/x/2 48264/Financial + Services/ECHR + its + Growing + Influence + on + Substantive + Competitio n + Law&email_access = on (accessed Jul. 30, 2014).
511. Bank of England, *Financial Policy Committee*, http://www.bankofengland.co.uk/financialsta bility/pages/fpc/default.aspx (accessed Jul. 30, 2014).
512. The *Financial Services and Markets Act 2000 (Regulated Activities) Order 2001*.
513. FCA, *Systems and reporting requirements*, http://www.fca.org.uk/firms/systems-reporting (accessed Jul. 30, 2014).
514. FCA, *What We Do*, http://www.fca.org.uk/about/what (accessed Jul. 30, 2014).

- making temporary product intervention rules, including the power to block an imminent product launch or to stop an existing product (FCA); and
- requiring firms to withdraw or amend misleading financial promotions (FCA).

While the FCA adopted the bulk of FSA responsibility, the FCA CEO, Mr. Martin Wheatley, has publicly stated that the FCA was created with a "completely different philosophy" to its predecessor. Noting the previous regulatory shortcomings that did not prevent harmful bank behavior, Mr. Wheatley stated that: "the focus is now on forward-looking regulation not backward looking, on compliance models not box ticking, and on an ethics of care not an ethics of obedience."[515]

[2]   Australia

Australia has developed an extensive bank regulatory system over time which, unlike the UK and United States, has experienced very few bank failures in recent decades.[516] The operations of Australian financial institutions are regulated by three bodies:

(1) The Reserve Bank of Australia (RBA) is responsible for overall systemic stability and the conduct of monetary policy pursuant to the *Reserve Bank Act 1959 (Cth)*.
(2) The Australian Securities and Investments Commission (ASIC) is primarily responsible for the performance of the financial system and financial actors and promoting investor and consumer confidence in the financial system pursuant to the *Australian Securities and Investments Commission Act 2001 (Cth)*; it also administers the *Corporations Act 2001 (Cth)*.
(3) The Australian Prudential Regulation Authority (APRA) was established by the *Australian Prudential Regulation Authority Act 1998 (Cth)* and is responsible for the prudential regulation of authorized deposit-taking institutions (ADIs), pension funds, and insurance companies pursuant to the *Banking Act 1959 (Cth)*, the *Financial Sector (Collection of Data) Act 2001 (Cth)*, and the *Legislation Amendment (Financial Claims Scheme and Other Measures) Act 2008 (Cth)*. APRA's supervisory framework requires financial entities to prudently manage their activities and risks. Unlike its UK and American equivalents, APRA does not guarantee depositors' funds; however it administers the Financial Claims Scheme for ADIs, which is the Australian Government's guarantee on deposits. It also assesses and maximizes the probability of ADI financial viability and their ability to meet obligations to depositors.[517]

---

515. Remarks made at the *Regulating Culture and the Manipulation of Financial Benchmarks and Currency Workshop*, Allens Linklaters and UNSW CLMR, Mar. 26, 2014, Sydney Australia, http://www.clmr.unsw.edu.au/workshops (accessed Jul. 30, 2014).
516. Kidwell et al., *supra* n. 76, at 408.
517. *Ibid.*, 420-444.

The *Banking Act 1959 (Cth)* is the key piece of banking legislation in Australia: it charters bank business and provides APRA with various powers, including the power to investigate non-compliant banks (section 13A), revoke ADI status (sections 9, 9A), and remove bank directors (section 23). APRA provides statements and guidance notes, which form the basis of prudential regulation in Australia with which banks must comply.

These three bodies, together with Treasury, form the Council of Financial Regulators, which aims to promote cooperation between the regulatory bodies and efficiency within the Australian financial system.[518]

[3]   United States

The American regime of bank regulation and supervision comprises multiple and independent agencies that regulate banks and also insurance and securities entities.

The US Federal Reserve System or "the Fed," is regarded as a "decentralized central bank" consisting of "a Board of Governors in Washington, D.C., 12 regional Federal Reserve Banks and their branches, and the Federal Open Market Committee."[519] The Fed manages monetary stability by gauging the prudential stability of the American banking system pursuant to the *Federal Reserve Act of 1913* (P.L. 63-43, 38 STAT. 251, 12 USC 221). Specifically, it supervises bank and financial holding companies, foreign bank operations, and state charter banks.[520]

Bank regulation and supervision is primarily codified in the *United States Code (USC) Title 12 Banks and Banking.* However, individual regulatory agencies (discussed below) have a significant portion of their responsibilities and powers set out separately in the *Code of Federal Regulations (CFR) Title 12 Banks and Banking* and their individual agency manuals.[521]

The first hallmark of the American banking regulatory framework is its dual nature. There are two formal systems of chartering (authorizing to operate) commercial banks: state banks are chartered by individual state charters; and national banks are chartered and prudentially supervised by the Office of Comptroller of Currency (OCC) pursuant to the *National Bank Act of 1864* (Chapter 106, 13 STAT. 99).[522] However, a national bank can convert its charter to a state charter and vice versa.[523]

This dual system is preserved in the *Dodd-Frank Wall Street Reform and Consumer Protection Act* (Pub.L. 111–203) (Dodd-Frank), which was passed into law on July 21, 2010. Dodd-Frank embodied massive reform of the financial regulatory regime by introducing, amongst other things: changes to the oversight and supervision of financial institutions; more stringent regulatory capital requirements for nonbank

---

518. *Ibid.*, 418.
519. Federal Reserve Bank of Atlanta, *Federal Reserve: Structure and Functions*, http://www.frbatl anta.org/pubs/frstructurefunctions/ (accessed Jul. 30, 2014).
520. See e.g. 12 USC § 1841 (bank holding companies) and 12 USC § 1841 (foreign banks).
521. See Federal Deposit Insurance Corporation, *Important Banking Laws*, http://www.fdic.gov/r egulations/laws/important/ (accessed Jul. 30, 2014).
522. See also 12 USC § 1; 12 USC, ss. 27(a)-(b); CFR s.4.2.
523. 12 USC, Chapter 2, § 214(a).

financial companies or codified the same for bank holding companies; significant changes in the regulation of derivatives, credit ratings, corporate governance, executive compensation and incentive-based compensation practices, and the securitization market.[524]

More specifically, Dodd-Frank changed the pre-GFC regulatory structure by merging and removing some financial regulatory authorities (Table 5.1) and creating a host of new agencies (Table 5.2).

In addition to the OCC, other agencies have significant prudential supervisory responsibilities in the United States:

(1) The Federal Deposit Insurance Corporation (FDIC), which administers the deposit insurance fund and is the primary regulator of state chartered banks and "thrifts" (also known as savings and loan associations that can take deposits and originate home mortgages).[525] National banks are required to be members of the Fed and the FDIC, the latter giving them access to the public deposit insurance fund.[526]

(2) Twelve regional Federal Reserve Banks that are separately incorporated, each with its own board of directors. Reserve Banks generate their own income, which comes mainly from interest on government securities acquired through open market operations. Reserve Banks monitor national and international economic conditions and provide information on their districts that is used in formulating monetary policy. Reserve Banks serve as "lender of last resort" to ADIs by holding reserve balances. They also examine and supervise certain types of ADIs and provide payment services to them and the Treasury.[527]

(3) The Board of Governors of the Federal Reserve System (the Federal Reserve Board) is the main governing body of the Fed. It oversees the twelve regional Federal Reserve Banks; assists the implementation of monetary policy; and issues rules under Dodd-Frank sometimes in conjunction with other government agencies.[528]

The Federal Financial Institutions Examination Council (the Council) attempts to facilitate uniformity and consistency between the supervisory and enforcement practices of these regulatory agencies.[529]

---

524. Federal Deposit Insurance Corporation, *Important Banking Laws*, http://www.fdic.gov/regulations/laws/important/ (accessed Jul. 30, 2014).
525. 12 USC § 1811 (a) (deposit insurance fund); 12 USC § 1815 (state chartered banks).
526. 12 USC § 282, § 1814(b).
527. Federal Reserve Bank of Atlanta, *The Federal Reserve*, http://www.frbatlanta.org/pubs/frstructurefunctions/structure.cfm#banks (accessed Jul. 30, 2014).
528. Board of Governors of the Federal Reserve System, *Implementing the Dodd-Frank Act: The Federal Reserve Board's Role*, http://www.federalreserve.gov/newsevents/reform_milestones.htm (accessed Jul. 30, 2014).
529. Federal Register / Vol. 62, No. 34 / Thursday, February 20, 1997 / Notices, *Revised Policy Statement on Interagency Coordination of Formal Corrective Action by the Federal Bank Regulatory Agencies*, 7782-7783, http://www.fdic.gov/regulations/laws/federal/rpsi.pdf (accessed Jul. 30, 2014).

Table 5.1   The US Financial Regulatory Framework after Dodd-Frank

Pre-2009	Enacting Legislation	Changes after Dodd-Frank Wall Street Reform and Consumer Protection Act (Pub.L. 111–203, H.R. 4173) (Dodd-Frank)
Dual system of chartering commercial banks (state/national)		Preserved in 12 USC Chapter 53 Subchapter III, § 5401 – Purposes.
The Office of Thrift Supervision (OTS)	*Financial Institutions Reform, Recovery, and Enforcement Act of 1989* (Pub.L. 101–73, 03 Stat. 183, H.R. 1278)	Abolished. All OTS functions, powers, authorities, rights and duties transferred to the Federal Reserve, the OCC or the FDIC.
Office of Comptroller of Currency (OCC)	*National Bank Act of 1864* (Chapter 106, 13 STAT. 99).	The OCC's powers have expanded as it takes on functions that were previously the responsibility of the OTS. Amongst other things, it now regulates national banks and federal thrifts of all sizes, and has all rulemaking authority relating to thrifts.[530]
The Federal Reserve System (the Fed)	*Federal Reserve Act of 1913* (ch. 6, 38 Stat. 251, enacted December 23, 1913, 12 USC. ch. 3)	The Act increased the powers of the Federal Reserve Board.[531] Dodd-Frank explicitly expanded the goals of the Fed to include financial stability in addition to the traditional goals of price stability and full employment.[532] Consequently, it has more responsibility for systemic risk assessment and regulation. The Fed can now regulate companies other than banks (such as insurance companies and investment firms) if they predominantly engage in financial

---

530. "Dodd-Frank: Title III – Transfer of Powers to the Comptroller of Currency, The Corporation and the Board of Governors," *Law Information Institute: Cornell University Law School,* http://www.law.cornell.edu/wex/dodd-frank_title_III.
531. P. Conti-Brown and S. Johnson, "Governing the Federal Reserve System after the Dodd-Frank Act," *Peterson Institute of International Economics Number PB13-25* (October 2013), http://www.piie.com/publications/pb/pb13-25.pdf; E. Schnidman, "Why the Federal Reserve is Dodd-Frank's Big Winner" (2011) *Harvard Business Law Review,* http://www.hblr.org/2011/06/why-the-federal-reserve-is-dodd-franks-big-winner/.
532. P. Wachtel and K. Schoenholtz, "Dodd-Frank and the Fed," *New York University: Leonard N. Stern School of Business* (Jul. 18, 2010), http://w4.stern.nyu.edu/blogs/regulatingwallstreet/2010/07/doddfrank-and-the-fed-kermit-s.html.

		activities and are selected for regulation by the Financial Stability Oversight Council (FSOC) based on an evaluation of their balance sheets, funding sources, and other risk-based criteria.[533] The Fed regulates thrift holding companies and subsidiaries of thrift holding companies, and has all rulemaking authority relating to thrift holding companies. The Fed continues to regulate State member banks.
The Federal Deposit Insurance Corporation (FDIC)	*Banking Act of 1933* (Pub.L. 73–66, 48 Stat. 162, enacted June 16, 1933).	Dodd-Frank assigned significant responsibility to the FDIC for writing and implementing new rules on regulatory reform. The FDIC's role under Dodd-Frank stems from the primary purposes of the Act: ending the "too big to fail" idea of banks, minimizing moral hazard, and mitigating systemic risk. The FDIC is responsible for implementing a number of initiatives under the Act,[534] some significant reforms include:[535] – strengthening and reforming the deposit insurance fund; – strengthening capital requirements; – creating risk retention rules for asset backed securities; – adopting rules with other federal banking agencies to curtail incentive based compensation. The FDIC will also regulate state thrifts of all sizes. The FDIC created the *Office of Complex Financial Institutions* (CFI) and *Division of Depositor and Consumer Protection* (DCP) to carry out responsibilities under Dodd-Frank:

533. W. Sweet, "Dodd-Frank Act Becomes Law," *The Harvard Law School Forum on Corporate Governance and Financial Regulation* (Jul. 21, 2010), http://blogs.law.harvard.edu/corpgov/2010/07/21/dodd-frank-act-becomes-law/.
534. Federal Deposit Insurance Corporation, *FDIC Initiatives under the Dodd-Frank Wall Street Reform and Consumer Protection*, http://www.fdic.gov/regulations/reform/initiatives.html.
535. S. Packer, "FDIC Reforms and Initiatives Under Dodd- Frank" (2010-2011) 30 *Review of Banking & Financial Law* 574, http://www.bu.edu/rbfl/files/2013/09/FDICReforms.pdf.

		– The CFI performs continuous review and oversight of bank holding companies with more than US$100billion in assets as well as non-bank financial companies designated as systemically important by the new Financial Stability Oversight Council. CFI will also be responsible for carrying out the FDIC's new authority under the Act to implement orderly liquidations of bank holding companies and non-bank financial companies that fail. (The FDIC has been able to seize and dismantle smaller banks since it was created in 1933).
		– The establishment of a new division dedicated to depositor and consumer protection will provide increased visibility to the FDIC's compliance examination and enforcement program. That program ensures that banks comply with a myriad of consumer protection and fair lending statutes and regulations.[536]
The Federal Financial Institutions Examination Council	*Public Law 95-630, 95th Congress, H.R. 14279: Financial Institutions Regulatory and Interest Rate Control Act of 1978*	Still responsible for facilitating uniformity and consistency.

Table 5.2   New US Financial Regulatory Institutions Created by Dodd-Frank

New Institutions	*Context and Function*
Financial Stability Oversight Council (FSOC)[537]	(i) Part of the US Treasury, the FSOC comprehensively monitors the stability of the nation's financial system and oversees financial institutions.[538]

---

536. Federal Deposit Insurance Corporation, "FDIC Announces Organizational Changes to Help Implement Recently Enacted Regulatory Reform by Congress," *FDIC Media Release* (Aug. 10, 2010), http://www.fdic.gov/news/news/press/2010/pr10184.html.
537. US Department of the Treasury, *Financial Stability Oversight Council*, http://www.treasury.gov/initiatives/fsoc/Pages/home.aspx.
538. *Ibid.*

	(ii) Voting members include representatives from, amongst others, Treasury, OCC, Bureau of Consumer Financial Protection, SEC, FDIC.
	(iii) Advisory members include representatives from, amongst others, the Office of Financial Research and the Federal Insurance Office.
Office of Financial Research[539]	(i) Serves the FSOC by conducting and sponsoring research related to financial stability, and by promoting best practice in risk management. Also part of Treasury.
Bureau of Consumer Financial Protection[540]	(i) Sits within and is funded by the Fed.
	(ii) Its purpose is to implement, examine and enforce compliance with federal consumer financial laws. It protects consumers from unscrupulous practices by financial entities such as mortgage lenders and credit card companies.[541]
	(iii) Core function: "We work to give consumers the information they need to understand the terms of their agreements with financial companies. We are working to make regulations and guidance as clear and streamlined as possible so providers of consumer financial products and services can follow the rules on their own."[542]
Federal Insurance Office[543]	(i) Sits within Treasury and serves as an advisory member of FSOC.
	(ii) It has authority to: – monitor all aspects of the insurance sector;

539. U.S Department of the Treasury, *Welcome to the Office of Financial Research*, http://www.treasury.gov/initiatives/ofr/Pages/default.aspx.
540. Consumer Financial Protection Bureau, *About Us*, http://www.consumerfinance.gov/the-bureau/.
541. L. Saunders, "The Role of the States under the Dodd-Frank Wall Street Reform and Consumer Protection act of 2010," *National Consumer Law Centre* (December 2010), http://www.nclc.org/images/pdf/legislation/dodd-frank-role-of-the-states.pdf.
542. Consumer Financial Protection Bureau, *supra* n. 540.
543. US Department of the Treasury, *About: Domestic Finance: Federal Insurance Office*, http://www.treasury.gov/about/organizational-structure/offices/Pages/Federal-Insurance.aspx.

| | (ii) – monitor the extent to which tradi-tionally underserved communities have access to affordable non-health insurance products; – represent the US on prudential aspects of international insurance matters. |
| Office of Credit Rating Agencies[544] | (i) Sits within the Securities and Exchange Commission (SEC). (ii) Oversees credit rating agencies: compliance-focused monitoring and examination. |

[B]   Corporate Law, Directors' Duties and Corporate Governance Norms

In addition to the specific banking rules, regulation and supervision with which banking business must comply, there are company laws that also regulate directors' conduct and decision-making more generally. Analysis of directors' fiduciary duties in the context of corporate law, corporate governance norms and climate-friendly bank practices is provided below in order to shed light on how corporate law and corporate governance norms may impact upon bank greening in the jurisdictions central to the case study.

[1]   United Kingdom

The 2009 UK Walker Report made clear that the "role of corporate governance is to protect and advance the interest of shareholders" by setting a company's strategic direction and appointing and monitoring management capable of achieving it.[545] Nonetheless, other stakeholders are to be considered in board decision-making, as enshrined in legislation. Pursuant to section 172 of the *Companies Act 2006* (c.46) (UK) (Companies Act), directors have a duty to "promote the success of the company for the benefit of its members as a whole" (section 172(1)). This duty is owed by directors to the company (section 170(1)). In discharging this duty, UK courts have suggested that directors consider the interests of current and future shareholders at common law.[546] Importantly, directors in the UK are legally required to consider a broader suite of considerations than just the bottom line and/or shareholders. While it can be argued

---

544. US Securities and Exchange Commission, *Office of Credit Ratings,* http://www.sec.gov/ocr#. U8dbub_rTWg.
545. D. Walker, *A Review of Corporate Governance in UK Banks and Other Financial Industry Entities: Final Recommendations: 26 November 2009* (The Walker Review Secretariat: London, 2009), 23.
546. See *Gaiman v. National Association for Mental Health* [1971] Ch 317, 330; *Brady* (1987) 3 BCC 535 (Norse LJ).

that this merely codifies the common law and good business sense to consider a range of interests when decision-making,[547] it is an innovation to place directors under a legal obligation to do so.[548]

Section 172(1) of the Act provides that directors must act in good faith to promote the success of the company "for the benefit of its members as a whole" by having regard to (amongst other matters):

- maintaining a reputation for high standards of business conduct (section 172(1)(e));
- operational impacts on the community and the environment (section 172(1)(d)); and
- likely long term consequences of a decision (section 172(1)(a)).

These three enumerated directives appear to legally mandate corporate greening and socially responsible behavior, particularly for heavy-polluting industries. Yet banks have no extractive or manufacturing activity, and their services and products do not create direct environmental impacts. Indeed, their relevance to climate change, as enunciated in Chapter 2, is significant but *indirect*. Thus, for banks, the first two directives above (sections 172(1)(d) and (e)), likely have little motivating effect on bank board decisions to assist climate change mitigation. As evidenced in the case study, most banks will consider that they are "maintaining a reputation for high standards of business conduct" by creating wealth. In other words that directive can be satisfied even when boards do not adopt climate-related strategies. Moreover, being required to consider community and environmental operational impacts is largely irrelevant for banks given the minimal direct socio-environmental impacts of their operations.

Therefore, the most likely lever in this legislation for greening banks (and other finance actors) is the directive that boards consider the likely consequences of their decisions in the long term. Arguably, this clause captures the indirect environmental and social consequences of bank activities, such as funding "dirty" clients and projects. Nonetheless, if ostracizing lucrative fossil fuel clients is likely to result in long term pecuniary and client service reputational damage to a bank vis-à-vis its competitors, then boards will need to consider *that* consequence too. As such, bank boards may still choose to adopt non-green strategies; even section 172(1)(a) does not guarantee greener decision-making.

Overall, the legal directives in section 172 of the Companies Act provide a potent example of how progressive legislation is sometimes pitted against ingrained cultural-cognitive norms and economic realities. Arguably the directives are best described as an important first-step to changing behavior; they are unlikely to have any immediate

---

547. R. Williams, "Enlightened Shareholder Value in UK Company Law" (2012) 35(1) *UNSW Law Journal* 360.
548. P.L. Davies, *Glower and Davies: The Principles of Modern Company Law* (Sweet and Maxwell: London, 8th ed, 2008).

impact on remedying bank directors' perceived and actual constraints on climate-friendly action.

[2]   Australia

Under section 180(1) of the *Corporations Act 2001* (Cth) (Australia) (Corporations Act), directors must act with the degree of care and diligence that a reasonable director would exercise. The parameters of section 180(1) are defined by section 180(2) which sets out what is commonly known as the business judgment rule, being a director's defense to an alleged breach of duty of care provided that they have acted in good faith, for proper purpose, impartially and honestly on credible information, and with a rational belief that the judgment was in the best interests of the corporation. Australian Courts are generally reluctant to interfere in matters that involve business judgment, especially where a range of decisions could have been made by a director in a particular situation.[549]

Pursuant to section 181 of the Corporations Act, Australian directors must act in good faith and for proper purpose in "the best interests of the corporation" (section 181(1)). However there is no explicit legislative direction or authoritative judicial statement to clarify this duty in the context of socio-environmental or broader stakeholder expectations. In contrast to the UK legislation, there are no enumerated directives to this end. Case law on the duty has almost solely focused on directors' exercise of power during internal contests for corporate control or other self-interested purposes.[550] The business judgment defense applies only to an alleged breach of duty of care (section 180(1) and (2)); it does not apply to a director's duty to act in the company's best interests under section 181(1).

Nonetheless, in *Provident International Corp v. International Leasing Corp Ltd* [1969] 1 NSWLR 424, 440, the court suggested that directors may consider the interests of future shareholders in discharging their duty to the company akin to the UK common law; and in *The Bell Group Ltd (in liq) v. Westpac Banking Corporation (No 9)* (2008) 39 WAR 1, 534, Owen J specified that the interests of shareholders and "the company" are correlative but not conflated.

Moreover, two of the most recent federal government inquiries on the discrete issue of CSR and Australian corporate law concluded that directors are not confined by law to short-term considerations, such as maximizing profit or share price returns, in their decision-making.[551] In other words, the best interests of the company include its long-term well-being. This raises a pivotal issue that was addressed by the Parliamentary Joint Committee on Corporations and Financial Services in 2006: the long-term

---

549. See Parliamentary Joint Committee on Corporations and Financial Services, *supra* n. 116, at 45.
550. see e.g. *Mills v. Mills* (1938) 60 CLR 150; *Whitehouse v. Carlton Hotel Pty Ltd* (1987) 162 CLR 285.
551. Corporations and Markets Advisory Committee. *The Social Responsibility of Corporations* (Australian Government: Canberra, 2006); Parliamentary Joint Committee on Corporations and Financial Services, *supra* n. 116.

interests of the corporation will not always accord with the short-term interests of shareholders. That is:

> [I]f directors make a decision... benefiting long term and future shareholders, but which results in a short term loss (and a short term decline in share value for current shareholders), then this decision will be in the interests of the company, but will be unwelcome news for shareholders who have taken a short term, perhaps speculative position.[552]

As such, that Committee concluded that the Corporations Act permits directors to "consider and act upon the legitimate interests of stakeholders [other than shareholders] to the extent that these interests are relevant to the corporation" regarding its long term viability "*even where they do not generate immediate profit.*"[553] Specifically, the Committee emphasized "enlightened self-interest" stating that "directors should act in a socially and environmentally responsible manner at least in part because such conduct is likely to lead to the long term growth of their enterprise."[554] For these and other reasons, that Committee and also the Corporations and Markets Advisory Committee opposed amending the Corporations Act to include directive provisions similar to the UK amendments.[555]

This has clear ramifications for climate-related bank efforts that are planted in the now at shareholders' expense but will likely reap reward in future in the form of enhanced client service reputation and social reputation for the bank as a company.

Buttressing some of the Committee's conclusions, a survey of 375 Australian directors from a variety of industries was conducted to explore directors' understandings of their legal obligations and how that affects their approach to stakeholders.[556] This study remains the most comprehensive of its type in Australia.[557] It found that no respondents believed the duty under section 181 required them to act only in the short-term interests of shareholders. Nearly all respondents (94.3 per cent) believed that corporate law regarding directors' duties was broad enough to permit consideration of the interests of both shareholders and stakeholders; with just over half (55 per cent) believing they needed to balance all stakeholders' interests. Yet this raises the key question of interest prioritization. In the survey, directors most commonly ranked the interests of shareholders as their number one priority and gave them the highest level of salience in terms of influence and ability to make demands on management (when compared to employees and creditors). So even though directors believe they have legal scope to consider and even balance stakeholders' interests, shareholders come

---

552. Parliamentary Joint Committee on Corporations and Financial Services, *supra* n. 116, at 44.
553. *Ibid.*, 52, emphasis added.
554. *Ibid.*, 52-53.
555. *Ibid.*, Recommendation 1 at xxi. See also Corporations and Markets Advisory Committee, *supra* n. 551.
556. M.E Anderson, M.A. Jones, S.D. Marshall, R. Mitchell and I. Ramsay, *Evaluating the Shareholder Primacy Theory: Evidence from a Survey of Australian Directors* (Legal Studies Research Report No 302, University of Melbourne, 2007).
557. S. Marshall and I. Ramsay, "Stakeholders and Directors' Duties: Law, Theory and Evidence" (2012) 35(1) *UNSW Law Journal* 291.

first in that balancing act. Most importantly for this discussion, "the environment" and "the community" were ranked second and third last respectively in terms of their perceived priority and salience by directors.

Arguably then, the legal environment in Australia can be interpreted as conducive to climate-friendly bank action, and may even permit measures that aim to benefit the company long-term but create short-term profit loss. In actuality however, this provides little comfort to banks concerned about shareholder or stock market backlash where greener efforts result in diminished short-term performance. This conjecture is supported by qualitative data in the case study; and is consistent with quantitative results in the aforementioned survey.

[3]   United States

American corporate law differs from UK and Australian law in two main respects. First, corporations in the United States are creatures of state not federal law, so the fiduciary duties of directors are defined by a company's state of incorporation.[558] Secondly, the primary fiduciary duties of "care" and "loyalty" are obligations at common law and remain largely uncodified in American legislation.

The fiduciary duty of care requires directors to make prudent decisions based on full and credible information.[559] Bernard S. Black opines that pursuant to this duty, directors are not obligated to make *sensible* decisions: "They only have to show up, pay attention, and make a decision that is not completely irrational."[560] The duty of loyalty obligates directors to act in good faith and in the best interests of the corporation and its shareholders.[561] Thus, unlike the UK and Australian jurisdictions, this duty is expressly owed to the corporation *and* its stockholders. Indeed, historically, United States' courts viewed company directors as trustees and individual shareholders as beneficiaries of that trust,[562] a metaphor that was even extended to shareholder value in *Dodge v. Woolsey* 59 US 331 (1855).[563]

Importantly, the *legal* requirement to make decisions in the best interests of shareholders provides the foundation of the "shareholder primacy" *norm* in American corporate culture.[564] Shareholder primacy means that the corporation's main purpose

---

558. See S.M. Bainbridge, "A Critique of the NYSE Director Independence Listing Standards" (2002) 30 *Securities Regulation Law Journal* 370; *Burks v. Lasker*, 441 US 471 (1979).

559. See B.S. Black, *The Principal Fiduciary Duties of Boards of Directors* (Presentation at Third Asian Roundtable on Corporate Governance, Singapore, Apr. 4, 2001), 6.

560. *Ibid.* See also R.S. Sprague and A.J. Lyttle, "Shareholder Primacy and the Business Judgment Rule: Arguments for Expanded Corporate Democracy" (2010-2011) 16 *Stanford Journal of Law, Business & Finance* 1, 8.

561. Black, *supra* n. 559, at 4-6.

562. see e.g. *Gray v. President* 3 Mass (3 Tyng) 364, 379 (1807).

563. See Sprague and Lyttle, *supra* n. 560.

564. *Ibid.*, 5. See also: D.G. Smith, "The Shareholder Primacy Norm" (2008) 23 *Journal of Corporate Law* 277, 278; S.M. Bainbridge, "In Defense of the Shareholder Wealth Maximization Norm: A Reply to Professor Green" (1993) 50 *Washington & Lee Law Review* 1423, 1423-25; K.B. Davis Jr, "Discretion of Corporate Management to Do Good at the Expense of Shareholder Gain: A Survey of, and Commentary on, the US Corporate Law" (1988) 13 *Canada-US Law Journal* 7, 8.

is to maximize shareholder *profit* and that "the powers of the directors are to be employed for that end" as per Michigan Supreme Court in *Dodge v. Ford Motor Co* 170 NW 668, 684 (Mich 1919), which may be the earliest judicial statement of the norm. In the near-century since the *Dodge* pronouncement, shareholder primacy has gained normative status as the "dominant force" in American corporate governance[565] and is embodied in the American Law Institute's 1994 *Principles of Corporate Governance.*[566]

Nonetheless, a majority of American states have now adopted constituency statutes that permit, but generally do not require, directors to consider the effects of decision-making upon stakeholders such as employees, customers, suppliers, creditors and local communities when deciding what is in the best interests of the corporation.[567] Some statutes are progressive, stating that directors are not required to give primacy to any particular set of interests and/or permitting profit sacrifice. Yet all statutes stipulate that directors will use their business judgment in determining the weight to be given to particular stakeholder interests.

Moreover, most publicly-traded American companies are incorporated in the state of Delaware, which does not have a constituency statute. The Delaware Court of Chancery has a long history of jurisprudence and precedent regarding directors' fiduciary duties.[568] Corporate Law commentators such as David Skeel have argued that Delaware's jurisprudential resolution of most corporate governance issues "tilt[s] to managers (and an occasional venting of stakeholder interests)" such that American corporate law is manager-centrist.[569]

This may be explained in part by the benefit of the Business Judgment Rule (BJR) for company directors when discharging their duties. The BJR is a presumption (rather than a defense) in American corporate law that in making a business decision the directors "acted on an informed basis, in good faith and in the honest belief that the action taken was in the best interests of the company."[570] The burden lies with plaintiffs to establish facts rebutting that presumption. Absent a showing of an abuse of discretion, conflicts of interest, or a grossly negligent procedure for becoming informed, a director's business judgment will be respected and not second-guessed by the courts with the benefit of hindsight.[571] As such, American courts do not penalize directors for making honest mistakes.

---

565. Sprague and Lyttle, *supra* n. 560.
566. L.L. Dallas, "The New Managerialism and Diversity on Corporate Boards of Directors" (2002) 76 *Tulane Law Review* 1363. *C.f.* L. Stout, *The Shareholder Value Myth: How Putting Share-holders First Harms Investors, Corporations, and the Public* (Berrett-Koehler: San Francisco, 2012).
567. See K. Hale, "Corporate Law and Stakeholders: Moving Beyond Stakeholder Statutes" (2003) 45 *Arizona Law Review* 823.
568. See B.S. Black, B.R. Cheffins, M. Gelter, H-J. Kim, R. Nolan, M.M. Siems, Legal Liability of Directors and Company Officials Part 1: Substantive Grounds for Liability (Report to the Russian Securities Agency) (2007) *Columbia Business Law Review* 614, 643.
569. Skeel, *supra* n. 254, at 990. See also Dallas, *supra* n. 566.
570. per Delaware Court of Chancery in *re Citigroup supra* n. 198, quoting from *Aronson v. Lewis* 473 A2d 805, 812 (Del 1984) overruled on other grounds by *Brehm v. Eisner*, 746 A.2d 244 (Del. 2000)).
571. *re Citigroup supra* n. 198; *Gantler v. Stephens* 965 A2d 695, 705-06 (Del 2009).

Arguably then, despite the shareholder primacy norm, the BJR allows American banks to adopt climate-friendly practices even if those decisions result in reduced shareholder returns. Boards need only act in good faith and with the requisite degree of care and loyalty when making a "green" decision. Indeed, analogous use of recent American jurisprudence in the context of stock downturn in the wake of the GFC can be cited to support this proposition. In re Citigroup, shareholders of Citigroup sued bank directors in 2007-2009 for financial losses suffered as a result of decisions to invest in sub-prime mortgage-related securities.[572] The Delaware Court of Chancery dismissed the claim on the basis that plaintiff shareholders had not rebutted the BJR presumption. Specifically, the Court re-stated the rationale of the BJR as follows:

> Through the business judgment rule, Delaware law encourages corporate fiduciaries to attempt to increase stockholder wealth by engaging in those risks that, in their business judgment, are *in the best interest of the corporation without the debilitating fear that they will be held personally liable if the company experiences losses.*"[573]

For similar reasons, plaintiff litigants were unsuccessful in suing Lehman in 2010 for flawed decision-making that included substantial investments in sub-prime mortgage related securities, which ultimately bankrupted the firm and decimated share value.[574] Moreover, citing Re Citigroup, the Delaware Court of Chancery in 2011 similarly dismissed a shareholders-plaintiffs' claim against Goldman Sachs' directors, which alleged that executive and employee compensation practices had incentivized risky behavior and that directors had breached their duty of care by not monitoring the business risk thereby created.[575] As in Re Citigroup, the Goldman Sachs decision stressed that taking on a high degree of risk is not a violation of the board's duty, even if it results in pecuniary loss.

Thus, using analogous and recent judicial precedent, it is arguable that bank directors have latitude within American corporate law to increase their climate-friendly endeavors. As long as managers can plausibly claim their actions intended long-term benefit for the bank as a company (in terms of reputation enhancement etc.) then it is nigh impossible for shareholders to challenge them in retrospect. However, as with the UK and Australia, it is improbable in reality that boards in the United States will prioritize green behavior above profit imperatives. Indeed, manager-centrism may have only enhanced "short-term decision making and window dressing to impress the stock market."[576] Despite a congenial legal environment for directorial discretion, banks in America must still make decisions in light of economic constraints and the

---

572. re Citigroup supra n. 198.
573. Ibid., at 139 (emphasis added).
574. see re Lehman Bros. Holdings Inc., No. 08-13555 UMP) at 4 (Bankr. S.D.N.Y. Mar. 11, 2010).
575. In re Goldman Sachs Grp., Inc. S'holder Litig., No. 5215-VCG, 2011 WL 4826104, at *23 (Del. Ch. Oct. 12, 2011). See also C.A. Hill and B.H. McDonnell, "Reconsidering Board Oversight Duties After the Financial Crisis" (2013) *University of Illinois Law Review* 859.
576. Dallas, supra n. 566, at 1363.

entrenched corporate governance *norm* of shareholder primacy. As depicted in my study, in practice they are not willing to risk adverse impacts on client service reputation and economic viability due to green endeavors that may amount to expensive short term "mistakes."

[4]   *Summation*

Data in the case study show that, consistent with corporate governance constraints (perceived and real) outlined above and the "it depends" literature outlined in Chapter 3, bank approaches to short-termism depend on the materiality of the business case. This was illustrated most clearly when bank respondents discussed the parameters of board support for green initiatives. It is arguable that even permissive law as an institution may be no match for deeply ingrained cultural-cognitive norms regarding shareholder wealth maximization, especially in light of current economic realities. Progressive law reform is an important first-step to codifying social norms and influencing corporate behavior but it is not a magic wand for creating macro normative change.

Dermine writes that the succession of financial crises over the last 30 years has inspired review of corporate governance in the banking sector. Although advocating a shareholder-based governance ,approach and acknowledging that financial markets reward short-term profits, he writes that "it is the responsibility of the bank's board to take care of long-term value creation, even if it means hurting reported revenue and the share price in the short term."[577] The case study data in this book show that we have some way to go yet before shareholder wealth maximization and long-termism are viewed *in practice* as compatible by bank boards.

Overall, an initiative will not get board support unless managers can show a clear and immediate business case. Arguably this is so for all new initiatives. Yet for climate-related initiatives in particular there is potential for: (a) disjunction between seeing how environmental-social risks and opportunities translate into fiscal ones, especially given that sustainability-related gains or losses can be difficult to quantify, and/or (b) a lag between green implementation and financial results. These limitations in business case logic have potential to impede *en masse* bank activity in climate-related spaces outside pension fund (long-term) investment.

§5.04   INCREMENTAL VERSUS TRANSFORMATIONAL CHANGE

[A]   "Organic" Mainstreaming via Isomorphism

So how is greener banking behavior likely to become mainstreamed voluntarily, that is, without government intervention? It is probable that such change would occur through

---

577. Dermine, *supra* n. 129, at 268.

isomorphism and mimetic behavior, that is, through institutional processes of organic corporate change whereby firms copy and adopt successful practices. As described in Chapter 3, institutional scholars posit that there are three mechanisms of isomorphic change, namely coercive, normative, and mimetic. The interview data revealed some elements of coercive isomorphism (when responding to community expectation and NGO pressure) and normative isomorphism (through professionalization in the banking industry). However, it was banks' propensity for mimetic isomorphism (whereby-organizations "model themselves after similar organizations in their field that they perceive to be more legitimate or successful")[578] that was evidenced most consistently throughout the data.The previous chapter demonstrated that banks are subject to peer pressure when adopting voluntary industry standards, especially when there is a lack of government regulation. Evidence of peer pressure and the mimetic nature of the banking industry supports the proposition that mainstreaming of leading bank practices will likely occur in the future via these institutional processes. Specifically, the data revealed that the high likelihood of "following the leader" in the banking industry is due to three main motivators: the competitive nature of the banking industry; the visibility of leading banks; and the various forms of uncertainty that come with climate change, not the least of which is regulatory uncertainty. In practice, leading banks are dealing with uncertainty by exploiting it: they are capitalizing on business opportunities created by immature markets while simultaneously honing risk mitigation strategies.

Mimetic isomorphism most often occurs when organizations are responding to uncertainty, which has been described as a "powerful force" that motivates imitation.[579] Institutional theory posits that, in uncertain environments, organizations will adopt the practices, rhetoric and symbols of more successful firms in order to be seen as "normal" and "appropriate" in order to survive. The risk/benefit approach of leading banks toward climate change can be viewed by the rest of industry as legitimate and compelling because it fits with business case imperatives and is consistent with the prevailing view of how banks see their role in society. In other words, risk/benefit language is redolent of the "organizational language" used by banks to demonstrate their rationality and legitimacy in the cultural context of capitalism. Indeed, risk/benefit discourse and economic efficiency are particular forms of rationality privileged by market economies. So, in institutional parlance, other banks will "garner more legitimacy if they can emulate or symbolically reproduce that rationality"[580] by adopting the climate-related risk/benefit strategies of leading banks.

Moreover, climate-related practices and products that have been created by leading banks provide good mimetic fodder for the rest of the industry. It is apparent from the data that leading banks tend to set standards together because none of them want to be the odd one out. If those standards/practices are perceived to be an industry

---

578. DiMaggio and Powell, *supra* n. 431, at 70.
579. *Ibid.*, 69.
580. Meyer and Rowan, *supra* n. 416, at 50.

standard or to provide competitive advantage as judged by the prevailing capitalist norm of economic efficiency then, suggest Jennings and Zanderbergen, adoption of environmentally responsible practices is likely.[581] In other words, as noted by Bebbington et al. "simply coming up with moral-based arguments for adopting [green] practices is unlikely to be convincing"[582] or to lead to mainstreaming in the industry.

Of course, leading banks may not intend to set standards for the rest of the industry or even *want* to be mimicked, especially if doing so diminishes first mover advantage. However, subjective intention is irrelevant to mimetic processes.[583] Wittingly or not, leading banks lend themselves to mimicry for two main reasons. First, they are loud and proud about their innovations through public disclosures on their websites and in industry publications, so their practices are readily accessible. Data in the case study confirmed that banks monitor each other's market ratings and inform themselves about competitor activities. Second, the banking industry is close-knit even though (or perhaps because) it is so competitive, which means that field cohesion is high and cross-pollination highly likely. Models can be diffused and copied through employee transfer and industry trade organizations.[584] It is evident from the case study that employees move frequently between banks; and high interaction in the field is indicated by, for example, industry conferences and Bankers Associations. Bansal and Roth's study confirms that organizations operating in highly cohesive fields adopt ecologically responsive practices that are established as legitimate in order to fit in.[585]

In short, industry-wide change will likely occur due to early-moving banks providing models of success that other banks can follow and copy. Their climate-related actions, together with soft law codes such as the Equator Principles, inculcate industry standards and taken-for-granted behaviors. It is probable that mainstreaming of leading practices in relation to climate change – despite inherent limitations to business case logic – will occur over time in the banking industry.

If mainstreaming occurs organically via institutional processes, then the critical question becomes this: is it prudent to put our faith in purely voluntary standard-setting by banks as the mode by which they assist timely transition to a low-carbon economy? There are two main concerns with doing so. First, how *far* can mainstreamed bank-created standards take us down a low-carbon path? Second, how *fast* can real change occur this way?

[B]   Going the Distance: Standard-Setting and Real Change

To address the first concern of whether purely voluntary action can take us far enough, we must ask "what exactly is being embedded as the industry norm?" The issue is that

---

581. D. Jennings and P. Zanderbergen, "Ecologically Sustainable Organizations: An Institutional Approach" (1995) 20(4) *Academy of Management Review* 1015.
582. J. Bebbington, C. Higgins and B. Frame, "Initiating Sustainable Development Reporting: Evidence From New Zealand" (2008) 22(4) *Accounting, Auditing & Accountability Journal* 588, 595.
583. DiMaggio and Powell, *supra* n. 431, at 69.
584. *Ibid.*
585. Bansal and Roth, *supra* n. 296, at 732.

if a low industry benchmark is set by leading banks then it may not be sufficient to create core or meaningful change to widespread corporate practices. Specifically, some scholars suggest that voluntary changes driven by business case imperatives result in low standard-setting; this is because such changes target only easily achieved cost efficiencies or "low hanging fruit."[586] Certainly, authors such as Steger and also Hoffman, who advocate the business case as driver for corporate greening, highlight how "easy wins" can be achieved through increased operational- or energy-efficiency measures rather than introducing environmental programs.[587] Indeed, Salzmann et al. have suggested that an increase in energy- or eco-efficiency is the basis for the environmental dimension of the business case.[588]

Whilst "efficiency" arguments can be viewed as an attractive and pragmatic selling point for corporate environmental uptake, they also seem to advocate "the 'no-brainers' of good (rather than corporate sustainability) management."[589] As noted by Bebbington, efficiency on its own is a necessary but insufficient condition for sustainability.[590] Spence further explains that "efficiency gains can potentially be undone by business growth" with the outcome that a business may create a larger overall environmental footprint despite making per unit eco-efficiency improvements.[591]

The concern is that "no brainers" can result in minimalist change becoming the corporate norm. That is, a low industry standard may equate to BAU with only peripheral changes. Literature on this issue, particularly in critical management studies and social-environmental accounting, contends that when businesses perpetuate BAU with peripheral change only, the magnitude of environmental issues and the level of corporate change actually required to address them is masked or deflected.

As noted in Chapter 3, some scholars have termed this phenomenon as corporate or managerial "capture" of the CSR agenda. It means that corporations may selectively choose elements of CSR that suit their business interests whilst appearing to be listening to criticism and enacting change. A swathe of critical accounting scholars suggest that doing so equates to legitimation of the status quo and, thus, resistance to desired change by "veering little from business as usual."[592]

Other scholars label this phenomenon as the "middle" or "third" way, which sits between "free-market fundamentalism and the green left" as a pragmatic and economic imperative for business.[593] These scholars view the demands of "the environment" and "development" as conflicting; they argue that business discourse that attempts to simplify this conflict actually obfuscates the complexities and tensions

---

586. Steger, *supra* n. 328, at 41. Bebbington et al., *supra* n. 582.
587. Steger, *supra* n. 328, at 42; Hoffman, *supra* n. 451, at 16.
588. Salzmann et al., *supra* n. 130, at 33.
589. *Ibid.*
590. J. Bebbington, "Sustainable Development: A Review of the International Development, Business and Accounting Literature" (2001) 25(2) *Accounting Forum* 128.
591. Spence, *supra* n. 318, at 870.
592. O'Dwyer, *supra* n. 291, at 527. See also: Milne et al., *supra* n. 317; Tinker et al., *supra* n. 318; Spence, *supra* n. 318.
593. Milne et al., *supra* n. 317, at 1230, 1236. See also: Tinker et al., *supra* n. 318; Spence, *supra* n. 318; P. Prasad and M. Elmes, "In the Name of the Practical: Unearthing the Hegemony of

inherent in far-reaching environmental-corporate change. That is, win-win discourse merely reinforces a continuing propensity by industry to damage the Earth while deflecting from doing so. This concern highlights the contested nature of win-win claims in the business strategy literature, as outlined in Chapter 3.

The bank case study shows that "sustainability" is predominantly conceived by bankers to mean self-sustaining instead of the longevity of the natural environment. There is a real concern that other finance actors are likely to adopt these prevailing conceptions in order to not be "presented as somehow missing out, not up with the play, and perhaps ultimately illegitimate."[594]

Yet there are two outcomes from creating or perpetuating the impression that meaningful change is occurring when actually it is not. First, corporate capture or middle-way framing of CSR discourse results in a missed opportunity to facilitate real solutions to pressing problems. In other words, it "crowds out and closes down alternative... principles and practices that might actually be better for conserving the Earth."[595] Second, it can result in subtle but decided corporate resistance to needed change.

If this is the case then mainstreaming of voluntary corporate green uptake motivated by business case logic is not so much a bridge to salvation as a tweak to the status quo. This is what I refer to as "BAU with a lemon twist."

To this end, NGOs and some academics have questioned the ingenuity of leading banks' commitment to a "core" green agenda. Criticism has been aimed at three main areas. First, new practices such as carbon trading and climate risk due diligence fall within the scope of current practice and can be viewed as simply logical as opposed to truly green. Secondly, the voluntary Principles may be sub-optimal for achieving environmental and sustainability objectives. Thirdly, funding continues for fossil fuel projects and at levels disproportionate to renewables and clean tech. These three areas of criticism are considered below in the context of the case study.

[1] Green Activities as an Extension of Current Practice?

In some banks, innovative and radical climate-related solutions and products had been born of business case logic with the potential to facilitate corporate GHG emissions reductions and societal benefit as suggested at the start of this book. For example, one bank had instituted an outright veto on lending to companies that practice mountain top removal coal mining. At another bank, a project financier described how he innovatively used hedges in energy/renewables projects to facilitate lucrative returns for his clients, which simultaneously increased investments in that space (because those projects were more attractive to clients) and enhanced his firm's client service

---

Pragmatics in the Discourse of Environmental Management" (2005) 42(4) *Journal of Management Studies* 845, 856; R.H. Gray and J. Bebbington, "Environmental Accounting, Managerialism and Sustainability: is the Planet Safe in the Hands of Business?" (2000) 1 *Advances in Environmental Accounting and Management* 1.

594. Milne et al., *supra* n. 317, at 1236.
595. *Ibid.* See also Prasad and Elmes, *supra* n. 593; M.A. Hajer, *The Politics of Environmental Discourse: Ecological Modernization and the Policy Process* (Clarendon Press: Oxford UK, 1997).

reputation (making it the "go to" bank in that space in the United States). Another bank had created a new set of climate change investment indices in response to the request of a very large pension fund client that re-classify industrial sectors into four climate-related themes. An analyst explained their impact: "The classification determines the asset allocation process. Fund managers have no translator between the way they see the world and the way the world is changing; we offer them a different view." (A2). In other words, the indices influence resource allocation into low-carbon investments and away from carbon-intense ones. Such a simple innovation has powerful flow on effects that can facilitate GHG emissions reductions through corporate networks in a ripple-out effect

However, the more common and widely-spread initiatives by leading banks were: activity in carbon trading; enhanced due diligence for climate risk; and signing onto the Equator Principles and/or Carbon Principles. Several NGO respondents asserted that such "new" practices fell within the scope of current bank practice and were therefore an extension of BAU as opposed to truly green. For example, one NGO respondent described the Carbon Principles as a "laughable version" of green activity:

> It is ridiculous to claim you are a climate leader when you are simply trying to assess a US power plant deal on the assumption that a carbon price will come in soon. That is such a no-brainer. It takes no environmental commitment; it's just logical. Another example is the measurement of GHG emissions in their loan portfolios in order to assess their own [carbon] liabilities. Again, it's not a radical or "green" idea. It is well within the scope of current thinking of banks. (NGO-D1).

An external consultant made a similar point regarding carbon trading: "Banks and bankers are good at talking about an ETS [emissions trading scheme]. They can just clip the ticket on it: it's another market and commodity that they can know better than anyone else." (Cons-B1).

Bankers tended to corroborate these perspectives. For example, transactional bankers involved in carbon trading noted that it was just another asset class to be exploited:

> [The bank] is active in carbon trading in the same way it is active in other trading areas such as oil, gas, electricity and coal trading desks. Therefore the carbon market is a natural addition to these desks. It makes no sense to trade in other commodities and not carbon. (E2)

For some bankers this was also true of signing onto soft law Principles:

> In my own view I don't really think the Carbon Principles impact upon the way we do business because we are already practicing a lot of what is presented in them regarding risk assessment and diligence levels. It wasn't a lot of effort to sign up and agree to all those things. (B2)

There is a rich critical literature on the issue of whether signing onto voluntary industry codes provides only the symbolic appearance of action without delivering substantive environmental and social outcomes.[596] In relation to banks adopting the voluntary Principles, most attention has been directed at the Equator Principles, which have been in force the longest. Criticisms of this voluntary code have focused mainly on two perceived shortcomings: limited coverage and a lack of accountability mechanisms. In relation to limited coverage, the Equator Principles apply only to project finance activities above $10million (since 2006). NGOs argue that this covers only "a small percentage of the environmental and social impacts of the financial sector as a whole"[597] and that corporate finance activities comprise a much larger part of bank balance sheets (NGO-A1; NGO-D1). As such, some commentators posit that regulating project finance represents "convenient 'low-hanging fruit'" for financial institutions and deflects attention from "other fundamental areas of concern" in banking practice.[598] In relation to accountability, NGOs and legitimacy scholars have also criticized the absence of clear accountability mechanisms in the Equator Principles at organizational or project levels, arguing that there is little opportunity to evaluate environmental results "on the ground" or assess whether "the essential machinery" of a bank has changed or improved.[599]

However, several commentators in recent years have argued that assessing effectiveness may encompass more than just measuring outcomes; that adoption of voluntary codes may facilitate processual benefits such as organizational learning and dissemination of best practices among adopting and non-adopting firms alike.[600] Specifically, some have suggested that the very act of adopting the Equator Principles "may be both evidence of and a catalyst for cultural change" within not only adopting banks but also lending syndicates.[601]

It was not the aim of this case study to investigate the effectiveness of the Equator Principles, Carbon Principles or Climate Principles in delivering improved environmental outcomes, nor whether banks are strictly complying with them. Instead, interview questions about these Principles were intended to grasp only the extent to which codes had become core to bank business and/or altered BAU.

---

596. See e.g. B.E. Ashforth and B.W. Gibbs, "The Double-Edge of Organizational Legitimation" (1990) 1(2) *Organizational Science* 177; M. Lenox and J. Nash, "Industry Self-Regulation and Adverse Selection: A Comparison Across Four Trade Association Programs" (2003) 12(6) *Business Strategy and the Environment* 343; P. Bansal and T. Hunter, "Strategic Explanations for the Early Adoption of ISO 14001" (2003) 46 *Journal of Business Ethics* 289; A. Prakash and M. Potoski, *The Voluntary Environmentalists: Green Clubs, ISO 14001, and Voluntary Environmental Regulations* (Cambridge University Press: Cambridge UK, 2006).
597. O'Sullivan and O'Dwyer, *supra* n. 166, at 581.
598. *Ibid.*, 555.
599. *Ibid.*, 576.
600. See e.g.: T.P. Lyon and J.W. Maxwell, "Environmental Public Voluntary Programs Reconsidered" (2007) 35(4) *Policy Studies Journal* 723; and N. Darnall and S. Sides, "Assessing the Performance of Voluntary Environmental Programs: Does Certification Matter?" (2008) 36(1) *Policy Studies Journal* 95.
601. Conley and Williams, *supra* n. 167, at 561. See also D.E. Rupp, C.A. Williams and R.V. Aguilera, "Increasing Corporate Social Responsibility through Stakeholder Value Internalization (and the Catalyzing Effect of New Governance): An Application of Organizational Justice, Self-Determination and Social Influence Theories," in M. Schminke (ed) *Managerial Ethics: Managing the Psychology of Morality* (Routledge: New York, 2010).

In relation to the Equator Principles, respondents at adopting banks affirmed that Equator Principles' standards were integral to ESG risk assessments for project finance, and usually coordinated through a Risk Committee. This finding accords with extant research that most adopting institutions standardize and benchmark their project finance procedures to World Bank standards and designate risk management personnel.[602] However, the case study data did not evidence that application of the Equator Principles was expanding to departments outside of project finance, such as commercial lending or underwriting.[603] Indeed, consistent with NGO concerns, the data suggested that internal procedures and standards (not the Equator Principles) continue to be used to assess corporate finance and other bank activities. Moreover, as evidenced in the previous chapter, there was no empirical support for the hope professed by some that the reflective processes required by the Equator Principles are deeply engaging "the moral and ethical sensibilities of employees and managers."[604]

I received mixed results in relation to the impact of the Carbon Principles on everyday bank practices, which apply only to coal clients/activities in the United States. Respondents at different banks stated that adopting the Carbon Principles had not changed BAU:

> We've always looked at the environmental consequences of new projects. All power plants will have a negative impact on the environment, even wind plants with birds and solar technology with turtles. That is already factored in and won't necessarily stop the deal. (C1).

However, a risk manager at a third adopting bank gave a different response:

> No bank was doing as in-depth a due diligence review prior to what is now required by the Carbon Principles. Some questions being asked – like "What's your company's carbon mitigation strategies?" – no bank was asking that question previously. (D1)

Only one leading bank had signed onto the Climate Principles. There was no mention at this bank of the effect of that code on their everyday business practices. So the level of penetration of the Climate Principles remained obscure. However, given that the principles were declared defunct by the Climate Group in 2012,[605] we may impute minimal penetration at the time of data collection.

Arguably, the main aim of the Equator Principles and the Carbon Principles is to mitigate risk via enhanced due diligence processes, being financial, regulatory and social reputation risk for signatories and environmental-social risk for local communities. There is no doubt that risk mitigation is a crucial component of the business case and an important first step in corporate change. Nonetheless, it is a first step only. As

---

602. E.g. A. Meyerstein, *On the Effectiveness of Global Private Regulation: The Implementation of the Equator Principles by Multinational Banks* (PRI Academic Conference, Sigtuna Sweden, Sep. 26-28, 2011).
603. Cf. Conley and Williams, *supra* n. 167, at 552.
604. *Ibid.*, 553. See also Rupp et al., *supra* n. 601; H. Spitzeck, "Organizational Moral Learning: What, If Anything, Do Corporations Learn from NGO Critique?" (2009) 88 *Journal of Business Ethics* 157.
605. See Chapter 2 *supra*.

evidenced earlier, banks became proactive and innovative when motivated to get a bigger slice of the client pie. In this way, grabbing opportunities (as opposed to mitigating risk) allowed banks to break with usual practice and leap two steps ahead of the curve instead of just one logical step further down the BAU path. In contradistinction, banks that were primarily driven by risk mitigation were reactive; their aim was to keep BAU running as smoothly as possible.

When a firm's goal is to minimize risks and costs then its aim is only to meet standards and not exceed them, known as satisfice.[606] Satisfice was most notably evidenced by bankers when the domestic climate-related law and policy context was highly uncertain. For example, although the Australian Senate had passed the *Clean Energy Act 2011* (Cth) comprising a number of climate-related initiatives including a carbon tax, there was much uncertainty around its implementation and even continued existence under a Coalition Government. Against this background, most Australian bank respondents revealed that there had been little change to BAU with the exception of enhanced due diligence processes for climate risks. They further noted that corporate clients were similarly reactive, stating that "you'd be hard pressed to find a top 100 ASX company that has its head around the carbon issue" (G1) and that clients were "in no mood to make $100million investment decisions in a climate of uncertainty." (G2). In contrast, climate law and policy was well-established in Europe, and in the United States federal incentives and a threatened carbon price existed at the time of interviews. In both of these jurisdictions, leading banks were motivated by opportunities more so than risk mitigation. It was in these jurisdictions that leading banks had developed innovative financial solutions for climate change mitigation.

Due to a desire to facilitate innovative change as opposed to satisfice, several NGOs had asserted that banks ought to go beyond risk mitigation via enhanced due diligence to a more rigorous process of preferential lending. Their reasoning was that preferential lending rewards low-carbon clients and penalizes GHG-intensive ones and is therefore an innovative and unambiguous facilitation of climate change mitigation:

> We need to see leadership from banks, for them to say "our commitment to environmental sustainability is such that we're taking a stand." Due diligence is quite the opposite: it's saying "we'll finance anything as long as it goes through a process that considers carbon." Banks should at least start pricing transactions to incentivize green outcomes and levy higher rates or more difficult terms on GHG-intense ones. (NGO-D1).

Another NGO respondent explained the broader intent:

> Are the climate-related acts of banks actually helping to mitigate climate change? It is obvious when banks make low-carbon investments. But when they just constrain or put overlays on what they already currently do regarding GHG intensive capital lending and financing then it's not enough... (NGO-A1).

These exhortations mirror scholarly urgings that banks can and ought to charge differentiated interest rates depending on the green performance of the borrower. For

---

606. Bansal and Roth, *supra* n. 296, at 728.

example, Marcel Jeucken writes that an inherent activity of banks is pricing risk; therefore interest rate differentiation based on lower environmental risks is a justified tool in greener bank practice.[607] In this way banks would have a quasi-regulatory role and "support the good-doers and offset the wrongdoers" in society.[608]

In reality, however, preferential lending is unlikely to occur of banks' own volition. No bank respondent in this study supported it. Some transactional bankers responded that their firm was acting sufficiently by virtue of having a renewables team and a sustainability department. Others asserted that preferential lending to incentivize clean energy activity was "not part of what we do on a daily basis" (C3) and too far outside BAU to be considered seriously. For example:

> The general philosophy of the bank is that we facilitate and give equal service to all clients. Giving preferential rates to renewable energy creates a difficult situation regarding why the bank is making certain choices. We need an economic rationale for doing this. It's not a bank's role to fudge the system. Get entrepreneurs or governments to incentivize. (B1)

Importantly, respondents made clear that lending conditions are not decided by imposing ethical values but by using risk metrics and good business sense on a case-by-case basis. For example:

> Sustainability is a strategic game. You achieve it through portfolio limits and pricing risk...If a coal station is giving us lots of business in Asia then we'll do the deal; and we might decline a renewable project because the client owns a leaking nuclear plant overseas. (G1).

Enhanced due diligence is preferable to bankers because it does not involve any perceived ethical or value judgment. As such, it is appropriate to the usual way banks do business:

> We don't bar sectors on the basis of legitimacy. It's a risk management process... We price risks associated with different industries...not because of some moral judgment. (F2).

Due diligence mechanisms do not belie a bank's intention to "choose" clients on a moral basis, which would compromise the bank's client service reputation and fee generation. Conversely, appearing to choose clients and projects on an economic basis is simply good business sense, which is appropriate and rational. Taking an "economic not emotional" (B1) approach to climate-related strategies indicates that a bank has good business sense, which is crucial to its client service reputation. So again we see the double-edged nature of client service reputation: it has the effect of motivating banks a little way down the "right" path, but also inhibits them going too far.

Yet the data also suggest that some leading banks *do* want to influence normative corporate change but they will pick the mode of influence on the basis of what is most palatable and appropriate to them. Indeed, enhanced due diligence may end up having

---

607. Jeucken, *supra* n. 79; M. Jeucken, *Sustainability in Finance: Banking on the Planet* (Eburon Delft: Netherlands, 2004); Lins et al., *supra* n. 385; Lundgren and Catasús, *supra* n. 164.
608. M Lundgren and Catasús, *supra* n. 164, at 190.

a normative and beneficial effect. Several respondents described enhanced due diligence as an indirect deterrence for clients to continue in high-polluting activities due to the extra hoops involved in seeking finance. In this way, they argued, banks can subtly and rationally "influence changes in economic activity." (B2). Nonetheless, this outcome is a happy corollary and not the main aim of climate risk due diligence.

[2]   New Green Initiatives in Competition with Established Non-Green Activities

A contested issue regarding the substantive nature of climate-related bank changes is the seemingly conflicted nature of their financing activities. Specifically, coal-fired plants account for nearly one half of man-made CO_2 emissions.[609] As such, NGOs argue that funding renewable and clean tech projects while continuing to fund fossil fuel projects is counterproductive to the aim of a low-carbon economy. For example, RAN has described banks that fund coal as:

> the ATMs for a dirty industry that is bad for health and bad for business. Coal is the ultimate subprime investment for the climate. We cannot solve climate change if banks continue to prop up this risky and outdated industry.[610]

For these reasons, an NGO respondent explained how his organization had commenced a project that endeavors to assess the carbon implications of bank lending and investment activities using the 2°C guardrail as a benchmark. Traditionally, banks have been benchmarked against each other. However this respondent contended that a comparative yardstick was not appropriate or sufficient in the context of climate change mitigation:

> Right now there is nowhere near a consistent approach by banks. It's like saying you're 15% pregnant: either you're doing climate change mitigation or you're not. They need to be saying "2°C is a challenge for us and we will tackle it." (NGO-A1)

NGOs argue that amounts of funding for fossil fuel projects far outweigh those for low-carbon initiatives, making the "green" efforts of banks tantamount to "greenwash." For example, RAN alleged that Citi's commitment in 2007 of US$50billion over 10 years to address global warming amounted "to less than 0.2 percent of the company's $2.2trillion in assets" and asked "What is Citi doing with the other 99.8 percent?...In 2006 Citi financed 200 times more money for dirty energy than it did for alternative energy."[611] Similar claims have been leveled at the big four Australian banks with Profundo evaluating in 2010 that they have invested over AU$5billion into

---

609. US Environmental Protection Agency, *Clean Energy: Air Emissions*, http://www.epa.gov/clea nenergy/energy-and-you/affect/air-emissions.html (accessed Apr. 17, 2014).
610. Per Amanda Starbuck quoted in Enews Park Forest, "US Banks Risk Public Health and Climate by Financing Coal," *enewspf.com* (May 1, 2012). See also RAN, BankTrack and Sierra Club, *Dirty Money: US Banks at the Bottom of the Class: Coal Report Card 2012* (RAN: San Francisco, 2012).
611. Rainforest Action Network (RAN), *Banks, Climate Change & the New Coal Rush* (RAN: San Francisco, undated), 5. See also H. Schücking, L. Kroll, Y. Louvel and R. Richter, *Bankrolling*

mining, burning, and transporting coal and only AU$0.78billion into renewable energy.[612] Greenpeace has distinguished between words and action on this point, asserting that "[m]ajor banks such as ANZ and Westpac have happily accepted sustainability awards while continuing to invest hundreds of millions of dollars into polluting coal stations."[613]

Bank respondents in this study reported a decline in financing new coal in the United States and Australia since 2009, citing the reasons as a mix of tougher EPA approvals, increased NGO pressure, and the prospect of a carbon price. Nonetheless, the data also reveal that leading banks in jurisdictions that (a) are rich in fossil fuels and/or (b) do not have well-established climate-related federal regulation in place are cautious in sending signals that could be interpreted as choosing nascent renewable energy client interests over lucrative fossil fuel ones.In these jurisdictions, coal entities comprise a large component of banks' client bases. For reasons of competitiveness and client service reputation, banks will not readily drop these clients. As stated by a risk manager: "we don't walk away from long-standing clients easily if an activity is not illegal." (D1). A transactional banker gave a more textured response:

> Because our clients are mixed [between fossil fuel and low-carbon], it's always dangerous to go too strong one way or the other. We are inherently conflicted due to our client mix. Banks usually don't like to advocate on any issue for that precise reason: they don't want to piss off clients. (F2)

Australian bank respondents lamented the lack of government incentives for renewable energy. Yet some also admitted that their bank would not lobby for such regulation for fear it would "irritate" its coal clients. One respondent related how a big coal company had contacted him to discuss switching across from a competitor bank that had proactively engaged with the government about pricing carbon. When I asked about this issue at the competitor bank, I was told that there is an "inherent conflict" for banks due to client service reputation. A respondent at that bank said: "You don't want to completely disadvantage yourself. But we thought we had more to gain in the long run [by engaging with government] so we did it." (F2). Nonetheless, that bank had not decided to redline coal clients, and respondents stated expressly that they had no intention of doing so.

[3]   *Built In versus Bolted On*

Chapter 3 described how change management scholars contend that sustainability-based thinking and behaviors must be embedded in everyday corporate activities in

---

Climate Change (urgewald, groundWork, Earthlife Africa Johannesburg and BankTrack, 2011); H. Schücking, *Banking on Coal* (urgewald, BankTrack, CEE Bankwatch Network and Polska Zielona Sieć, 2013).

612. Greenpeace, *Pillars of Pollution: How Australia's Big Four Banks Are Propping Up Pollution* (Greenpeace: Sydney, 2010), 5. See also Profundo, *Australian Banks Financing Coal and Renewable Energy: A Research Paper Prepared For Greenpeace Australia* (Profundo: Amsterdam, 2010).

613. Greenpeace, *supra* n. 612, at 3.

order to generate real organizational change. For example, Hess's study of why financial actors adopt green strategies found that "real changes should take place in the core business" if the business case as driver is to generate a positive sustainability impact.[614] Similarly, Doppelt argues that sustainability-based thinking and behaviors must be embedded in "everyday" activities in order to generate real organizational change.[615] In other words, green values and action must become "built in" to corporate life and not just "bolted on" to it in order to have far-reaching positive impact within the organization itself which in turn can facilitate broader societal benefit.

Specifically, Doppelt states that an organization needs to link "bonuses, promotions, new hiring, and succession planning to performance on sustainability" in order to motivate cultural change toward genuine triple bottom line sustainability.[616]

Yet the data showed that remuneration and KPIs are not so linked in leading banks. Transactional bankers revealed that their compensation depends on fee generation and how the individual, department, and bank perform financially each year. In some cases there was commission on deals. Yet, the "greenness" of a deal was not relevant to remuneration. As stated by one bank respondent: "whether I IPO climate friendly companies or dirty companies I still get paid." (E1). One interviewee asserted that her remuneration was linked to "carbon." Yet, after further probing, it became clear that this was due to her involvement in trading carbon as a commodity; it did not equate to a KPI for positive environmental performance or climate change mitigation.

Similarly, researchers and analysts' compensation was "linked to financial not environmental outcomes" (E2). These employees were judged on their level of external recognition via client and industry voting, which depended on how much their work had added value to client investment and decision-making processes. Finally, the KPIs of sustainability managers were described as "very subjective" (D2), being linked to "coalition building with externals" (D2) and "external recognition" via government ratings, NGO rankings, and "getting good sustainability ratings and the resulting reputational impact for the firm." (C2).

Overall, there was no evidence that banks had heeded Hart's exhortation for companies to "close the loop on their own rhetoric" by rewarding employees that move "the company *and the world* toward sustainability."[617]

Furthermore, different divisions within the one bank had different priorities and cultures. For example, sustainability managers and transactional bankers had disparate attitudes to social reputation, and conceptions of CSR (and related terms) varied between departments and individuals even within the same bank. Similarly, attention to climate change, let alone mitigation of it, did not appear consistent throughout the banks. In two leading banks only, it was evident that climate change had been elevated to the corporate governance level with senior management oversight and regular meetings of inter-departmental managers to discuss and coordinate climate strategies.

---

614. Hess, *supra* n. 384, at 16.
615. Doppelt, *supra* n. 383.
616. Doppelt, *supra* n. 382, at 3.
617. S. Hart, *Capitalism at the Crossroads: Aligning Business, Earth and Humanity* (Wharton School Publishing: Upper Saddle River, 2007), 231-32 (emphasis added).

As stated by an NGO respondent: "If climate change is not something the CEO reports on then it hasn't arrived at where it needs to be sitting." (NGO-A1).

Moreover, climate-related changes were not consistently nor widely-spread throughout the banks. No bank respondent could explain clearly whether attention to climate change was spread throughout their bank, whether it was consistent between departments, or whether there were future plans for diffusion of climate-related knowledge, practices or mitigation efforts throughout the firm. Most managers could only tell me how their department or division addressed climate change (or even just "carbon issues") and/or diffused that knowledge. In some cases climate-related consciousness seemed confined or siloed to specific departments only. For example, activities within institutional or investment banking divisions usually involved attention to climate change by focusing on carbon as a risk factor or commodity and clean tech as an investment opportunity. Yet this was not the case for activities in other divisions such as wealth management or retail banking.

One banker attempted to explain how it worked in practice:

> It's still early days. But to be honest, in Retail Banking, when a person walks into a branch and wants a home loan, we're not interested in the sea level elevation of the property or carbon emissions. In Corporate Lending, we'll check a company's credentials and if we like their business plan then we'll lend the money. In Institutional Banking, that's when we take into account the environmental practices, reputation management, and long-term sustainability [economic viability] of the business. Could we do more? Absolutely. But that's where we are. (G1)

On this last point, banks and NGOs seem to agree: even leading banks could be doing more to ensure that their climate-related practices are core to everyday business and facilitate real corporate progress toward a low-carbon economy as opposed to tweaking BAU. For example, in addition to enhanced due diligence processes and activity in carbon and clean tech markets, banks could be linking employee KPIs to climate mitigation efforts; providing preferential lending rates; redlining the most GHG-intense clients and projects; ensuring that attention to climate change and mitigation efforts is present in all departments and under the umbrella of board oversight; and even assessing bank activities by how far they move us toward the $2\,°C$ guardrail on global warming.

In the words of an NGO representative: "banks shouldn't think this is their contribution to saving the world... What we are looking for is a little bit of ambition and heroism" (NGO-D1). Yet the honest response from leading banks, as evidenced in the data, is that they see their core business as servicing clients and making money, *not* saving the world. Pursuit of commercial success can have flow-on effects that benefit the planet, but these are happy corollaries.

In short: business case logic is both a motivator and a barrier to corporate green uptake; and client service reputation as a driver is a double-edged sword. As stated by one banker: "being a good corporate citizen – as an individual – is encouraged [by the bank]... But at the end of the day we are not paid to be nice people, we are paid to be successful." (B1) And, as evidenced in the data, "success" is measured by business case standards. Whether by hard or soft means, an initiative must translate into making money and not losing money for the bank, and employee remuneration is tied to this imperative.

[4] Market Size

Furthermore, and given the importance of business case logic to bank behavior, it is relevant at this point to acknowledge current limitations to the size of renewable energy and clean tech markets. Having regard to the risk-opportunity spectrum in the previous chapter, we know that banks and their clients are actively looking (or beginning to look) for opportunities associated with climate change as much as they are trying to mitigate risks. As such, the "opportunity" component of the business case is an increasingly important motivator for corporate green uptake.

Renewable energy and clean tech spaces present such opportunities for banks. These industries are lucrative and they have long-term potential for significant investment opportunities. For these reasons, most respondents stated that being in the renewables space is core to their bank's business and "pretty mainstream now in the [bank] industry generally" (B1).

However, renewable energy and clean tech markets are relatively nascent and considerably smaller in size than traditional energy markets. This has obvious repercussions for bank activity in both low-carbon and fossil fuel spaces:

> We're currently doing doubles and singles not multiples and home runs [in the renewables space] because although the market is currently $100billion in size that is still not a huge market for a big investment bank. (C3)

The comparatively small size of alternative energy markets ensures that banks will not make radical changes that have a chilling effect on fossil fuel clients. There is no logical reason for doing so if there is not a guaranteed or large enough return to warrant or entice such a switch. In other words, it would not make "good business sense."

For an industry driven by profit and client service, the current size of the renewables market is not conducive to banks' ostracizing, let alone excising, fossil fuel clients and funding only alternative energy initiatives. Renewables and clean tech markets are not yet significant enough to drive radical changes to bank practices. Yet, ironically, private finance is required to bring low-carbon energy solutions to scale. The answer is that these markets need to be expanded in order to attract more capital; and this needs to happen now, not decades in the future. The question is "how?" My suggestions are detailed in the next chapter.

In sum, there are grounds for genuine concern about how far purely voluntary bank changes can take us toward a low-carbon existence. Genuine innovations do exist but they are few. The more prevalent actions, such as carbon trading and enhanced risk assessments, fit readily into the category of "BAU with a lemon twist." Specifically, client service reputation is a double-edged sword. It drives banks to adopt green practices that can enhance and protect client service in order to enhance and protect profits. Yet, for this very reason banks will only make rational and conservative changes that do not compromise their client base or potential for profit maximization.

[C]   Timeliness: The Imperatives of Expeditious and Radical Change

The previous section showed that there are grounds for concern about how *far* purely voluntary bank changes can take us toward a low-carbon existence. Even if we assume the best, whereby bank industry standards are set high and resulting change is substantial not peripheral, we still need to contemplate the second concern about whether such action can create real change *fast* enough. To do this we need to examine types of corporate change and re-visit the urgency of global warming.

Within the fields of organizational change and organizational development, scholars consider corporate change as either "incremental," being small alterations to existing practice, or "fundamental," being transformational change. In the context of sustainable development, incremental corporate change can be defined by what it is not. It does *not* include "radical changes in strategy, structure, capability or organizational realignment."[618] Indeed incremental change "for the most part impacts on the organization's day-to-day operational processes," which includes "the drive to efficiency."[619] As such, business strategy literature that advocates increased operational- and eco-efficiency is actually advocating incremental change only.

In contrast, transformational corporate change is fundamental, radical and deep. Quinn writes that "[d]eep change differs from incremental change in that it requires new ways of thinking and behaving. It is change that is major in scope, discontinuous with the past and generally irreversible."[620] Levy and Merry describe this kind of change as involving a paradigmatic shift.[621] In other words, fundamental or transformational change goes well beyond "BAU with a lemon twist."

There are circumstances in which incremental corporate change is appropriate and successful, for example, creating "new strategic opportunities, more minor culture shifts, capability development or changes in the workforce skills mix."[622] Some might even argue that incremental change is better than no change at all, or that grabbing low hanging fruit is a good starting point for a change management process to move in the right direction. However, the magnitude and urgency of global warming requires fundamental and transformational change to address it. Returning to some statistics from Chapter 1, the imperative of GHG emissions reduction is to cap global warming at 2°C within the next few years in order to stop catastrophic change. Yet GHG emissions are now occurring three times faster than predicted by the Intergovernmental Panel on Climate Change in 2007 and we are already moving outside the climatic parameters

---

618. D. Dunphy, A. Griffiths and S. Benn, *Organizational Change For Corporate Sustainability: A Guide For Leaders and Change Agents of the Future* (Routledge: London, 2003), 230.
619. *Ibid.*, 228.
620. R. Quinn, *Deep Change: Discovering the Leader Within* (Jossey-Bass, San Francisco, 1996), 3.
621. A. Levy and U. Merry, *Organizational Transformation* (Praeger, New York, 1986). See also S. Waage and J. Torok, "Organizational Change for Sustainability," in S Waage (ed) *Ants, Galileo and Ghandi: Designing the Future of Business Through Nature, Genius and Compassion* (Greenleaf: Sheffield, 2003), 215; D. Anderson and L. Anderson, *Beyond Change Management: Advanced Strategies for Today's Transformational Leaders* (Jossey-Bass: San Francisco, 2001); G. Barczak, C. Smith and D. Wilemon, "Managing Large-Scale Organization Change" (1987) *Organizational Dynamics* 23.
622. Dunphy et al., *supra* n. 618, at 229.

required for healthy social and economic development. In this context, there can be no meaningful discussion of incremental mitigation of climate change. Notions that 4°C warming is close enough to 2°C or that getting it right in 30 years is better than not doing it at all are nonsensical. Climate change mitigation requires radical action.

Yet the banking industry has never been known for its radicalism. Indeed, the data evidence that even leading banks regard changes in corporate practice and the switch to a low-carbon global economy as gradual processes. As stated by one bank respondent: "The planet does need coal fired energy...We can't pull the plug due to our shareholders but we can work strategically to change over time." (C2) Similar phrases used by bankers such as "we can't turn off the tap altogether" (E1) and "we can't just switch off coal" (G3) indicate that sudden and radical voluntary change is unthinkable to even leading banks right now.

Illustrative of this point, transactional bankers in all jurisdictions made clear their intent to keep servicing coal clients today on the basis that there might be new technology that makes coal "clean" tomorrow. A wait-and-see approach makes good business sense to banks:

> Technology will keep developing. I believe that over time technologies like CCS [carbon capture and storage] will become cheaper, in 20-40 years' time, we can have a kind of "clean" coal. There is so much coal in this country that it is not very smart to ignore it. (F1)

This would indeed be win-win for all concerned: coal clients, banks, the environment, and communities that rely on secure and cheap coal-fired energy. Indeed, banks link a "wait-and-see" approach to the needs and expectations of society, which gives their perspective legitimacy and rationality. For example, most transactional bankers claimed that a bank's social responsibility is to help keep the world's lights on by continuing to support fossil fuel clients and projects. Moreover, some respondents went further by stating that financial support for non-green clients will *facilitate* green solutions. For example:

> Fossil fuel companies are the most able to push change right now because they have the resources and wherewithal to do so. We are seeing companies in fossil fuels that are investing in emerging tech. To not help or enable those companies would set back the whole clean tech market. (B2)

Yet critics describe a "wait and see" approach as a ruse to protract BAU. It is now known that large-scale deployment of CCS cannot be realized, if at all, until long after 2020,[623] which will be too late to assist with the 2°C target. Some commentators have asserted that humans have the capability to develop innovative technologies to redress

---

623. International Energy Agency, *Technology Roadmaps: Carbon Capture and Storage* (OECD/IEA, Paris, 2009), 4; also Carbon Sequestration Leadership Forum, *Press Release: IEA/CSLF Report to the Muskoka G8 Leaders' Summit* (IEA/PRESS, Paris, Jun. 14, 2010).

a range of human-induced crises, including global warming.[624] To these claims, Shu and Bazerman respond that:

> little concrete evidence exists that a new technology will solve the problem in time. In fact, such claims make the task of the climatologist all the harder... [and] serve as an ongoing excuse for the failure to act today.[625]

The fact is that we already have the technology to reduce GHG emissions. Required now are investment strategies, in all jurisdictions, to grow existing renewables and clean tech markets expeditiously so as to be competitive with fossil fuels within a few years, not a few decades.

The data reveal that leading banks have an evolutionary not revolutionary attitude to moving away from fossil fuels and toward a low-carbon existence. Yet this corporate attitude mirrors macro complexities and failures. The world is currently dependent on fossil fuels and most governments have not acted decisively to change that. A good test of political will arose during my first round of interviews. At the time of interviews in the United States and UK, the 2010 BP *Deepwater Horizon* oil rig disaster in the Gulf of Mexico was in its fourth week. That disaster is the world's largest accidental oil spill into marine waters with an estimated 35,000 to 60,000 barrels of oil released per day.[626] It flowed unabated for three months; it caused environmental and social carnage, which was captured vividly by international media; it created such public outcry that BP's chief executive resigned. It was a perfect opportunity for governments around the world, particularly those reliant on oil, to instigate changes to the status quo. However, and as two American bank respondents dourly pointed out, despite media and government proclamations on the need to switch away from fossil fuels, the Obama Administration did not bring a Bill before Congress to change the energy structure in the United States. Other respondents noted that public outrage caused by the spill had not slowed down oil consumption: "the ordinary person is horrified by the pictures on the news, but they are still gonna get into their car the next day and drive it." (B2).

Importantly, banks do not view themselves as *de facto* regulators. The data showed no support for an extended view of corporate citizenship whereby business is willing to fulfill a similar role to government in voluntarily solving social problems. One of the reasons for the rejection of CSR by Levitt and also Friedman was the view that corporations should not be regarded as governments, with Levitt in particular asserting that "government's job is not business, and business's job is not government."[627] Certainly, concerns about democratic accountability arise where that is not

---

624. E.g. S.D. Levitt and S.J. Dubner, *Freakonomics: A Rogue Economist Explores the Hidden Side of Everything* (Harper Perennial: New York, 2009); J. Bennett, *Little Green Lies: An Exposé of Twelve Environmental Myths* (Connor Court Publishing: Ballan Australia, 2012).
625. L. Shu and M. Bazerman, *Cognitive Barriers to Environmental Action: Problems and Solutions* (Harvard Business School Working Paper 11-046, 2010), 8.
626. C. Robertson and C. Krauss, "Gulf Spill is the Largest of its Kind, Scientists Say," *The New York Times* (Aug. 2, 2010).
627. Levitt, *supra* n. 490, at 47; M. Friedman, "The Social Responsibility of Business is to Increase Profits," *New York Times Magazine* (Sep. 13, 1970), 32.

the case,[628] and these concerns were voiced by leading banks in the case study. For example, a CSR manager related to me how he had dealt with NGO calls for banks to cease funding coal:

> I said to them: You want four banks to make public policy? That is, we've got money so we'll make the decisions; we don't care if we're a democracy with regulations in place? I said you're asking us to use financial muscle to achieve what *you* want us to achieve. (G4)

Data in this case study make clear that banks will not fill governmental gaps if to do so means risking client service reputation and fee generation and/or compromising their perceived legitimate role in society.

§5.05 CONCLUDING REMARKS

The prevailing mind-set of leading banks comprises the following elements: (a) enhance and protect client service reputation in order to enhance and protect competitive edge and profits; and (b) make logical and conservative changes, not radical or irrational ones, especially in an uncertain climate. This conclusion is based on empirical evidence around three themes. First, business case logic is intrinsically bounded whereby the "materiality" of the business case comprises an impediment to corporate environmental uptake. Secondly, there is a cautious preference by banks to extend BAU and not act radically or entrepreneurially, which demonstrates the inherently conflicted and limited nature of the business case. Innovative solutions certainly exist, but they are not yet the norm. Thirdly, climate-related changes driven by the business case are bolted on (siloed or peripheral) not built in (or core) to bank organizational functioning. Overall, the result of the examination is decidedly "BAU with a lemon twist."

A central finding of the case study is that client service reputation – a critical ingredient of the business case – is a double-edged sword. It drives banks to adopt green practices that can enhance and protect client service in order to enhance and protect their own profits. Yet, concomitantly and ironically, this means that banks will only make rational and conservative changes to not compromise their client base or potential for profit maximization.

Banks' current mind-set, combined with the thesis that isomorphism arises from cognitive acceptance of persistent norms, indicates that it may take some time for "enlightened" climate-related strategies to become mainstreamed in the banking industry. Similarly, it can be deduced that bank industry change through purely voluntary means will be incremental and not transformational.

In conclusion, the business case drives voluntary corporate change but, simultaneously, impedes change that is far-reaching and expeditious. Accordingly, in the context of stemming dangerous global warming within the next few years, if we relied *solely* on voluntary corporate mainstreaming motivated by business case logic and a

---

628. Matten and Crane, *supra* n. 301, at 176.

wait-and-see approach then we would miss a vital opportunity. Importantly, we would miss the opportunity to capture and leverage the full potential of the banking industry and other private finance actors to facilitate climate change mitigation and the timely shift to a low-carbon global economy. Discussion of this opportunity, and how best to capture it via regulatory modalities, is the basis of the next chapter.

CHAPTER 6

Empirically Informed Regulation

§6.01 SUMMARY OF EMPIRICAL FINDINGS

Moving to a low-carbon and climate-resilient global economy requires the financial input and facilitation of private finance actors. Yet little is understood about the relationship between private finance actors and climate change or its interplay with regulatory context. Indeed, the dearth of empirical investigation in this area has meant that bankers, lawyers and policy-makers are not across each other's domains or even speaking the same language. Without a realistic understanding of the climate-related potential, motivations and limitations of private finance actors, we cannot hope to address climate change sufficiently or expeditiously.

As such, the heart of this book comprised data and findings from an empirical case study of early-moving or leading transnational banks in order to provide "real life" learnings regarding the broader finance sector and climate change. This is one of the first studies to do so.

Qualitative evidence from the case study addressed the question of why, in their own words, seven transnational banks had adopted climate-related practices despite regulatory uncertainty on the issue. Interview-based data were gathered in 2010-2011 from managers who headed up units responsible for climate-related bank practices at leading banks in the United States, Europe and Australia. For triangulation purposes, interviews were also conducted with external respondents such as independent consultants and relevant NGOs.

In summary, the data revealed that the overarching driver of climate-related initiatives by early moving banks is business case logic. The business case comprises two facets. First, climate risks – credit, investment, litigation, reputation, regulatory – are minimized. Second, profits are enhanced: directly via fee generation; and indirectly via competitive edge. The common, ulterior motivation for banks is simple: make money and do not lose money.

Arguably, this is an unsurprising result from corporate players in jurisdictions with a predominant shareholder primacy norm. Yet, as summarized below, the case study progresses our knowledge and understanding of levers and limits in the corporate climate finance space by revealing deeper "real life" complexities regarding:

- corporate reputation as a powerful motivator of bank behavior;
- how regulatory environments shape corporate decision-making;
- the limited role of CSR as an actual driver of beneficial corporate change; and
- the limits of corporate voluntarism for broader societal benefit.

[A]   A New Taxonomy of "Corporate Reputation" as Driver

This book creates a new taxonomy of "corporate reputation," which comprises two elements: a newly classified "client service reputation" in addition to the well-established "social reputation." These elements act as twinned drivers of corporate financial greening. In the literature, the terms "social reputation" and "corporate reputation" are often used interchangeably, conflating a company's reputation with its social standing. Yet the case study reveals a more complex and nuanced understanding of corporate reputation in business practice. It comprises not only social license but also a reputation for good business sense in delivering excellent service to help large corporate clients flourish; what I termed "client service reputation." The existence and effects of client service reputation on corporate greening have been largely unexplored to date.

For all bank respondents, regardless of their jurisdiction or unit, client service reputation was a prime motivator to create climate-related products and services and to enter new markets. In nearly all of the banks studied, large corporate and/or wealthy individual clients had approached them to seek solutions not only for mitigating regulatory risk associated with extant or threatened carbon pricing, but even more so to capitalize on opportunities created by new markets and amenable government policy. By giving innovative responses to these requests, banks helped corporate clients to survive and thrive in an increasingly carbon-constrained world.

And by helping clients, banks help themselves. Providing responsive and innovative client service creates two important benefits for firms. The first consequence is direct: it generates fees. These fees are significant, coming from large corporate and/or affluent individual clients and they create profit for the bank and increase value for shareholders. In other words, fee-generation is a hard or measurable item that comprises an unambiguous business case. The second consequence is indirect and harder to quantify, yet it ties into the first. Excellent and innovative client service creates a reputation for the bank as the "go to" bank in the space, which gives it competitive advantage. This is crucial because the banking industry is highly competitive with banks fighting for a fixed size of the client purse. The corollary is that banks work to win clients from other banks whilst satisfying their own; an endeavor that was assisted greatly by banks' climate-related innovations.

Conversely, regard for social reputation and concomitant NGO pressure as a driver for climate-related bank behavior was variable. Its meaning and manifestation for banks is consistent with the literature: it constitutes both the risk to a company if it is rejected by communities and the benefit to a company if it is socially acceptable. However the salience of social reputation as a driver of bank behavior was contingent. It had most traction:

(1) with banks that have a direct relationship with civil society, that is, private and retail banks that have individual or "mom and dad" account holders; not investment banks that have only corporate clients;

(2) with sustainability/CSR managers within a bank, as opposed to transactional bankers and analysts; and

(3) in a regulatory context where:

    (a) there is a lack of government intervention or industry standards and/or the regulator is weak; and

    (b) NGO campaigning is voluble and hostile.

While trust and good image management were crucial to both social reputation and client service reputation, they manifested differently. In relation to client service reputation, trust manifested as "good business sense." The aim was to help corporate clients (and therefore banks) to keep up with and/or get ahead of the curve. Consequently, banks market themselves as corporate partners with good business sense, which preserves and enhances their client service reputation. Importantly, banks can be agnostic about climate change in this context; the "green" driver is the greenback not environmental benefit. In contrast, trust in the context of social reputation manifested as "good conscience." To this end, marketing and branding strategies aim to create an emotional response in the hearts and minds of civil society account holders. It is here that concepts such as "CSR," "sustainability" and "good corporate citizenship" become most relevant to bank branding.

Corporate reputation is often regarded as a "soft" issue that is intangible, unmeasurable, and therefore under-researched as relevant to a business case. Yet it is clear from this case study that reputational components can be disaggregated in order to better understand how (soft) reputation is as crucial as (hard) fee generation in driving corporate behavior and innovation.

[B] Climate Change as Risk or Opportunity? The Relevance of Regulatory Context

Interplay between the two types of reputation, regulatory context, and bank behavior became clear by examining banks' perspectives of climate change as a risk or an opportunity. In large part, perspectives were jurisdiction-specific and shaped by regulatory context, which included not only state interventions (a carbon price, financial incentives for renewables, a RET, or even direct coercive legislation) but also social pressure from NGOs and mass media.

The more sophisticated the state interventions, the more likely that banks will perceive climate change as an opportunity and leverage regulatory incentives to enhance both their client service reputation and social reputation. Against such a regulatory backdrop banks can focus on profit enhancement. This was illustrated by European leading banks, which were situated at the "opportunity" end of the climate risk-to-opportunity spectrum.

In contrast, the weaker or less certain the state interventions, then:

(1) the more important NGO activity and voluntary industry standards become to mobilize better corporate behavior; and
(2) the more likely that banks will see climate change as a risk.

In this regulatory context banks will focus on strategies for downside prevention, particularly for protecting their social reputation. This was evidenced by Australian leading banks, which were situated at the "risk" end of the spectrum.

Importantly, banks that are primarily driven by risk mitigation are reactive, particularly to social pressure. Their aim is to keep BAU running as smoothly as possible; and they are less likely to be innovative in addressing climate-related issues.

The interesting point here is that a threatened carbon price (not even an actual one) when combined with financial incentives motivated banks to view climate change as more of an opportunity than a risk. Indeed, comparing data from the United States with those from Australia, it is clear that inclusion of financial incentives in government regulation is crucial to the greening of private finance actor behavior.

[C]   CSR as a Non-driver

The case study revealed that CSR considerations are subordinate drivers for green activity and in some cases non-existent. The findings suggest that CSR is a limited explanatory tool for what drives actual corporate environmental-social uptake. In so doing, the case study disconfirms the CSR literature which posits that ethical conceptions of "doing the right thing" can and do drive corporate greening.

There are two points to make here. First, banks do not care about climate change; people in banks care about climate change. The data suggest that a firm needs a CEO or other high-ranking senior manager to be a charismatic "green leader" by setting the tone for the bank, which then shapes bank policy. Equally, however, attention to climate change needs to be diffused throughout all bank units before the bank (and not just the CEO) can truly call itself "green." It is also possible that passionate individual managers might create new climate-related products and services that fill a market gap. Yet the data also showed that managerial ethical discretion is curtailed by perceived and actual structural constraints.

Secondly, CSR is conceived heterogeneously – when it is considered at all. Just under half of all bank respondents did not mention CSR; for these banks and/or individuals it is simply not a motivating force for *anything*. Of the respondents that did mention CSR, it had varying labels and definitions, meaning different things to different people and departments even within the same bank. These different conceptualizations had created occasional friction between sustainability managers and transactional bankers in the same bank, showing that CSR is still a contested concept *within* firms.

Notably, the terms used most often by bank respondents were "CSR," "good corporate citizen," and "sustainability." Definitions of these terms were often specific to each bank and even to individual managers, and often inconsistent with the literature. For example, the overwhelming conception of "sustainability" by respondents equated to "economic longevity of the bank." This is in stark contrast to the UNEP Finance Statement on Sustainability to which all banks bar one in the case study were signatories and which provides that economic imperatives must be *balanced* with human welfare and a healthy environment.[629]

Moreover, a number of respondents stated that voluntarily taking radical steps toward a low-carbon economy by excising fossil fuel clients would be socially *irresponsible* as it would jeopardize fuel security. They had tied their conception of CSR to the perceived needs and wants of society, which gave it (and them) legitimacy albeit with counter-intuitive results for climate change mitigation.

Finally, corporate conceptions of "CSR" (and related terms) incorporated social as well as environmental elements. The implication is that even if a bank considers itself to be a socially responsible corporation, and behaves accordingly, that does not guarantee greener decision-making. For example, while climate or environmental issues were central to the CSR agenda at some banks, philanthropy and/or economic empowerment were central to the CSR activities at other banks. Strikingly, no respondent used the term "corporate environmental responsibility" when describing bank CSR activities or philosophies.

In summary, this study found that CSR action is:

(1) at best, heterogeneously conceived; at worst, irrelevant;
(2) often dependent on the personal preferences of a senior officer such as the CEO;
(3) reactive to external circumstances, such as the decimation of bank balance sheets and reputations due to the GFC; and
(4) not a motivating force for change on its own.

Overall then, the diffuse and heterogeneous nature of CSR means that it cannot lead the charge to sustainability; it can create marginal change only. Accordingly, the findings support the conclusions of previous scholars who contend that although CSR exists, it is not central to business strategy or market forces and therefore of limited significance in driving meaningful change.[630]

---

629. See Chapter 1 *supra*.
630. E.g. Vogel, *supra* n. 338; Richardson, *supra* n. 116.

[D]   The Limited Nature of Voluntarism for Mainstreaming

If there is no ethical rudder by which business can steer a course then rational economic behavior is their only compass. And yet, from the findings in Chapter 5 we see that change wrought by purely voluntary action based on business case logic is inherently limited. Overall, radical or entrepreneurial voluntarism was not evidenced in the data due to inherent limitations in business case logic and current size of alternative energy markets.

Nonetheless, without government intervention, mainstreaming of purely voluntary leading bank practices would probably occur over time via mimetic processes. The data suggest that the banking industry follows its leaders due to: their high visibility; peer pressure; the competitive nature of the banking industry; and the uncertain regulatory and business environments that come with climate change.

However, the data also raised concerns about how far and fast purely voluntary corporate behavior can move us down a low-carbon path. Arguably, low benchmark-setting by leaders embeds an industry standard that is equivalent to "BAU with a lemon twist" while deflecting attention from the level of corporate change actually required to address dangerous global warming. Certainly, the data show that even leading banks could be doing more to ensure that their climate-related practices are core to everyday business and facilitate real corporate progress toward a low-carbon economy. For example, in addition to enhanced due diligence processes and activity in carbon and clean tech markets, banks could also be linking employee KPIs to climate mitigation efforts, providing preferential lending rates, and redlining the most GHG-intense clients and projects. Yet the data reveal that banks will not do so voluntarily. The reason is that business case logic drives greener behaviors but, concomitantly and ironically, it impedes change that is far-reaching and expeditious. It is both a motivator *and* a barrier to corporate green uptake.

Central to this is the finding that client service reputation – a critical ingredient of the business case – is a double-edged sword. On the one hand it mobilizes banks to innovate new products and to become active in green spaces due to the desire for competitive advantage and enhanced fee generation. Perversely however, client service reputation impedes entrepreneurialism. Even leading banks are not willing to sacrifice lucrative clients on the green altar. So, while client service reputation is a strong motivator for banks to voluntarily drive in a greener direction, it is a simultaneous brake on driving too far down that path.

[E]   Reflections

When reflecting on the empirical evidence, it becomes apparent that banks' self-perceptions, their obsession with client service reputation, and the current size of low-carbon markets are all too limiting to incite voluntary financial entrepreneurialism. Accordingly, the "network change potential" of the banking industry and other private finance actors to facilitate exponential corporate GHG emissions reductions as postulated in Chapter 2 cannot be realized without something more.

At this point we may blame the banking industry for not being radical enough, or denigrate business case logic for its limitations, or even curse capitalism just for being itself. The contention in this book however, is that these are the realities with which we need to work in market economies. Proponents of CSR have tended to focus on the "ought" and not the "is" of corporate greening. Data in this study show that the question of why firms *actually* go green can be separated from a normative discussion about why they *ought* to go green. Appreciating these real life aspects is crucial to understanding how beneficial corporate change might best occur.

So the question now becomes: how best to work with what we've got in order to get what we want?

§6.02 GENERALIZABILITY OF FINDINGS TO OTHER PRIVATE FINANCE ACTORS

Before proceeding, some preliminary questions about generalizability of the findings need to be answered. Yin writes that "the description and analysis of a single case often suggest implications about a more general phenomenon."[631] Can the findings in this case study be generalized to other populations and therefore be more broadly useful? In other words, are the business case based motivations and limitations of banks unique to them?

On the one hand, the banking industry is a quintessential and unconstrained exemplar of capitalism. Through the inner workings of this industry we see capitalism in its most naked form. Banks are the poster child for capitalism; they are not just actors in a capitalist paradigm, they personify it. This sense of fiscal urgency and privilege, and the business case logic driving it, permeated all of the interviews.

On the other hand, it can be argued that these predilections are not unique within a market economy. Indeed, when considering empirical studies with cross-sectoral samples, such as Spence's investigation of CSR in UK companies or O'Dwyer's research into CSR conceptions by Irish managers,[632] it becomes clear that the banking industry is *not* the only industry to privilege business case logic. It may simply be considered the brightest star in the capitalist constellation.

Specifically, the case study findings and implications regarding the levers and limits of voluntary "green" corporate action can be generalized to other private finance actors such as institutional investors and insurers. Large pension funds and financiers for energy companies, infrastructure and transport all operate pursuant to business case imperatives. Arguably they are incentivized and disincentivized in ways similar to banks. Specifically, institutional investors seek to mitigate risk and capitalize on opportunities associated with climate change due to their fiduciary obligation to members. Indeed, Richardson has highlighted the lure of a countervailing business case for institutional investors to support GHG-polluting companies.[633] Moreover, a 2010 New Energy World Network survey that targeted pension funds, insurance

---

631. Yin, *supra* n. 472, at 144.
632. Spence,*supra* n. 318; O'Dwyer, *supra* n. 291.
633. Richardson, *supra* n. 116, at 507.

companies and asset management firms found that "[v]irtually all institutional investors believe generating returns to be the overriding objective to any commitments they
make in the green sector."[634] That is, generating profit and mitigating risk was the
principal aim of green investment; environmental and social responsibilities were
secondary. Neuhoff et al. argue that "a profound change in the portfolio of investors"
must occur if we are to move to a low-carbon economy.[635] Yet, the business case is both
a driver and a barrier to socially responsible investment by institutional investors,
including insurers, in ways similar to that of banks.

§6.03   LEGAL IMPLICATIONS OF FINDINGS: THE CASE FOR REGULATORY INTERVENTION

While the findings in this book might be disappointing for CSR protagonists, they have
important policy implications. If NGO or government activity directed at the finance
sector attempts to nurture a "warm cuddly glow" then it is unlikely to have any impact.
Such an approach has no resonance with, and hence no motivating effect upon, that
sector. This realization is especially timely given the frustration that civil society and
regulators may feel toward banks and other finance actors for their role in the GFC and
subsequent scandals, and any need by constituents to somehow make these actors
care.

Working with the "is" of the empirical findings and moving to the "ought" of
what is desired, this book contends that practitioners and policy-makers and even civil
society can direct energy and resources to points of actual leverage with banks and
other private finance actors: that is, profit motive, risk/opportunity, and reputation.

In this way, private finance actors can be harnessed and best utilized because the
regulatory mechanisms will be meaningful to them. What harnesses their business
case imperatives to make money and not lose money? What leverages their self-
perception of their role and purpose in society? The data reveal that finance actors view
themselves as economic leaders not environmental stewards. Nonetheless, their role in
perpetuating or mitigating environmental harm through indirect economic behaviors is
undeniable. Thus, it is vital to accurately leverage their strengths as economic leaders
in order to harness their potential as environmental stewards.

In short, the contention of this book is that the business case can provide an
available and actionable modality of transformational change if it is harnessed by
clever government regulation.

The CSR literature is not overtly interested in the interaction between voluntary
corporate action and government regulation.[636] Indeed, much of the point of CSR
scholarship is to envisage business regulation devoid of or beyond state intervention

---

634. New Energy World Network/AltAssets, *Investing in Green Private Equity and Venture Capital Funds – Survey of Institutional Investors 2010* (New Energy World Network, 2010).
635. Neuhoff et al., *supra* n. 399, at 10.
636. Notable exceptions include: Vogel, *supra* n. 338; Gunningham et al., *supra* n. 390; and J-P Gond, N. Kang and J. Moon, "The Government of Self-Regulation: On the Comparative Dynamics of Corporate Social Responsibility" (2011) 40(4) *Economy and Society* 640.

through, for example, industry self-regulation and/or modes of social control. Yet the findings from the case study provide fertile ground for theorizing how government intervention and corporate voluntarism can interact and mutually enhance and reinforce each other as opposed to one being a substitute or surrogate for the other.

A number of independent studies have confirmed that the technological capacity already exists for a timely transition to a low-carbon, renewable-based electricity sector that can meet the entire electricity demand at national, regional and even global levels.[637] Technological capacity is not the issue. The problem is achieving timely scale up; and a key obstacle is mobilization and infusion of significant capital.[638]

An important finding from the case study data is that the comparatively small size of alternative energy markets ensures that banks will not make radical changes that have a chilling effect on fossil fuel clients. There is no logical reason for doing so if there is not a guaranteed or large enough return compared to the risk to warrant or entice such a switch. The renewables market is not yet significant enough to drive radical changes within an industry that is driven by profit and client service reputation. Yet, ironically, private finance is required to bring low-carbon energy solutions to scale. These markets need to expand in order to attract more capital; and they need to attract more capital in order to expand.

The question is how? The answer is that government intervention is required. It needs to be certain and specific to the issues of climate change, clean energy, and carbon emissions. It must tangibly influence the decision-making of private finance actors in order to overcome inherent limitations to both business case logic and CSR variability.

As discussed in Chapter 1, focus is now on national and regional climate law and policy due to regulatory uncertainty at the international level. This is a good thing from a financial perspective. Domestic policy is the key determinant of whether and under what conditions financiers and investors will deploy capital in energy efficiency, clean tech and renewable energy.[639] These sub-global efforts are not only desirable but also indispensable.[640]

The empirical case study demonstrated that national/regional government policy can change the behavior of financiers and investors. Specifically, government interventions, both in their substance and certainty, modify the risks and returns faced by

---

637. For studies at the national level, see e.g. P. Willson et al., *Powering the Future – Mapping our Low-carbon Path to 2050* (2009); M. Wright and P. Hearps, *Australian Sustainable Energy: Zero Carbon Australia Stationary Energy Plan* (2010). For the regional level see e.g. European Climate Foundation, *Roadmap 2050 – A practical guide to a prosperous, low-carbon Europe* (2010); PriceWaterhouseCoopers, *100% Renewable Electricity: A Roadmap to 2050 for Europe and North Africa* (2010). At the global level, see e.g. M.Z. Jacobson and M.A. Delucchi, "Providing all Global Energy with Wind, Water, and Solar Power, Part I: Technologies, Energy Resources, Quantities and Areas of Infrastructure, and Materials" (2011) 39 *Energy Policy* 1154, 1164.

638. Mormann, *supra* n. 38, at 686.

639. R. Sullivan, *Investment-grade climate change policy: financing the transition to the low-carbon economy* (Institutional Investor Group on Climate Change, the Investor Network on Climate Risk, the Investor Group on Climate Change Australia/New Zealand and the United Nations Environment Programme Finance Initiative, 2011), 15.

640. Farber, *supra* n. 24.

private finance actors as well as the information and processes they use in their decision-making. This, in turn, influences whether and to what extent private sector finance flows to climate change mitigation and adaptation endeavors.

The trick is to leverage the resources at the necessary scale. To this end, public policy for private finance plays a crucial dual role: first, policy can establish the incentive frameworks needed to catalyze high levels of private investment in mitigation and adaptation activities, and second, it can generate public resources for needs which private flows may address only imperfectly.[641]

Given the heterogeneity of projects and investors, there is no one-size-fits-all prescription for designing effective climate finance policy.[642] Broadly speaking, federal- and regional-level government intervention can take two forms in order to motivate private finance to address climate change. The first is direct and coercive ("command-and-control") regulation, which is considered high interventionist. The second form is indirect or steering ("nudging") regulation, which is low interventionist. Both of these regulatory modalities are explored in detail below.

§6.04   DIRECT AND COERCIVE GOVERNMENT REGULATION

[A]   Attributes and Benefits of Direct Coercion

Legislation that directly regulates the finance sector appears to be *en vogue* in the aftermath of the GFC. Sweeping new regulatory schemes have been legislated in the United States, Europe and Australia with a view to improving financial stability and sustainability via prescriptive measures.[643] For example, the *Dodd-Frank Wall Street Reform and Consumer Protection Act* was signed into American law on July 21, 2010 with the objective to "promote the financial stability of the United States by improving accountability and transparency in the financial system" which includes ending "too big to fail," protecting American taxpayers by ending bailouts, and protecting consumers from "abusive" financial services practices. Similarly, the European Commission proposed nearly 30 sets of rules between 2010 and 2014 aiming "to ensure all financial actors, products and markets are appropriately regulated and efficiently supervised."[644] These efforts culminated in finalization of the Banking Union legislative reform package on April 15, 2014, which promotes proactive bank risk management and protection of taxpayers.[645]

---

641. M. Brahmbhatt, *Mobilizing Climate Finance: A Paper prepared at the request of G20 Finance Ministers* (World Bank Group: Washington DC, 2011).
642. See e.g. empirical study by International Energy Agency, Deploying *Renewables: Principles for Effective Policies* (2008), 85-87, http://www.iea.org/textbase/nppdf/free/2008/deployingrenewables2008.pdf ; Mormann, *supra* n. 38, at 696-724.
643. See Chapters 1 and 5 *supra*.
644. European Commission, *A comprehensive EU response to the financial crisis: substantial progress towards a strong financial framework for Europe and a banking union for the eurozone* (European Commission – MEMO/14/244 28/03/2014), http://europa.eu/rapid/press-release_MEMO-14-244_en.htm.
645. European Commission, "Commissioner Barnier welcomes trilogue agreement on the framework for bank recovery and resolution" (Memorandum, Memo/13/1140, Dec. 12, 2013);

These reforms reflect a push towards increased prudential regulation, crisis management and market efficiency. Clearly, they are directed at the economic activities of private finance actors – interest rate hikes, salaries of CEOs, activities of "too big to fail" – as opposed to their environmental or climate-related practices. Nonetheless, the US Community Reinvestment Act sets a precedent for financial regulation that seeks to achieve social objectives. As such, it is plausible that coercive regulation could be similarly legislated to mandate how private finance actors *must* facilitate climate change mitigation and/or adaptation. The obvious caveat is that this regulatory approach is highly interventionist and therefore carries attendant political challenges.

The advantage of direct and coercive regulation of the finance sector is that a high minimum-standard of green behavior can be set externally. This ensures that no actor is going it alone and thereby risking its precious client base by compromising client service reputation. It also ensures that "BAU with a lemon twist" is not *de rigueur*.

What might direct and coercive climate finance legislation contain?

[B] Examples of Direct and Coercive Climate Finance Regulation

[1] Prescriptive and Proscriptive Legislation

In order to achieve the objectives of using financial actors to assist climate change mitigation and/or adaptation, direct and coercive legislation would need to prescribe not only how private finance actors must alter their own practices but also how they must influence the practices of other corporate actors.

Legislation could stipulate that financiers must lend at preferential rates for infrastructure adaptation projects; or proscribe support for listed carbon-intense projects or industries by institutional investors. Indeed, using the US Community Reinvestment Act as inspiration, subjects of coercive climate finance regulation could be examined and rated on their compliance by federal financial agencies. An unsatisfactory rating would trigger sanctions such as government denial of business expansion with obvious ramifications for corporate reputation.

[2] Taxation

Alternatively, direct legislation could prescribe new taxation measures. New taxes on the finance sector have been proposed in international fora as a way to raise money for public climate finance.[646] These taxes typically take two forms: first, a broad-based financial transactions tax (FTT) levied on the value of a wide range of financial transactions; and secondly, a financial activities tax (FAT) levied on the profits of

---

European Commission, "Finalising the Banking Union: European Parliament backs Commission's proposals (Single Resolution Mechanism, Bank Recovery and Resolution Directive, and Deposit Guarantee Schemes Directive) (Press Release, Statement/14/119, Apr. 15, 2014).
646. Brahmbhatt, *supra* n. 641, at 24.

financial institutions.[647] In international circles, the FTT has acquired greater political acceptability whereas the FAT has garnered greater support from tax policy specialists.

At the national level, a FAT could be levied on defined private finance institutions to tithe a percentage of their annual profits, which is then dedicated to specified green initiatives. Interestingly, while this may be distasteful for finance actors, it may be popular amongst citizens post-GFC. For example, a 2010 Australia Institute poll found that 81 percent of Australians wanted government consideration of a bank industry "super profits" tax to tithe annual bank profits over a certain threshold.[648] Finance from this type of tax is usually funneled into a national sovereign fund; in the case of a "climate finance tax" however, proceeds would need to be sequestered in a separate fund for prescribed climate change mitigation and adaptation initiatives.

[C]   Criticisms of Coercive Legislation

Despite the potential benefits of coercive legislation for achieving desired outcomes, in recent decades there has been voluble criticism of top-down, command-and-control or, in the United States, "new deal" regulation. Most notably, scholars of new governance theory have described traditional regulation as "sanctioned,"[649] "hierarchical,"[650] "inefficient, ineffective, and undemocratic"[651] due to its coercive nature. Specifically, some American scholars have opined that the advent of command and control environmental regulation in the 1970s created a counter-productive adversarial relationship between government and industry on "green" issues.[652]

New governance theory emerged from a myriad of alternative regulatory theories with epithets such as "soft law,"[653] "responsive regulation,"[654] "regulatory capitalism,"[655] "regulatory pluralism,"[656] "collaborative

---

647. International Monetary Fund (IMF), *A Fair and Substantial Contribution by the Financial Sector, Final Report for the G-20* (International Monetary Fund: Washington DC, 2010).
648. The Australia Institute, *Media Release: Australians Want Bank Super Profits Tax On the Agenda* (The Australia Institute: Canberra, 2010).
649. O. Lobel, "The Renew Deal: The Fall of Regulation and the Rise of Governance in Contemporary Legal Thought" (2004) 89 *Minnesota Law Review* 342, 344.
650. J.M. Solomon, "Book Review Essay: Law and Governance in the 21st Century Regulatory State" (2007-2008)86 *Texas Law Review* 819, 820.
651. J. Freeman, "Collaborative Governance in the Administrative State" (1997) 45 *UCLA Law Review* 1, 3.
652. A.J. Hoffman and M.J. Ventresca, "The Institutional Framing of Policy Debates: Economics Versus the Environment" (1999) 42 *American Behavioral Scientist* 1368, 1375.
653. See e.g. D.M. Trubek and L.G. Trubek, "Hard and Soft Law in the Construction of Social Europe: The Role of the Open Method of Coordination" (2005) 11 *European Law Journal* 343.
654. See e.g. I. Ayers and J. Braithwaite, *Responsive Regulation: Transcending the Deregulation Debate* (Oxford University Press: New York, 1992).
655. See e.g. D. Levi-Faur, "The Global Diffusion of Regulatory Capitalism" (2005) 598(1) *The ANNALS of the American Academy of Political and Social Science* 12; J. Braithwaite, *Regulatory Capitalism: How It Works, Ideas for Making It Work Better* (Edward Elgar: Cheltenham UK, 2008).
656. See e.g. N. Gunningham and D. Sinclair "Regulatory Pluralism: Designing Policy Mixes for Environmental Protection" (1999) 21 *Law and Policy* 49; J. Black, "Decentring Regulation: Understanding The Role of Regulation and Self-Regulation in a 'Post-Regulatory' World" (2001) 54(1) *Current Legal Problems* 103; and C. Coglianese and E. Mendelson,

governance,"[657] and "neoliberal governmentality."[658] Jason M. Solomon defines new governance regulation as:

> an umbrella term covering a kind of interaction between the state, regulated entities, and other stakeholders... [It has] a number of desiderata – public participation, data provision, transparency, benchmarking, sharing of best practices, fora for deliberation on ends and means, and autonomy and flexibility for those subject to regulation.[659]

According to new governance theory, the democratic state is moving toward a "post-regulatory" model, captured in the linguistic shift from "government" to "governance."[660] The essence of this transition is a "fermenting" of top-down government regulation[661] in favor of diffused regulatory power amongst multi-levelled networks and players.[662] These include non-state actors, such as corporations and NGOs.[663]

Some commentators have pointed to the heightened role of markets and market actors in this regulatory transition. For example, Christian Vannier describes it as "a shift in governance from a direct supervisory role [for the state] to an indirect role premised on the organizational norms of the free market."[664] Benjamin Cashore similarly highlights that these networks often use market forces to advance social and environmental goals.[665]

In this way new governance theory is apt for evaluating direct regulation that coerces private finance actors to assist the social objective of climate change mitigation. In undertaking this evaluation, it becomes apparent that direct regulation (on its own at least) is ill-equipped to deal with an emerging and dynamic area like corporate climate finance.

For example, climate finance regulation will need to be responsive to new information gained through implementation. In other words, "[l]earning by doing, embracing flexible responses to changing conditions and new knowledge, and revising to improve solutions"[666] need to become guiding lights for effective climate finance

---

Meta-Regulation and Self-Regulation (University of Pennsylvania Institute for Law & Economics Research Paper No. 12-06, 2010).

657. See e.g. Freeman, *supra* n. 651.
658. See e.g. C. Vannier, "Audit Culture and Grassroots Participation in Rural Haitian Development" (2010) 33 *Political and Legal Anthropology Review* 282.
659. Solomon, *supra* n. 650, at 834.
660. See Lobel, *supra* n. 649; and G. de Búrca and J. Scott (eds), *Law and New Governance in the EU and the US* (Hart Publishing: Oxford and Portland, Oregon, 2006).
661. S. Burris, M. Kempa and C. Shearing, "Changes in Governance: A Cross-Disciplinary Review of Current Scholarship" (2008) 41 *Akron Law Review* 1, 1.
662. R. Shamir, "Between Self-Regulation and the Alien Tort Claims Act: On the Contested Concept of Corporate Social Responsibility" (2004) 38 *Law and Society Review* 635, 659.
663. C. Scott, "Regulation in the Age of Governance: The Rise of the Post Regulatory State" in J Jordana and D Levi-Faur (eds) *The Politics of Regulation: Institutions and Regulatory Reforms for the Age of Governance* (Edward Elgar, Cheltenham UK, 2004); AM. Slaughter, "Global Government Networks, Global Information Agencies, and Disaggregated Democracy" (2003) 24 *Michigan Journal of International Law* 1041.
664. Vannier, *supra* n. 658, at 284.
665. B. Cashore, "Legitimacy and the Privatization of Environmental Governance: How Non-State Market-Driven (NSMD) Governance Systems Gain Rule-Making Authority" (2002) 15 *Governance: An International Journal of Policy, Administration, and Institutions* 503.
666. K.A. Strausser, *Myths and Realities of Business Environmentalism: Good Works, Good Business or Greenwash?* (Edward Elgar: Cheltenham UK, 2011), 74.

policy. In contrast, lists of funding activities (whether proscribed or permitted) can become outdated and remain "on the books" long past their use-by date due to inertia and/or scarce state resources. Corporate climate finance regulation will need to allow for adaptation and adjustment given the complexity and newness of this area.

Further, if regulators lack expertise to make rules for a highly specialized industry such as finance and investment, then inefficiencies will manifest in implementation and enforcement. New governance theory posits that collaborative processes help to "craft better policy ideas and implement them more effectively."[667] In other words, complex technical policy areas (like climate finance) require input from experts in the field, including those very entities that are the subject of the regulation (namely banks and institutional investors). Without collaboration, the text and implementation of technical legislation is likely to be inaccurate. If that occurs then direct and coercive regulation will not achieve its goals.

Moreover, new governance scholars argue that collaborative policy is more alive to problem solving than coercive regulation; and it is therefore more adaptable to changing economic, environmental and social demands. In particular, some scholars opine that command-and-control regulation "supports a belief that government regulators and industry decision makers cannot find solutions that offer mutual gain."[668] Therefore, mutually beneficial problem solving is less likely to occur within a context of coercive regulation.[669] In the context of addressing climate change, this situation would create a missed opportunity: the problem is urgent and the finance sector has the expertise, resources, and self-interest to assist mitigation. Using new governance theory, it appears that policy-making that is less coercive and more collaborative encourages parties to be engaged, and to take ownership of the problem and the need for a solution.[670] These aspects are key to effective law and policy on corporate climate finance.

Finally, coercive regulation forces private finance actors to become government-directed instruments of environmental protection. In market economies it is not surprising that they prefer to not be so regulated.

Thaler and Sunstein opine that "[t]he sheer complexity of modern life, and the astounding pace of technological and global change, undermine arguments for rigid mandates or for dogmatic laissez-faire."[671] Certainly, as there are limitations with purely market-driven changes as exemplified by the business case as driver in Chapter 5, so too there are limitations with sole reliance upon direct and coercive government regulation as highlighted by new governance thinking above.

---

667. *Ibid.*, 66.
668. Hoffman and Ventresca, *supra* n. 652, 1376.
669. O. Amir and O. Lobel, "Stumble, Predict, Nudge: How Behavioral Economics Informs Law and Policy" (2009) 108 *Columbia Law Review* 2098.
670. Strausser, *supra* n. 666, at 74.
671. R. Thaler and C. Sunstein, *Nudge: Improving Decisions About Health, Wealth, and Happiness* (Yale University Press: New Haven CT, 2008), 253.

§6.05 INDIRECT AND "NUDGING" GOVERNMENT REGULATION

Amir and Lobel assert that "policy can be improved dramatically by expanding the tools of regulation."[672] That is, for better problem solving in modern life, the regulatory toolbox needs to contain a screwdriver or even magic glue, not just a bunch of hammers. In this way, "economic efficiency and democratic legitimacy can be mutually reinforcing" and might redress the pervasion of both regulatory and market failures.[673]

What does that mean exactly? In sum, the empirical findings inform regulatory endeavors that:

- harness private finance actors' limited voluntarism and drive it further;
- establish conditions that support a business case for renewable energy and clean-tech that surpasses a competing one for fossil fuels to bring low-carbon technology to scale;
- create a policy environment in which private finance actors can choose renewables/clean tech clients and eschew fossil fuel clients without compromising their client service reputations; and
- capitalize on new understandings of "soft" factors like corporate reputation and risk/opportunity as drivers of socially beneficial corporate behavior.

This approach does not propose to cure the defects of voluntarism or "un-make" inherent limitations in business case logic. Instead, it acknowledges the defects and limitations, and then harnesses them. In this way, I am arguing that private finance actors can be harnessed *accurately* and thus *fruitfully* for societal good.

In hypothesizing why environmental law has failed to substantially curb degradation and unsustainable conduct, Richardson writes that "[e]nvironmental law is hampered where it ignores or misunderstands the behavioral tendencies of the actors it seeks to influence."[674] The key to better environmental outcomes lies in a better understanding of human and also corporate motivations. Hence, the purpose of the empirical case study was to understand the "is" with a view to then crafting empirically informed regulation[675] to assist timely attainment of the "ought." I cannot overemphasize the importance and novelty of this process in the corporate climate finance space. In short, climate finance regulation will need to take account of financial regulatees' behavioral shortcomings as well as their strengths in order to achieve its goals.

---

672. Amir and Lobel, *supra* n. 669, at 2127.
673. Lobel, *supra* n. 649, at 344.
674. B.J. Richardson, *A Damp Squib: Environmental Law from a Human Evolutionary Perspective* (Osgoode Hall Law School Research Paper Series, Research Paper No. 08/2011, 2011), 6.
675. I borrow this phrase from C. Sunstein, "Empirically Informed Regulation" (2011) 78 *The University of Chicago Law Review* 1349.

[A]   What's in a Nudge?

Research in the fields of cognitive psychology and behavioral economics focuses on the minds of decision-makers. The literature has documented numerous biases inherent in human decision-making,[676] many of which are responsible for faulty decisions that contribute to poor environmental outcomes, including inaction on climate change.[677] For example, Shu and Bazerman focus on three cognitive biases that have particular relevance for behaviors that negatively impact the environment: first, over-discounting the future and over-weighting the present; second, having positive illusions that a problem is not severe enough to warrant action; and third, behaving egocentrically and expecting others to do more to solve a problem.[678]

In retrospect, we can see that all three biases were evidenced in the empirical case study.

Cognitive biases stem from bounded rationality. Bounded rationality "refers to the obvious fact that human cognitive abilities are not infinite."[679] Humans have limited brain power, limited time, and imperfect memories.[680] Thus, when making decisions, we tend to take mental shortcuts and apply rules of thumb, which result in biased or flawed judgments.[681] Importantly, these shortcuts and rules of thumb are predictable.[682] Accordingly, human behavior departs spectacularly but systematically from neoclassical economic models of "rational" behavior.

In the words of Cass Sunstein, that finding has "transformed our understanding of regulation and its likely consequences."[683] It has motivated contemplation about the

---

676. See e.g. A. Tversky and D. Kahneman, "Judgment under uncertainty: heuristics and biases," in DJ Levitin, (ed.) *Foundations of Cognitive Psychology: Core Readings* (MIT Press: Cambridge, MA, 2002), 585; C. Jolls, C.R. Sunstein and R. Thaler, "A Behavioral Approach to Law and Economics" (1998) 50(5) *Stanford Law Review*, 1471; M.H. Bazerman and D. Moore, *Judgment in Managerial Decision Making* (Wiley: New York, 2008).

677. See e.g. M.H. Bazerman and A.J. Hoffman, "Sources of environmentally destructive behavior: individual, organizational, and institutional perspectives" (1999) 21 *Research in Organizational Behavior* 39; M.H. Bazerman, J. Baron and K. Shonk, *You Can't Enlarge the Pie: Six Barriers to Effective Government* (Basic Books: New York, 2001); R. Gifford, "Psychology's Essential Role in Alleviating the Impacts of Climate Change" (2008) 49(4) *Canadian Psychology* 273; Shu and Bazerman, *supra* n. 625; A.J. Hoffman, "Climate change as a cultural and behavioral issue: addressing barriers and implementing solutions" (2010) 39(4) *Organizational Dynamics* 295; M.H. Bazerman, "Climate Change as a Predictable Surprise" (2006) 77 *Climatic Change* 179; C. Sunstein, "Nudges.gov: Behavioral Economics and Regulation" (Forthoming) *Oxford Handbook of Behavioral Economics and the Law* (Feb. 16, 2013) *available at* http://dx.doi.org/10.2139/ssrn.2220022.

678. Shu and Bazerman, *supra* n. 625.

679. Jolls et al., *supra* n. 676; C.R. Sunstein and R. Thaler, "A Behavioral Approach to Law and Economics" (1998) 50(5) *Stanford Law Review*, 1471, 1477. See also H.A. Simon, "A Behavioral Model of Rational Choice" (1955) 69 *Quarterly Journal of Economics* 99.

680. D. Kahneman and A. Tversky, "On the Reality of Cognitive Illusions" (1996) 103(3) *Psychological Review* 582.

681. See e.g. et al., Jolls et al., *supra* n. 676; Tversky and Kahneman, *supra* n. 676; Shu and Bazerman, *supra* n. 625.

682. Tversky and Kahneman, *supra* n. 676.

683. Sunstein, *supra* n. 677, at 1.

actual operation and potential improvement of legal systems and policymaking.[684] More specifically, it has prompted enquiry of how policy-makers might harness predictably flawed decision-making to achieve better "real world" outcomes.

The most influential manifestation of this endeavor is Richard Thaler and Cass Sunstein's work on *Nudge* theory.[685] Thaler and Sunstein argue for self-conscious efforts by governments and private entities to influence people's behavior in order to improve their lives. Nudge is both a verb and an acronym. The acronym comprises the tools with which to combat limited human decision-making: iNcentives, Understanding mapping, Defaults, Giving feedback, Expecting errors, and Structuring complex choices. By using nudges, "choice architects" can "steer" choices to make people's lives "longer, healthier, and better."[686] A choice architect can be a government regulator or policy-maker, a company director, an employer, or anyone that "has the responsibility for organizing the context in which people make decisions."[687] In other words, architects influence or steer best outcomes by nudging people in the right direction. Thaler and Sunstein clarify that "[n]udges are not mandates."[688]

Given the human focus of nudge theory, how can it be applied to the case of private finance actors and climate change whilst remaining true to the theoretical precepts of a behavioral economics (as opposed to a rational economic) approach? I propose three main motivations.

The first motivation highlights the human stain indelibly inked into the frameworks of private finance actors. Data from the case study showed that banks copy each other at a cultural-cognitive level – and therefore without much thought – for reasons of legitimacy; and they may not always do what is "right" or "best" of their own volition given their eschewal of ethical motivations and their obsession with client service reputation. Moreover, previous studies have shown that corporations are also subject to bounded rationality whereby businesses have failed to adopt proactive environmental policies even when it is in their economic interests to do so; while the reasons for inaction are multifarious, they include the attitudes and biases of human managers.[689] In other words, decisions of private finance actors are made by individual managers who are subject to cognitive biases within perceived and even actual structural constraints. Despite the legal fiction of a corporation's status as a "person" and separate legal entity, company decisions are made by human beings.

This leads to the second point: the corporate cognitive element is an important regulatory soft spot. The case study showed clearly the importance of client service reputation and, to a lesser extent, social reputation to private finance actors. Their brand, image and reputation are assets that they value and want to maintain to a high

---

684. See e.g. Jolls et al., *supra* n. 676; Amir and Lobel, *supra* n. 669; J.J. Rachlinski and C.R. Farina, "Cognitive Psychology and Optimal Government Design" (2002) 87(2) *Cornell Law Review* 549; Richardson, *supra* n. 674.
685. Thaler and Sunstein, *supra* n. 671.
686. *Ibid.*
687. *Ibid.*, 3, 83.
688. *Ibid.*, 6.
689. Gunningham et al., *supra* n. 390.

standard. Reputation is one of the characteristics to which nudge theory plays. Adapting nudge theory to "supply side" commercial enterprises, Adam Oliver argues that policy-makers should aspire to "budge" profit-oriented industry to serve the social good more effectively by leveraging their social reputation.[690] I argue that this logic can be extrapolated to client service reputation for private finance actors.

Thirdly, government intervention cannot be avoided even in free and functioning markets[691] and markets alone cannot achieve the best environmental or social outcomes.[692] Indeed, it is well-accepted that markets contribute to environmental problems because incentives are not aligned properly and environmental degradation is treated as an externality.[693] Sunstein has noted that, in addition to standard accounts of market failures, "we are now in a position to identify a series of *behavioral market failures*, and these do appear to justify regulatory controls."[694] One approach is for government to realign poorly-aligned incentives and to "put the right incentives on the right people."[695] From a behavioral economics perspective, incentive-based initiatives exploit "present bias" or "hyperbolic discounting" by offering an incentive to offset the upfront unpleasantness of doing a task now that will have longer-term benefits. In the context of addressing climate change, the right incentives are economic, behavioral and cognitive.

This can be illustrated by elaborating upon an example provided in *Nudge*. Let us equate the climate field to a cafeteria. In this cafeteria healthy carrots are placed at eye-level and under bright light (renewables/clean tech incentives) and unhealthy candy bars sit on the floor where it is harder to reach, dirty, and a little embarrassing if one is seen handling them (carbon-emitting disincentives). In this way, the cognitive (rather than economic) cost of choosing the candy bar has increased, which acts as a disincentive for choosing it.[696] By exploiting the dichotomous reputational soft spot of private finance actors, well-designed nudging regulation can:

- increase both cognitive and economic *costs* for private finance actors that associate with high GHG-polluting actors and projects; and
- increase both cognitive and economic *opportunities* for those that support and facilitate renewable and clean tech solutions.

How does the cafeteria scenario relate to learnings from the case study? Pricing carbon penalizes GHG-intensive corporations and projects and externalizes the cost of carbon emissions. This is textbook economics. Yet, the case study data show that a carbon price also has the *consequential* effect of disincentivizing *financial* support for those industries/projects by affecting bank risk assessment, financing, lending and investment practices. This is an important finding for policy-makers: it shows the concentric

---

690. A. Oliver, "From Nudging to Budging: Using Behavioural Economics to Inform Public Sector Policy" (2013) 42(4) *Journal of Social Policy* 685, 698.
691. Thaler and Sunstein, *supra* n. 671; Amir and Lobel, *supra* n. 669.
692. Vogel, *supra* n. 338; Gunningham et al., *supra* n. 390.
693. Thaler and Sunstein, *supra* n. 671, at 184-5.
694. Sunstein, *supra* n. 677, at 9 (emphasis in original).
695. Thaler and Sunstein, *supra* n. 671, at 97.
696. *Ibid.*, 8.

circles of intermediary actors influenced by economic regulation due to non-economic or cognitive reasons. As explained in Chapter 1, climate risk is broader than credit and investment risk; it also includes litigation and reputational risk for private finance actors that associate with "dirty" clients and projects. As we now know, this corporate cognitive or reputational element is very important for private finance actors. By exploiting it, government regulators and policy-makers can increase consumption of healthy options and decrease consumption of unhealthy ones by strategically re-arranging the cafeteria without necessarily changing the menu. The case study data bring to life the inner workings of how soft considerations bear on private finance actor decision-making; a realization that is *not* found in economics or legal text books.

Yet the American and Australian political environments have not been amenable to a carbon price in recent years. From a regulatory perspective, it is important to note that human-exacerbated climate change has been conflated with pricing carbon in political and media debates in these two jurisdictions. Yet they are *not* the same thing. The first is a phenomenon accepted by the vast majority of the world's scientists; the other is one of a range of possible policy responses to that phenomenon. This conflation has only served to fuel political ambitions and confound public discourse.

It is critical at this juncture to de-conflate these elements and to emphasize the gamut of policy responses, which include creating opportunities and benefits and not just inflicting costs.

On this point, the empirical data from the case study showed that the *threat* of a carbon price *combined with* financial regulatory incentives motivated banks to view climate change as more of an opportunity than a risk. This is fascinating and almost counter-intuitive. Nonetheless, by comparing the American data regarding regulatory context to those from Australia, it is clear that including financial incentives in a climate policy framework is crucial to mobilizing private finance in low-carbon spaces. Once again, this is due to both economic and cognitive reasons. Leading banks in the United States leveraged government incentives to increase their client service reputation through the creation of innovative products and services. They were motivated by opportunities more so than risk mitigation despite the fact that no federal-level carbon price existed. In contrast, Australian leading banks were driven by downside prevention. They were not inspired to innovate and wanted to keep BAU running as smoothly as possible despite the recent implementation of a carbon tax.

Thus, it is clear from the evidence that pricing carbon is not the only imperative. Incentivizing regulation is also required.[697] That is, government policy that incentivizes investment in and uptake of renewables, clean tech and, just as critically, energy storage technology[698] by private finance actors and their clients. This may take the

---

697. M. Bowman, "Nudging Effective Climate Policy Design" (2011) 35(2/3/4) *International Journal of Global Energy Issues: Special Issue – Carbon Markets: An International Perspective* 242.

698. See J. Intrator, E. Elkind, A. Abele, S. Weissman, M. Sawchuk, E. Bartlett, *2020 Strategic Analysis of Energy Storage in California* (University of California Berkeley School of Law, University of California, Los Angeles, University of California, San Diego: November 2011), available at http://www.energy.ca.gov/2011publications/CEC-500-2011-047/CEC-500-2011-047.pdf.

form of hard economic incentives such as FITs, tax credits, and/or grants; it may also take the form of behavioral and cognitive incentives such as securities listing guidances on material climate risks, or company rankings regarding GHG emissions reductions. Indeed, the data reveal that cognitive incentives can influence corporate behavior for economic reasons and vice versa. Importantly, incentivizing regulation is a low interventionist option for governments and therefore more politically palatable than coercive regulation. Thaler and Sunstein opine that "some of the best nudges use markets; good choice architecture includes close attention to incentives."[699]

Importantly, nudging regulation overcomes the limitations inherent to business case logic that were evidenced in Chapter 5, particularly regarding the delayed materiality of a green business case. Thaler and Sunstein note that decision-makers "may most need a good nudge for choices that have delayed effects."[700] In other words, nudging regulation can help private finance actors to map long-term costs/benefits thus mitigating the impetus of short-term profit maximization. Secondly, regulation that incentivizes renewables and disincentivizes high-carbon activities helps to undermine a countervailing business case for continued financial support for GHG-intensive options. This is particularly necessary in fossil fuel rich jurisdictions such as the United States and Australia. For example, one bank respondent noted how incentivizing regulation created necessary demand:

> …[W]e don't need renewable energy power because we have so much cheap coal. So we need regulation to make a demand for renewable energy… Right now coal is a cheap and reliable energy source for people. We need competing policy for solar. Without that, [renewable energy] plants will never get done voluntarily. (G1).

Overall, well-designed nudging regulation that focuses on the profit and reputation-driven nature of "supply side" commercial enterprises[701] births several important consequences.[702] First, it can facilitate the finance necessary to start up and deploy clean technology on a scale sufficient to transition expeditiously to a low carbon global economy. The case study data show that when government regulation addresses market risk then banks will shift their attention accordingly. As stated by one banker:

> Some [clean] technologies are still not economic. They need to be competitive with very cheap coal plants. Regulation can be helpful, it can pick up the true cost of energy, it can help to bridge the gap between starting up and getting returns" (C3).

Attracting capital into energy storage technologies and renewable energy and clean tech markets will facilitate deployment at scale. Enlarging these markets is critical for the transition to a less carbon-dependent world. As evidenced in the previous chapter, this is not going to happen sufficiently or expeditiously by itself.

---

699. Thaler and Sunstein, *supra* n. 671, at 253.
700. *Ibid.*, 76-7.
701. Oliver, *supra* n. 690, at 697.
702. M. Bowman, "Corporate Care and Climate Change: Implications for Bank Practice and Government Policy in the United States and Australia" (2013) 19(1) *Stanford Journal of Law, Business & Finance* 1, 35-36.

Secondly, nudging regulation (both incentivizing and disincentivizing) may encourage innovation by private finance actors because it capitalizes on the "go to" motivation born of client service reputation. This is evidenced in the data. For example, innovative investment indices were created by a bank in response to a pension fund's concern about portfolio risk from a carbon price. Similarly, bankers in the energy sector boasted how they created competitive edge for their bank by marrying hedging practices with tax credits for renewables to ensure lucrative returns for their clients and themselves.

This in turn leads to the third important consequence of nudging regulation: it can realize the private finance sector's network change potential. By harnessing the finance sector's profit motive, nudging regulation also harnesses that sector's ability to enroll other corporate actors in climate change mitigation efforts. Specifically, through day-to-day activities as risk managers, financiers, lenders, investors and advisers, private finance actors have potential to influence choices by other actors in the financial and corporate chain. This includes large pension funds as well as polluting entities. For example, the climate investment indices cited above have influenced allocations by fund managers; and the banking advice based on tax breaks lured client capital into low-carbon spaces.

Finally, for maximum effect, the cognitive element of incentivizing regulation means that policy nudges can dovetail with "social nudges."[703] In the case of private finance actors, social nudges include: positive feedback from the market due to high profits and also from clients in the form of repeat business and/or industry awards; and negative feedback from NGOs in the form of name and shame campaigns and also from stockholders in the form of shareholder activism. In this way, social nudges leverage not only the social reputation but also the client service reputation of private finance actors for greener behavior.

As such, clever climate finance law and policy leverages both hard and soft aspects of the business case as revealed through empirical investigation. Such regulation makes it more difficult for finance actors to choose poor alternatives from the choice set, as opposed to eliminating those alternatives completely. It nudges the right choice by them.

[B]   Critiques of "Nudging" Regulation

Drawing on nudge economics and new governance theories for inspiration, well-designed law and policy for corporate climate finance can harness the private finance sector's profit motivation, corporate reputation, expertise, and network capacity to assist GHG emissions reductions. Specifically, indirect government regulation can nudge private finance actors in the "right" direction via incentives and disincentives.

However, it must be acknowledged that there are limitations to this type of regulation and some voluble criticisms of nudge theory itself, which need to be considered when advocating for it.

---

703. Thaler and Sunstein, *supra* n. 671, at 54, 191.

*[1]   Incentives Can Have Unintended and Unexpected Perverse
        Consequences*

Several authors have noted that although flexible regulation can be environmentally effective and cost-efficient, it is not flawless.[704] In particular, Bennear and Coglianese highlight how market-based regulatory approaches can be "susceptible to the law of unintended consequences" due in part to their very flexibility.[705] They provide the example of tradable permits in pollution reduction schemes whereby permits can create "so-called hot spots" if they are concentrated in one area, and "banking" of permits can create early achievement of environmental goals but at the expense of lower environmental standards at a later date.

The Australian AU$1.5billion Solar Flagships Program provides another example. In 2009 the Federal Government launched this Program with intent to "support the construction and demonstration of large-scale, grid connected solar power stations in Australia [that could operate] within a competitive electricity market."[706] Two projects were selected at tender; however, neither project commenced because they could not secure a power purchase agreement (PPA) with an energy retailer due in part to the low price and over-supply of RECs,[707] which in turn meant they could not secure private financing. Under a PPA the generated electricity is purchased (usually by the retailer) for a set price and period of time, which provides investor certainty by ensuring the project's viability; without it, financiers will not get on board.

Nonetheless, it is equally important to consider where and why incentivizing regulation has succeeded. McConnell points to the success of FITs as the alternative to a grants scheme, stating that in the same three-and-a-half year period during which the Australian federal solar program produced nothing, the German FIT delivered over one hundred times the capacity of one proposed Australian project.[708] Although FITs are not flawless, the German experience illustrates how an efficient and programmatic FIT system can manage costs by adjusting tariffs as PV prices fall, instead of making ad hoc adjustments that create boom/bust cycles and investor insecurity.[709]

Moreover, the UK GIB is an innovation in this category of intervention. As noted in Chapter 1, it is a government-supported corporation that leverages private finance for energy efficient and low-carbon solutions in the UK. GIB investments are assessed for "green impact" prior to approval and undergo continuous monitoring during the life

---

704. E.g., Gunningham et al., *supra* n. 390; Gunningham and Sinclair, *supra* n. 656; L.S. Bennear and C. Coglianese, *Flexible Environmental Regulation* (University of Pennsylvania Institute for Law & Economics Research Paper No. 12-03, 2012).
705. Bennear and Coglianese, *supra* n. 704, at 11, 20.
706. Per Federal Department of Energy, Resources and Tourism at http://www.ret.gov.au/energy /clean/sfp/Pages/sfp.aspx viewed May 29, 2012.
707. D. McConnell, "Not Dead Yet: Flagship 'Collapse' Only Part of Australia's Solar Story," *The Conversation* (Feb. 10, 2012).
708. *Ibid.*
709. See generally International Energy Agency, *supra* n. 642; Mormann, *supra* n. 38.

of the project. The measure of green impact is assessed by reference to one or more specific measures or "green purposes":[710]

- GHG emissions reduction;
- efficiency in the use of natural resources;
- protection or enhancement of the natural environment and/or biodiversity;
- promotion of environmental sustainability.

In terms of encouraging private finance, the success of the GIB is demonstrated by its 1:3 leverage: by mid-2014, the GIB had mobilized £4.8billion and directly committed £1.3billion into eligible projects within the UK.[711] Importantly, the GIB acknowledges that it is leading the way in "a relatively new area of banking"; and thus, it has purposely adopted a new governance style of approach "based on 'learning through doing' and underpinned by a commitment to continuous improvement."[712]

Whilst there have been no manifestly perverse outcomes arising from the UK GIB, German FITs or American tax credits for renewables, it is certainly important to monitor the effects on-the-ground of this form of regulation. Doing so is consistent with a new governance model of collaborative and responsive regulation. The point is that regulation needs to have strong empirical foundations, which occurs "through careful analysis in advance and through retrospective review of what works and what does not."[713] As posited previously, climate finance regulation will need to be adaptive, particularly when new information comes to light during the effort to implement solutions.

[2]   Governments May Not Know What is "Best" or "Right"

The potential for perversity feeds into the second concern about nudging regulation: that a government may not know what is "best" or "right." Indeed, a key ingredient of Thaler and Sunstein's nudge theory is the assumption that government knows best. In particular, they advocate a "new movement" called "libertarian paternalism."[714] By constraining individual choice, paternalism on its own provides one of the main objections to direct and coercive regulation in market economies. Thus, Thaler and Sunstein argue that by providing regulatory nudges within a *libertarian* paternalism construct, governments and other decision-makers choose the architecture (paternalism) within which individuals can then make choices (libertarian).

However, the concept of libertarian paternalism is not without criticism. The main critique is whether a person's response can truly be a choice if it has been elicited covertly; and whether covert governmental messaging is appropriate or desirable in a

---

710. Green Investment Bank, *Green Impact*, http://www.greeninvestmentbank.com/green-impact/ (accessed Aug. 30, 2014).
711. Green Investment Bank, *Summary of Transactions*, 1, http://www.greeninvestmentbank.com /media/25380/gib_ar_transactions_250714.pdf accessed Aug. 7, 2014.
712. Green Investment Bank, *Green Impact*, http://www.greeninvestmentbank.com/green-impact/ (accessed Aug. 30, 2014).
713. Sunstein, *supra* n. 675, at 1350.
714. Thaler and Sunstein, *supra* n. 671, at 4-6.

democracy.[715] Moreover, Orly and Lobel describe libertarian paternalism as an "ideologically loaded marketing formula" that does not explain as accurately as new governance theory why improvements are made to safety, health or financial security.[716] Alternatively, Raghuram Rajan has asserted that the notion of choice in libertarian paternalism is illusory because "the paternalism [is] largely unconstrained" and this opens the door for "paternalistic mistakes."[717]

Political history teaches us that it is *not* axiomatic that "government knows best." Indeed, critical evaluation of the Australian federal approach to climate policy in the past decade shows that there is plenty of room for government officials to get it wrong. The climate policy framework implemented in Australia in 2011 was recommended at that time to investor networks around the world as "investment grade" and apt to provide "investors with real confidence when investing in areas such as renewable energy."[718] The framework included a carbon price, a RET, compulsory reporting of intensive corporate GHG emissions, and the Clean Energy Finance Corporation which bore resemblance to the UK GIB. However, in less than three years, the legal landscape had altered radically. A change in government and political attitude to climate change in 2014 heralded regulatory revision and the key policy platforms were dissembled. As a result, investor enthusiasm shifted elsewhere despite Australia's natural abundance of renewables.

European experiences similarly illustrate the crucial role of governments in ensuring stable policy environments that support green markets. Deutsche Bank writes: "As investors, we essentially look for Transparency, Longevity and Certainty (TLC) in assessing the potential success of policies."[719] In other words, incentivizing and disincentivizing regulation must be long term and stable in order to create investment certainty, which then motivates private finance and business activity in renewable and clean tech sectors. This also means that a carbon price (whether in the form of an ETS or tax) must be set high enough to ensure markets can flourish whilst carbon emissions are actually reduced. Moreover, regulation cannot be repealed suddenly or retroactively. For example, due to a tariff deficit in the electricity sector, Spain abolished FITs with immediate effect for newly installed renewable energy generation in 2012/2013.[720] Such action was described by bankers in the case study as "potentially disastrous" for the financing of renewables. It also opens a government to potential legal action by operators and investors.

There is no doubt that markets and/or demand can change. If a government is adopting a truly adaptive approach and regulatory revision is warranted, then I urge

---

715. See e.g. Oliver, *supra* n. 690, at 687.
716. Amir and Lobel, *supra* n. 669, at 2125-2126, 2132.
717. R. Rajan, "It's the 'Nudge,' Stupid, the Key to Steady the Nerves," *The Saturday Age* (Apr. 14, 2012) 12, 13.
718. Sullivan, *supra* n. 639, at 15.
719. Deutsche Bank, *supra* n. 232.
720. Spanish government, Royal Decree-Law 1/2012 and Royal Decree-Law 9/2013, published in the Spanish Official State Gazette. See also J. Alcauza, "Spain kills Feed-in Tariff for renewable energy," *CSP World* (Jul. 13, 2013), http://www.csp-world.com/news/20130713/001121/spain-kills-feed-tariff-renewable-energy.

regulators to make changes prospectively and to communicate those changes transparently with clear timeframes and set criteria,[721] preferably in consultation with industry participants.

Sunstein's approach is succinct: "No one denies that nudges can go wrong. If they do, the challenge is to get them right."[722] He further notes that chosen rules might be harmful in a situation where choice architects are "uninformed or self-interested."[723]

On this latter point, Thaler and Sunstein acknowledge that governments are comprised of intrinsically flawed human decision-makers.[724] Politicians are subject to the usual suite of human cognitive biases: they suffer from present bias due in part to short-term electoral cycles in democracies and they are also susceptible to powerful lobby interests.[725] Indeed, some studies have examined the effects of bounded rationality on policy-making and argue that government regulators themselves can be nudged to "do the right thing."[726]

This is an apt proposition for optimal climate finance policy-making. Regulators need to see empirical work such as the banking case study to steer their own choices for informed and effective policy-making.[727]

Finally, it is relevant to note that governments around the world are now experimenting with nudge-type interventions and some have even created units dedicated to this endeavor. For example, in the UK the Cameron government created the Behavioural Insights Team in the Cabinet Office (colloquially known as the Nudge Unit) to improve public policy[728] and the FCA is interweaving behavioral economics insights into financial market regulation in order to protect consumers.[729] In the United States, consumer protection via disclosure requirements can be found in a gamut of legislation from Dodd-Frank to the *Patient Protection and Affordable Care Act (2010)*.[730] In Australia, ASIC has adopted behavioral insights about financial decision-making into financial literacy strategies.[731]

---

721. Sullivan, *supra* n. 639, at 4.
722. C. Sunstein, "There's a backlash against nudging – but it was never meant to solve every problem," *The Guardian* (Apr. 25, 2014), http://www.theguardian.com/commentisfree/2014 /apr/24/nudge-backlash-free-society-dignity-coercion.
723. C.R. Sunstein and L.A. Reisch, "Automatically Green: Behavioral Economics and Environmental Protection" (Preliminary draft 9/19/2013), *SSRN-id2245657*, 2013), 20.
724. Thaler and Sunstein, *supra* n. 671, at 10.
725. Oliver, *supra* n. 690, at 697; Sunstein, *supra* n. 722.
726. See e.g. Rachlinski and Farina, *supra* n. 684; J.C. Cooper and W.E. Kovacic, *Behavioral Economics: Implications for Regulatory Behaviour* (Jul. 21, 2011), http://ssrn.com/abtract = 1 892078 (accessed Aug. 12, 2012).
727. See Bowman, *supra* n. 697.
728. See https://www.gov.uk/government/organisations/behavioural-insights-team.
729. Financial Conduct Authority, *Applying Behavioural Economics at the Financial Conduct Authority* (April 2013): http://www.fca.org.uk/static/documents/occasional-papers/occasion al-paper-1.pdf.
730. *Dodd–Frank Wall Street Reform and Consumer Protection Act* (Pub.L. 111–203, H.R. 4173) "Title IX – Investor Protections and Improvements to the Regulation of Securities"; and the *Patient Protection and Affordable Care Act* (Pub.L.111-148) (2010) SEC.1103. "Immediate Information that Allows Consumers to Identify Affordable Coverage Options."
731. See e.g. Australian Securities and Investments Commission, *Keynote Address: Speech to ISIE and IOSCO Global Investor Education Conference* (Jun. 21, 2013), http://www.asic.gov.au/as ic/asic.nsf/byheadline/Global + investor + education + conference + - + Keynote + address?ope

The fascinating and relevant commonality throughout these initiatives is the inculcation of empirical findings on soft issues into hard financial and economic regulation. It is no stretch to extrapolate this species of regulatory innovation to corporate climate finance law and policy-making.

§6.06   THE NECESSITY OF A CLIMATE FINANCE "REGULATORY MIX"

Nudge protagonists opine that replacing traditional regulation with nudges can result in better governance;[732] yet new governance scholars do not necessarily agree. For example, Strausser asserts that there is no serious discussion in the new governance literature of replacing "old government."[733] Instead, the real enterprise of "business environmentalism" is to work out "the most effective combination" of the two modalities.[734]

This is true also for corporate climate finance, and the broader climate endeavor. In reality, lone applications of regulatory nudging or coercion are no panacea for the climate crisis. In order to address an issue as complex and far-reaching as climate change, a regulatory mix is required to mobilize the private finance sector and other relevant actors. We need to bring the full regulatory arsenal to ensure expeditious transition to a low-carbon existence.

An effective climate finance regulatory mix will comprise various combinations of:

- emissions reduction targets;
- renewable and energy efficiency targets;
- prescriptive and proscriptive regulation (for finance actors and other corporates);
- incentivizing and disincentivizing regulation;
- climate-related corporate disclosure requirements and securities listing guidances;
- enhanced and centralized climate change information repositories;
- industry self-regulation;
- civil society and NGO interventions.

In addition, this chapter has evidenced the following three points for policy-makers:

(1) There is no one-size-fits-all design prescription for effective climate finance policy given the heterogeneity of projects and investors, so a regulatory framework will need to be responsive, adaptive and contain provision for retrospective analysis and prospective revision.

---

nDocument; Australian Securities and Investments Commission, *Report 230: Financial Literacy and Behavioural Change* (March 2011) http://www.financialliteracy.gov.au/media/218309/fi nancial-literacy-and-behavioural-change.pdf.
732. Thaler and Sunstein, *supra* n. 671, at 14.
733. Strausser, *supra* n. 666, at 80-81. See also Scott, *supra* n. 663, at 146.
734. Strausser, *supra* n. 666, at 80-81.

(2) As far as possible, both regulatory implementation (ex post) and revision (ex ante) need to be empirically informed in order to be effective.

(3) For any climate finance policy initiative, the guiding litmus test is disarmingly simple: how does it harness hard or soft aspects of the business case to promote societal benefit?

With this last point in mind, the next chapter broadens the regulatory focus to consider entrepreneurial climate finance modalities and actors.

CHAPTER 7
Re-setting the Regulatory Sights

§7.01   OVERVIEW

Given the centrality of the business case to private sector financial activity, this chapter considers risk/return trade-offs of that business case across a range of relevant private finance actors. In so doing, it broadens the regulatory sights beyond conservative or risk-averse financial actors such as banks and pension funds to include capital providers with greater risk tolerances such as venture capital funds and private equity firms. Specifically, this chapter presents and analyzes three entrepreneurial investment modalities to augment renewable energy and clean tech: tax equity partnership structures; climate and green bonds; and Evergreen buy-out strategies.

It makes sense to start this chapter with a high-level overview of the risk/return trade-off as conventionally presented within corporate finance theory, highlighting the well-known insights of the capital asset pricing model (CAPM). Although it is not the unique asset pricing approach to the topic, the CAPM is far and away the most recognized approach by chief financial officers, investment managers, financial economists and courts.[735] That is not to say the CAPM is universally embraced; the appropriateness of its assumptions were scrutinized both before and after the GFC, and some have described the CAPM as "a very mediocre model" in terms of accuracy, particularly across long-term horizons.[736] Moreover, "smart beta" strategies that acknowledge the complexity of markets in terms of multiple sources of risks and risk premiums are increasing, which weight assets by stock characteristics other than market capitalization, such as volatility and company fundamentals.[737] Nonetheless,

---

735. I. Welch, *Corporate Finance* (3d ed, UCLA, 2014), 242.
736. I. Welch, *The CAPM: Mostly Cookbook With Explanations (Welch, Chapter 09)*, Jan. 1, 2012 (UCLA Anderson School), 45. See also I. Welch, *supra* n. 735, at 243-245.
737. See e.g. J. Hsu, *Smart Beta: The Best of Both Worlds* (Research Affiliates, Newport Beach CA, 2013); Towers Watson, Using Smart Beta In Equities (Towers Watson, London, 2014); P. Bertrand and V. Lapointe, *Smart Beta Strategies: The Social Responsibility of Investment*

the CAPM remains a critical benchmark model, and the predominant paradigm by which the cost of capital is calculated.[738] As such, prior to considering specific entrepreneurial investment modalities, it is worthwhile walking through the steps of risk/return thinking employed by banks and other finance actors in order to understand key determinants of how financing decisions are made.

§7.02   THE CAPITAL ASSET PRICING MODEL AND RISK/RETURN REQUIREMENTS FOR DIFFERENT PRIVATE FINANCE ACTORS

The core ingredients of the CAPM are a universe of risky financial investments (or "assets") that a population of risk-averse investors hold in their individual investment portfolios. The model assumes these investors will be attracted to high expected returns in their respective portfolios, but dislike variability as measured by the overall variance of their portfolios' returns. Indeed, under the assumptions of the CAPM, investors care only about expected returns and variance; they are insensitive to other measures of return or variability. Consequently, when assessing whether to invest in (or even whether to short-sell) a financial asset, an investor in the CAPM model cares about how that investment contributes to the expected return and variance of her overall portfolio. This is due to two critical CAPM assumptions: investors can take either "long" or "short" positions in any financial asset without cost; and there exists one financial asset in the economy that is riskless, generating a risk-free return with absolute certainty.

In assessing the contribution of a candidate asset within her portfolio, a CAPM investor will subdivide the asset's total risk into two categories: unsystematic risk and systematic risk. Unsystematic (or "idiosyncratic") risk is unique to an asset and not others, potentially including a labor strike or an occupational health and safety disaster that will affect a particular industry or company. Such unsystematic risks are of little moment to the CAPM investor, since simply diversifying one's portfolio can reduce the effects of idiosyncratic risk to negligible magnitudes.

In contrast, systematic risk is common to all financial assets, such as the risk of a macro-economic slowdown or a hyperinflation. Such risks cannot be diversified away, and consequently investors require compensation (beyond the risk-free rate of return) in order to bear systematic risk. It is this type of risk, then, that truly concerns most investors. Systematic risk is measured with *beta* (β) which measures how closely an asset's returns are associated with the returns of the market as a whole. The higher the beta of the asset, the greater the systematic risk associated with that asset. In short, beta is a measure of individual stock risk relative to the overall volatility of the stock market. In technical terms, β is equal to the covariance of the asset's return with that of the market portfolio, divided by the variance of the market portfolio. Thus, the β of the market portfolio itself must always equal 1.0.

Universes Does Matter (Working Paper, CERGAM, IAE Aix-en-Provence and Aix-Marseille University), April 2013, ssrn.com/abstract = 2254578, 2.
738. Welch, *supra* n. 735, at 242.

So, if a β of 1.0 is assigned to the market as a whole, then an asset with a β of 2.0 will be regarded as twice as aggressive because it carries twice the systematic risk than the market, which means its inclusion in a portfolio will magnify the effects of market swings on the returns of that portfolio. Conversely, if an asset has a beta of 0.5 relative to the market beta of 1.0, it clearly carries less systematic risk than the market and will be considered defensive, which will likely subdue the effects of market swings on portfolio return.[739] That is, for an investor who holds a portfolio similar to the market portfolio, investing in an asset with a low market β reduces overall portfolio risk: it is "less toxic" and therefore preferred by investors (all else held constant).[740]

Of course, not all else is constant. Importantly for investors, the CAPM predicts that the amount of systematic risk will push up the expected return that investors will demand from that asset. In fact, according to the CAPM, the relationship between systematic risk (measured by β) and required return is linear in nature, as reflected in Figure 7.1.

Figure 7.1   Risk/Return Trade-Off: The CAPM and SML

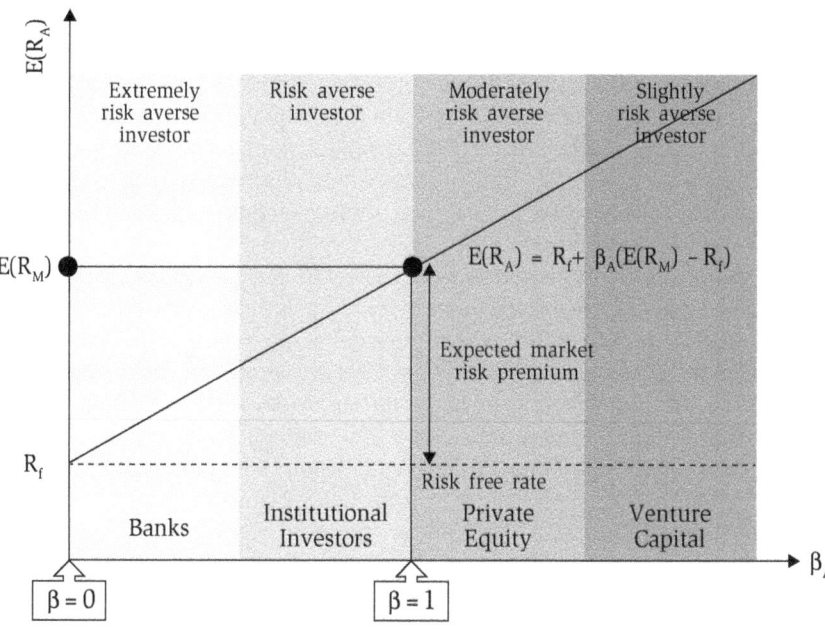

Figure 7.1 depicts the CAPM which postulates the relationship between systematic risk and the expected return from an asset/project. The linear relationship depicted is frequently referred to as the security market line (SML). The slope of the SML

---

739. Kidwell et al., *supra* n. 76, at 306.
740. Welch, *supra* n. 735, at 220.

represents the average return reward demanded for each incremental unit of systematic risk that the asset represents; the greater the value of β, the larger the gap (or "spread") investors will require over the risk free rate R_f. Specifically, the model predicts that this spread will tend to be a multiple of the spread exhibited by the market as a whole,[741] or , as depicted in the following expression:

$$E(R_A) - R_f = \beta_A \times E(R_M) - R_f$$

In this expression, $E(R_A)$ denotes the expected return on the asset, R_f denotes the risk-free return, $E(R_M)$ denotes the expected return on the market and β_A denotes the asset's beta.

This equation is merely the algebraic expression of the line from Figure 7.1 and it shows that the expected spread on a risky asset or project is the beta factor multiplied by the spread on the market portfolio. Alternatively, the expected return on an asset *equals* the return on a risk-free security (such as a government bond) plus a reward (or risk premium) for bearing the asset's systematic risk. The upshot is that the higher the beta of the asset or project, the higher the required/expected rate of return by investors.

What does the CAPM mean for private finance actors? As depicted in Figure 7.1, banks and institutional investors tend to be regarded as relatively risk-averse investors, which would cause them to stake out a place on the SML that involves relatively low systematic risk and correspondingly lower returns. In contrast, private equity and venture capital entities are conventionally thought to inhabit the riskier portions of the SML (as the SML heads northwards of $E(R_M)$) while simultaneously sitting at the higher end of systematic risk (β). The upshot of this observation is that "high-β" investments are more likely to be the province of private equity and venture capital investors, and not traditional banks.

Thus, risk/return theory is relevant to renewable and low-carbon technology investments and financing by private finance actors. It is clear from the discussion above that higher beta assets will appeal to capital providers with higher risk tolerances. Clean tech and renewable energy projects are considered high risk in their early phases due to a combination of high start-up costs, high technological and operational risks, significant regulatory risks, and unpredictable returns.[742] As a consequence the required risk/return trade-off may simply be prohibitive for more conservative finance actors, deterring them from financing at the scale required to make renewable energy and clean tech cheaper and more accessible than traditional energy sources.

Moreover, risk/return theory buttresses the empirical findings in this book that government policy intervention is required if renewable and clean tech projects are to be attractive to conservative finance actors such as banks. As demonstrated in Chapter 5, these actors will be considering alternative uses of their equity and debt capital, including investment in traditional (fossil fuel) forms of energy.

---

741. Welch, *supra* n. 736, at 11.
742. Sullivan, *supra* n. 639, at 12.

Accordingly, viewing the previous chapter through a risk/return CAPM lens, in order to engage banks and other conservative finance actors in sector-wide green financial activity, government intervention is required to make:

- renewable/clean tech investments more attractive by *reducing* β by covering or mitigating start-up and/or operating costs, for example, through grants, tax credits and/or GIB-type leveraging;
- "dirty" energy investments less attractive by *increasing* β through, for example, regulatory uncertainty and pricing carbon.

On the debt side, policy-makers can take concrete steps to harness the business case via regulation as detailed in the previous chapter and via climate bonds as detailed below. Yet innovative equity structures can also help raise equity capital in renewable and clean tech projects. Focusing on the equity side carries two implications. First, it permits the regulatory sights to be re-set toward capital providers with greater risk tolerances, namely venture capital (VC) funds and private equity (PE) firms; and secondly, it obviates the need for government intervention in the form of special bank regulations.

Different sources of finance come into play at different stages of low-carbon development and deployment. The later stages of manufacturing scale-up and commercial roll out can be financed by public equity markets as well as debt and project finance. Prior to that, technology development is generally financed by VC funds and PE firms, which are prepared to take a portfolio approach to risk analogous to the intuitions of the CAPM.[743]

VC funds tend to focus on early or growth stage technology companies and projects. They are keen to invest in new technology and new markets and have a high risk appetite to do it. They acknowledge a high risk of failure in every venture and therefore require a rate of return many multiples of the original investment (between 50 – 500 percent).[744] Venture capital is often provided prior to a company's IPO, which starts the listing process. Given that a company's beta is determined from historical data, the beta for an unlisted firm will be unknown. Nonetheless, the use of proxies in this space tends to assist with risk/return calculations. VC investment horizons are around four to seven years.

PE firms are interested in companies and projects with more mature technology, including: those preparing to raise capital on public stock exchanges pre-IPO; demonstrator companies; or under-performing public companies. They have a medium risk appetite and therefore a higher expected rate of return around 25 percent.[745] They generally expect to exit their investment and make their returns within three to five years.

Each of these sources of capital may be responsive to governmental policies meant to encourage investment in renewable energy and clean tech, albeit perhaps in

743. S. Patel, *Climate Finance: Engaging the Private Sector* (IFC: Washington DC, 2011), 4.
744. S. Justice, *Private Financing of Renewable Energy: A Guide for Policymakers* (UNEP, SEFI, Bloomberg New Energy Finance: Chatham House, 2009), 6, 8.
745. *Ibid.*

different ways. A few promising avenues for attracting all of them (in some degree) draw on tax policy and climate bond structures – topics to which I now turn.

§7.03   TAX EQUITY PARTNERSHIP STRUCTURES AND CORPORATE CLIMATE FINANCE

In addition to depreciation benefits, previous chapters identified two types of tax credits that incentivize renewable energy (RE) projects: PTCs that provide an inflation-indexed tax credit for every kilowatt hour of energy produced over a certain time period (say 20 years); and ITCs that provide a one-off tax credit (say 30 percent) of a project's capital costs. Developers and VC/PE Sponsors often do not have sufficient taxable income to qualify for or utilize these tax benefits. Thus they require tax equity investors to provide equity investment in return for a RE project's tax benefits and some of its cash flows.[746]

Figure 7.2   Tax Equity Partnership Structure

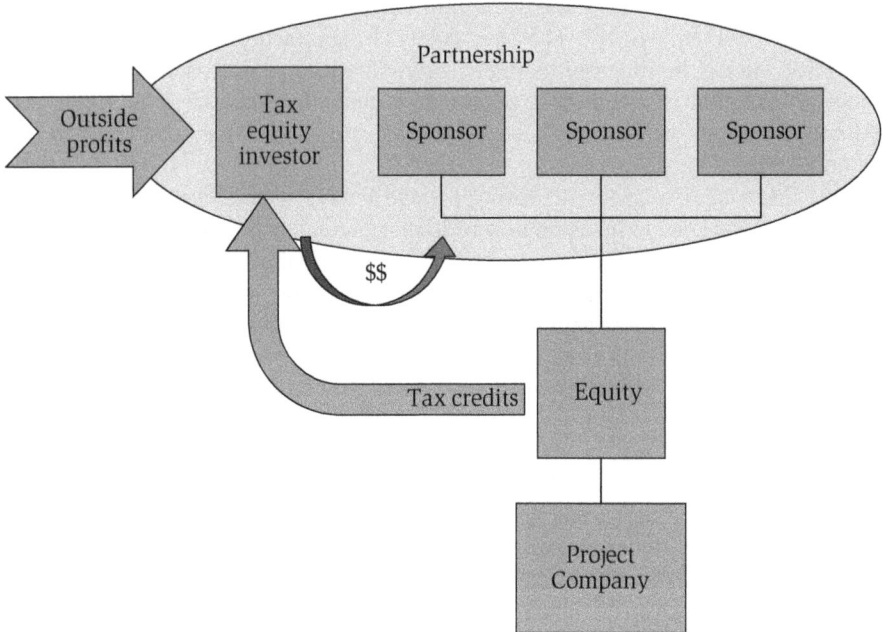

746. M. Meyers, B. Davidson and C. Gladbach, "Bridging the Tax, Equity Funding Gap," *Project Finance International: Renewables Report May 2012* (Thomson Reuters, 2012), 6, http://www .orrick.com/Events-and-Publications/Documents/4804.pdf.

Previously in the United States, the cash grant program under Section 1603 of the *American Recovery and Reinvestment Act* had successfully filled the role played by tax equity. The combined tax equity plus grant financings aggregate in each of 2010 and 2011 exceeded that for tax equity financing alone in 2007 which was considered a high-water mark year for tax equity (approximately US$6.1billion).[747] With the expiration of that grant in the United States and the absence of it in other jurisdictions, tax equity investment remains a prime strategy for RE project investment, providing that tax credits are in existence.

One form of innovative equity structure that can raise capital in RE and clean tech projects is the tax equity partnership structure depicted in Figure 7.2 above. This structure requires two main corporate inputs. First, it needs any combination of RE project developers and/or VC/PE sponsors to inject some cash equity and operationalize the project. These entities earn their return on investment through cash flows, a minimum allocation of tax benefits, and long-term ownership.[748]

Secondly, in order to finance the project's high up-front capital expenditures, it requires a deep-pocketed tax equity investor; or an entity with access to a client base that can absorb the tax credits and offer a tax credit syndication product to appeal to enough investors to create the deep pocket. The tax equity investor earns the required rate of return on the asset/project by receiving a majority allocation of tax credits and taxable losses/income plus some distributable cash.[749] That is, the tax equity investor can take advantage of tax deductions on capital losses from the upfront investment, which shields the operating revenues from taxes. It also has sufficient taxable income to monetize the tax benefits. Under a "partnership flip" the tax equity investor will exit the project when the developer/sponsor exercises its fair market value purchase option.[750]

Importantly, the partnership can be both owner of the assets and producer of the energy, which means that the tax equity investor can claim PTCs created by the project. Indeed, in the United States, partnership special allocation rules have been utilized to specially allocate tax incentives to an investor.[751] Providing that factual thresholds are met for this commercial arrangement to constitute a legal partnership,[752] the tax equity structure will generally be viewed as a *bona fide* partnership for tax purposes.

A prime advantage of this structure is that it allows very large non-financial players that have significant tax liabilities to get involved as tax equity investors, such as utilities, retailers, and technology companies. This is important because, as described previously, conservative finance actors will not buy equity if it is deemed too risky or the returns too low; and large financial institutions that might normally take on

---

747. *Ibid.*
748. G. Hecimovich and T. Stevens, *Introduction to Tax Equity Structures* (Deloitte, 2012), 9, http://www.deloitte.com/assets/Dcom-UnitedStates/Local%20Assets/Documents/Energy_u s_er/us_er_AESem2012_1_1_2_1IntTaxEquity_101012.pdf.
749. *Ibid.*, 8.
750. *Ibid.*
751. *Ibid.*, 7.
752. See e.g. : *Historic Boardwalk Hall, LLC v. Commissioner*, 136 T.C. 1 (2011) (United States); the *Limited Liability Partnerships Act 2000* (UK); sub-national Partnership Acts (Australia).

equity, such as investment banks, have been adversely impacted by the GFC and macroeconomic factors, which has reduced their risk appetite and available finance for RE projects. Ironically, bank reticence in this space due to the GFC has diluted a formerly "elite club" of tax equity investors comprised by the largest and most sophisticated financial firms,[753] thus enabling entry by other large but non-financial corporate entities.

Specifically, a tax equity investor must have "fairly predictable and reliable taxable income and large balance sheets."[754] The usual floor for tax equity investments is US$25-30million with the "sweet spot" generally twice or thrice that amount due to high deal complexity and transaction costs.[755] This requirement is satisfied by many Fortune 500 companies with Bloomberg New Energy Finance reporting that the 500 largest public companies paid more than US$137billion in taxes in 2010-2011.[756] Arguably, there is significant capacity within current markets for non-financial corporations to take on RE tax equity investor roles.[757]

To this end, the tax equity partnership structure provides a vehicle for large companies like Wal-Mart, Facebook, Walt Disney Company or San Diego Gas & Electric to park their profits and passively reap tax advantages. Google's fast-rising status as tax equity investor is a good example: in 2010 it entered the RE market and by 2014 had invested US1.4billion in RE production (over half in solar PV) and received commensurate tax breaks.[758]

Moreover, accelerated asset depreciation reduces the tax base of capital-intensive non-financial companies, which enhances the attraction for them of tax equity investment.

A final benefit of this tax equity partnership structure is that RE projects can be difficult to stage with a single VC/PE sponsor. This structure addresses that problem by attracting passive equity investments from multiple operating companies, VC funds and PE firms.

However, there are also several barriers to the proliferation of RE tax equity partnership structures. The most obvious barrier is regulatory uncertainty in the form of a "frustrating cycle of on-again/off-again tax policy incentives"[759] as evidenced by American energy policy and economic stimulus for the period 2007-2017. Regulatory risk remains the hardest type of risk for firms to mitigate and manage in this space. Yet despite regulatory uncertainty at both national and international levels, participants in

---

753. F. Mormann, "Beyond Tax Credits – Smarter Tax Policy for a Cleaner, More Democratic Energy Future" (2014) 31 *Yale Journal on Regulation* 303.
754. Meyers et al., *supra* n. 746, at 8.
755. *Ibid.*, 9.
756. E. Goossens, "US Clean Energy Needs Private Funding as Stimulus Wanes," *Bloomberg.com* (Nov. 22, 2011), http://www.bloomberg.com/news/2011-11-21/u-s-clean-energy-needs-priv ate-funding-as-stimulus-wanes.html.
757. Meyers et al., *supra* n. 746, at 8.
758. B. Womack, "Google Reaps Tax Breaks in $1.4billion Clean Energy Bet," *Bloomberg.com* (Mar. 10, 2014), http://www.bloomberg.com/news/2014-03-10/google-reaps-tax-breaks-in-1-4-billi on-clean-energy-bet.html.
759. Meyers et al., *supra* n. 746, at 9.

RE and clean tech markets expect the climate finance market to be robust even in the face of regulatory and liquidity fluctuations.[760]

Secondly, tax equity structures are complex. The main financing structures used to employ tax equity for RE projects are the partnership flip, the sale-leaseback, and the inverted-structure or lease pass-through.[761] These structures are complicated and tend to be administered by legal and financial players in the market that have developed high-level expertise in these types of transactions. Accordingly, new entrants need to conquer a steep learning curve in order to competently administer or participate in these structures.[762] Nonetheless, the successful entry by players such as Google can provide some confidence for other Fortune 500 companies.

A third and related barrier is the monetary thresholds for tax equity investing due to high transaction costs. These thresholds tend to preclude smaller projects and investors from the space.[763] Yet there is the possibility of expanding the investor base by reducing transaction costs through standardized transaction documents. Using the loan syndication market as a precedent, Meyers et al. suggest that industry participants such as energy finance lawyers and financial institutions could draft and peer-review standardized transaction documents for subsequent review by RE trade associations prior to release into the market.[764] Arguably this process could occur not only within domestic jurisdictions but also at an international level, perhaps under the auspices of the UNEPFI.

§7.04 CLIMATE BONDS AND GREEN BONDS

Climate bonds and green bonds are conventional debt-financing instruments that operate like a normal bond. They are a fixed-income long-term financial instrument issued to raise finance for a specific purpose. In the case of climate bonds their purpose is to facilitate GHG emissions reduction and/or increased resilience to the inevitable impacts of climate change. A climate bond is certified as such by the Climate Bonds Initiative,[765] a specialist NGO launched in 2009 by the international Network for Sustainable Financial Markets and supported by the Carbon Disclosure Project. For a green bond, the purpose is to generate sustainable outcomes more generally. For example, the Ile-de-France regional council issued an environmentally and socially

---

760. *Ibid.*
761. See e.g. T. Lowder, "How Could Securitization Debt Fit with Tax Equity in the Solar Financial Landscape? Pt. I," *National Renewable Energy Laboratory* (Apr. 10, 2013), https://financere.n rel.gov/finance/content/how-could-securitization-debt-fit-tax-equity-solar-financial-landscap e-pt-i; Hecimovich et al., *supra* n. 748; Mintz Levin & GTM Research, *Greenpaper: Renewable Energy Project in the US: 2010-2013 Overview and Future Outlook* (Mintz, Levin, Cohn, Ferris, Glovsky and Popeo, P.C, 2012), 10.
762. Meyers et al., *supra* n. 746, at 9. See also Lowder, *supra* n. 761.
763. S. Corneli, *Clean Energy and Tax Reform: How Tax Policy can Help Renewable Energy Contribute to Economic Growth, Energy Security and a Balanced Budget* (US Partnership for Renewable Energy Finance, 2012), 13.
764. Meyers et al., *supra* n. 746, at 7.
765. Climate Bonds Initiative, "Climate Bond Certified," *climatebonds.net*, http://www.climatebo nds.net/standards.

responsible bond in 2012 for clean energy, retrofitting schools and low-energy social housing; it was over-subscribed by pension funds and insurers in just 30 minutes.[766] The proceeds of climate and green bonds are often set aside in a separate account for exclusive investment in eligible climate-related or sustainability projects.

Akin to traditional bonds, climate and green bonds are risk-weighted and credit rated based on the creditworthiness of the issuer. Bond issuers can be federal or state governments, municipal councils, transnational banks or other corporations. Federal government bonds are considered akin to risk free; state and municipal government bonds are not considered risk free; and corporate bonds carry the highest risk of project default unless the corporation is state-backed or state-owned. The issuer guarantees to repay the bond plus interest over a specific period of time and the interest may also be tax-free. As such, climate/green bonds potentially provide investors (or buyers) with two kinds of income: (1) current income from periodic interest payments paid over the life of the bond; and (2) capital gains via repayment of the bond (being a fixed amount of principal) at maturity.[767]

The appeal of climate and green bonds is three-fold. First, they are attractive to conservative investors because they are a lower-risk option. The risk of delivery is borne by the bond issuer, which contrasts with equity investment where the investor bears a large part of the risk. Secondly, bond returns are more stable than share returns and therefore provide excellent diversification in a portfolio. That is, adding bonds to a portfolio can reduce portfolio risk without reducing required return, which satisfies the CAPM and business case imperatives of private finance actors. Thirdly, these bonds mobilize institutional capital to achieve economies of scale because they are best suited for large-scale investments and projects. US$300million is a minimum bond investment due to high transaction costs. This matches the investment and debt financing appetite of institutional investors, especially insurers and pension funds, as they often require a sizeable threshold of at least US$250million. The culmination of these three factors makes climate and green bonds particularly well-suited to mobilizing capital for the long-term infrastructure required to build a low-carbon and climate-resilient global economy.[768]

Indeed, utilizing climate and green bonds to facilitate the timely transition to a global low-carbon economy is clever policy. Returning to statistics at the start of this book, around US$15.2trillion of additional finance for both developed and developing nations needs to be mobilized for global GHG emissions mitigation, which is an exponential increase on the current annual investment of US$160billion.[769] These are significant amounts. Yet in 2010 alone, more than US$6trillion in new mainstream

---

766. Sustainable Business, "French Green Bond Oversubscribed in Half Hour," *SustainableBusiness*.*com* (May 23, 2012), http://www.sustainablebusiness.com/index.cfm/go/news.display/id/2 3726.
767. L. Gitman, M. Joehnk, S. Smart, R. Juchau, D. Ross and S. Wright, *Fundamentals of Investing* (Pearson Australia, Frenchs Forest, 2011), 321.
768. Climate Bonds Initiative, *Bonds and Climate Change: The State of the Market in 2013* (Climate Bonds Initiative and HSBC Climate Change Centre of Excellence, June 2013), 2, http://www. climatebonds.net/files/uploads/2013/08/Bonds_Climate_Change_2013_A4.pdf.
769. Green Climate Fund, *Business Model Framework: Private Sector Facility, GCF/B.04/07*, Jun. 12, 2013.

bonds were issued and funds under management reached US$105trillion.[770] More specifically, from 2010 to 2012 the climate-themed bond market grew from approximately US$48billion to US$74billion outstanding, 79 percent of which was issued by state-owned companies or implicitly backed by governments through guarantees.[771] These figures are encouraging given the nascent nature of the market with the first security issued by the European Investment Bank in May 2007.[772]

Nonetheless, these figures also demonstrate that climate and green bonds require government and policy support:

(1) At the front end, governments can reduce the risks of bond financing by: issuing or guaranteeing bonds directly; signing purchase contracts that allow companies to issue corporate bonds; providing tax exemptions for periodic interest payments; and/or permitting state-owned companies to issue the bonds.
– China is leading the way in this respect. China has the largest stock of climate-themed bonds and has issued policy to encourage corporate bond issuance for energy saving and environmental protection.[773] By July 2014, the amount of issued climate bonds in China reached US$127billion with 92 per cent of bonds issued by China Railway Corp (the previous Ministry of Railways) and the remainder issued by renewable energy manufacturers and nuclear operators.[774] China Railway Corp won approval from the National Development and Reform Commission to issue as much as 150billion yuan (US$24.1billion) of bond debt.[775] The plan for the vast majority of these bonds is to expedite construction of green infrastructure, including railways. Globally, low carbon transport (notably rail) accounts for 75 percent of climate bonds, followed by clean energy and then climate finance.[776]

(2) Moreover, a bond must be repaid out of savings (such as savings made from energy efficiency measures) or revenues (which are usually generated as a result of the capital investments). So policy mechanisms are required to drive income streams.

---

770. Climate Bonds Institute, "Climate Bonds Move Forward," *www.sustainablebusiness.com* (Mar. 16, 2012).
771. Climate Bonds Initiative, *Bonds and Climate Change: The State of the Market in 2013* (Climate Bonds Initiative and HSBC Climate Change Centre of Excellence, June 2013), 3, http://www.climatebonds.net/files/uploads/2013/08/Bonds_Climate_Change_2013_A4.pdf.
772. S. Bakewell, "Green Bond Bankers in Japan, Sweden Beat US to $7billion," *Bloomberg.com* (Jan. 25, 2012).
773. Robins, *supra* n. 153, at 15.
774. "Report: US$346 bln market for green bonds," *China Development Gateway* (Jul. 15, 2014), http://en.chinagate.cn/2014-07/15/content_32953628.htm.
775. "China Railway Said to get approval for 150billion Yuan of Bonds," *Bloomberg.com* (Mar. 21, 2014), http://www.bloomberg.com/news/2014-03-21/china-railway-said-to-get-approval-for-150-billion-yuan-of-bonds.html.
776. Climate Bonds Initiative, *Bonds and Climate Change: The State of the Market in 2013* (Climate Bonds Initiative and HSBC Climate Change Centre of Excellence, June 2013), 3, http://www.climatebonds.net/files/uploads/2013/08/Bonds_Climate_Change_2013_A4.pdf.

- This can occur through direct legislation. For example, in the United States the *American Recovery and Reinvestment Act of 2009* increased the amount of funds available to issue new clean renewable energy bonds from US$800million to US$2.4billion for electricity generated from renewable sources.[777] Similarly, the Act increased the amount of funds available to issue qualified energy conservation bonds from US$800million to US$3.2billion to finance governmental programs to reduce GHG emissions and for other conservation purposes.[778]

- Another policy approach is indirect incentivizing regulation such as tax credits and FITs as detailed in the previous chapter. For example, the Climate Bond Initiative's experience with RE bonds in Europe has shown that a wind farm can repay its bond using revenues from its energy production guaranteed under a FIT.[779]

(3) Finally, government policy is required to enable a bond issuer to make returns when savings are not coming directly to that issuer. For example, when a municipal council issues an energy efficiency bond, it is the customers who make the cost-savings not the council; so the council must increase property rates to raise revenue to repay the bond. Accordingly, legal frameworks that permit rate rises by councils for these purposes need to be in place prior to the bond issuance.

In an encouraging development, the Green Bonds Principles were created in January 2014 as a set of voluntary guidelines for the issuance of green bonds by corporate and public sector entities in order to deploy capital to "projects and activities that promote climate or other environmental sustainability purposes."[780] Led by Bank of America Merrill Lynch, Citi, Crédit Agricole, and JP Morgan, the Principles were signed by 13 commercial and investment banks and are designed to:

> provide issuers guidance on the key components involved in launching a credible Green Bond; they aid investors by ensuring availability of information necessary to evaluate the environmental impact of their Green Bond investments; and they assist underwriters by moving the market towards standard disclosures which will facilitate transactions.[781]

The Principles list categories of eligible projects for the use of bond proceeds, including: renewable energy, energy efficiency, sustainable land use, clean transportation, clean water, and biodiversity conservation. With a view to strengthening the environmental integrity of investments, the Appendix to the Principles also provides links to benchmarks for use of proceeds. These benchmarks have already been developed in

---

777. *American Recovery and Reinvestment Act of 2009, supra* n. 34, at §1111.
778. *Ibid.*, at §1112.
779. See http://www.climatebonds.net/.
780. *Green Bond Principles, 2014: Voluntary Process Guidelines for Issuing Green Bonds* (Jan. 13, 2014), 2, http://www.ceres.org/resources/reports/green-bond-principles-2014-voluntary-process-guidelines-for-issuing-green-bonds/view.
781. *Ibid.*, 1.

conjunction with green bond issuers, NGOs and other stakeholders, and adopted by multilateral entities such as the IFC.

Nonetheless, BankTrack has critiqued the Principles as lacking "real commitments" to science-based definitions of a "green" purpose or to independent verification of sustainability information and use of proceeds by issuers.[782] In order to increase transparency, the Principles urge issuers to report annually or semi-annually via newsletters, websites and/or financial reports on what investments have been made from green bond proceeds.[783]

Despite initial criticisms of the Principles, they are an important first step in facilitating *en masse* bond deployment to provide environmental and climate-related benefits. A set of robust Principles will likely embed a coherent approach to climate and green bonds, which will mitigate perceptions of risk and complexity by issuers and investors. The upshot could be significant expansion of fixed income capital allocation to green and clean tech projects by both risk-averse and risk-tolerant private finance actors.

As issuer and buyer know-how increases amongst corporate and public sector entities transnationally, this growing "green debt" market is enabling institutional investors to make fiduciary duty-aligned investments to help mitigate climate change (such as renewable energy installations, low-carbon tech, reforestation) and increase climate adaptation (such as watershed management, flood protection, wild weather resilient infrastructure). In other words, climate and green bonds will become increasingly important financial mechanisms through which to harness the deep and liquid fixed income capital pool to accelerate a low-carbon global economy.

§7.05 EVERGREEN BUY-OUT STRATEGIES

Evergreen buy-out strategies are nascent and largely theoretical to date. It remains to be seen whether they get market traction. Nonetheless, they represent an innovative alternative to conventional equity investing for a low-carbon economy and are therefore worthwhile exploring in this chapter.

Evergreen buy-out strategies are predicated on the evergreen direct investing (EDI) model. This model is designed for pension funds and other large institutional investors to make direct investments in companies and thus bypass capital markets. In this way, returns are not dependent on the ultimate sale of the investee enterprise as determined by public market valuations ("exit-on-sale driven"). Instead value is derived through long-term cash flows in order to meet the investor requirements for payouts to beneficiaries.[784] The theory is that, through a long-term investment approach that emphasizes direct engagement between investors and enterprises, EDI investors can have actual influence on the ESG goals, performance and corporate

---

782. BankTrack, "BankTrack Calls for Strengthening of Green Bond Principles," *www.banktrack.org* (Apr. 14, 2014).
783. Green Bond Principles, *supra* n. 780, at 5.
784. Capital Institute, "Evergreen Direct Investing: Co-Creating the Regenerative Economy," *capitalinstitute.org* (September 2013), 7-8, http://capitalinstitute.org/edi_fg.pdf.

governance of investee enterprises and thus "help engineer a robust, dynamic, adaptive, and sustainable enterprise."[785]

Building on this model, the Evergreen buy-out strategy can be conceptualized as a form of creative destruction from the inside-out. In a nutshell the strategy involves buying out a fossil fuel or other polluting company, recovering the invested capital over a defined period, then winding up the company or changing its business internally. It comprises three main components:[786]

(1) The transaction involves negotiating a merger of the corporation into a new private company shell at an agreed price per share that is accepted by the Board of Directors and approved by a majority of shareholders pursuant to relevant company law. Following the merger the corporation would be converted into a Limited Liability Company co-owned by corporation management (Enterprise management) and the institutional investors who sponsored the transaction (Investors). Investors and Enterprise management negotiate an operating agreement that splits their rights and responsibilities, contractually addressing financial returns as well as non-financial outcomes such as improved ESG governance and/or climate-related outputs. The agreed strategy would then be executed by Enterprise management.

(2) Invested funds are recovered incrementally and over time as cash flows are generated within the enterprise. The cash flow is shared or split between the Investor and Enterprise management. Splits are front-loaded in favor of the Investor to increase certainty of funds recovery and the realization of fiduciary returns. The remaining portion is controlled by Enterprise management for employee compensation, profit sharing or reinvestment in the business. As pre-agreed investment return milestones are passed, the share of cash flow to the Enterprise increases in order to motivate management for achieving milestones on or ahead of schedule. Once the final investment return milestone is passed, the sharing of cash flow can continue as "tail profits" indefinitely or until the Enterprise is wound up. If the goal is to break the cycle of reinvestment in fossil fuels then agreed covenants may restrict investment discretion of Enterprise management.

(3) Recovered funds and tail profits are returned to the Investor's portfolio to satisfy their fiduciary obligations and be redeployed into other investments as determined by the Investor. In this way, reinvestment can be for climate-friendly ends, such as the construction of renewable energy infrastructure, or acquisition of forestlands as carbon sinks. Conversely, the Investor can internally change the corporation's business model to become climate-friendly.

---

785. *Ibid.*, 23.
786. Evergreen Investment Architects, *Case Study: James River Coal* (Evergreen Investment Architects, undated); and Capital Institute, *supra* n. 784, at 9.

Evergreen buy-outs of fossil fuel enterprises may be apt for two main reasons. First, it is almost impossible for fossil fuel businesses to be or become truly green. Several of the largest fossil fuel companies invest in renewable energy/clean tech and have established strong corporate governance frameworks, which is laudable. Yet the fact remains that their *raison d'être* is to exploit and disseminate fossil fuel resources. In the context of investigating a market for virtue back in 2005, David Vogel used BP to exemplify how climate-related gains are peripheral when compared to core business activities.[787] He noted that the significance of BP's contribution to the production of GHG emissions is not through its own pollution; rather it is "primarily from the consumption of the fossil fuels it markets and whose sales continue to form the basis of its business."[788] Tweaking at the edges of that core business does not produce significant climate-related benefit.

Secondly, trends demonstrate that publicly-traded oil, gas and coal companies are increasingly not meeting market demand for growth in share price. This has led industry experts and media commentators to label fossil fuel investments as "stranded assets" and even the next "subprime."[789] For example, Morgan Stanley reported that the biggest European oil groups had spent US$161billion on operations and dividends in 2013 but generated only US$121billion in cash flow; and American investment in shale gas ventures is barely breaking even due to a supply glut.[790] The reason for potential mass stranding is threefold. First, low-hanging fruit has already been picked, after which access to resources becomes more difficult and therefore more expensive. Secondly, climate-related changes in government policy have created a less-certain future for fossil fuel investments. The International Energy Agency reported that two-thirds of the earth's known reserves of oil, gas, and coal must remain underground and unburned if a binding international climate treaty limits global GHG emissions to the 2°C guardrail, which equates to a loss of US$28trillion of gross revenues across fossil fuel industries over the next 20 years.[791] Even without an international agreement, there are significant ramifications for coal companies due to recent federal government interventions in the United States and China particularly.[792] Thirdly, the increasing competitiveness of established renewable energy such as solar and wind

---

787. Vogel, *supra* n. 338, at 125-127.
788. *Ibid.*, 127.
789. See e.g. HSBC, *supra* n. 153; Robins, *supra* n. 153; Standard & Poor's, *supra* n. 152; M. Lewis, *Stranded Assets, Fossilised Revenues*, KeplerCheuvreux (Apr. 24, 2014), http://www.keplerch euvreux.com/pdf/research/EG_EG_253208.pdf; R. Sullivan, *Climate Change: Implications for Investors and Financial Institutions: Key Findings from the Intergovernmental Panel on Climate Change Fifth Assessment Report* (University of Cambridge, UNFCCC, UNEPFI, 2014), 12; A. Evans-Pritchard, "Fossil Industry is the Subprime Danger of This Cycle," *The Telegraph*, Jul. 9, 2014, http://www.telegraph.co.uk/finance/comment/ambroseevans_pritchard/10957292/Fo ssil-industry-is-the-subprime-danger-of-this-cycle.html.
790. Evans-Pritchard, *supra* n. 789.
791. International Energy Agency (IEA), *World Energy Outlook 2012: Executive Summary* (OECD/ IEA, 2012), 3, http://www.iea.org/publications/freepublications/publication/english.pdf.
792. See Chapter 1 *supra* and Chapter 8 *infra*.

power, as well as expeditious developments in energy/battery storage, will eventually erode the viability and necessity of fossil fuel development.[793]

For any of these reasons, fossil fuel executives and buyers/utilities may find an Evergreen buy-out opportunity attractive.

Thus, in theory, Evergreen buy-out strategies are a self-fulfilling and virtuous cycle: the more buy-outs of small and medium-sized fossil fuel enterprises that occur, the more expeditiously a widespread stranded asset scenario is realized at which point even the largest fossil fuel entities become potential targets. Moreover, as the new green corporate entities (remodeled from the inside) become established and generate positive cash flows, there may be a decline in the demand and supply for coal and other fossil fuels; and an increase in the generation and deployment of low-carbon energy solutions. For these reasons, proponents argue that Evergreen can facilitate the long-term transition to a low-carbon global economy.

Nonetheless, there are some concerns to address before this theory gets market traction. First, fossil fuel buy-outs in the context of increasingly stranded assets may not actually provide the return required by fiduciary entities. Proponents argue that an Evergreen approach is most suitable where market capitalization and/or share prices in a company have fallen significantly.[794] Evergreen buy-outs are designed to work without requiring growth in share price. If the market capital for a company is too high relative to its cash flows then the investor's fiduciary purpose cannot be achieved through an Evergreen buy-out strategy.[795] It is preferable for the market capital or share price to drop so the buy-out can occur at a lower cost to the investors and invested capital can be recovered more quickly. The counter-argument is that if the Investor cannot recover invested funds to meet their fiduciary obligations then they will not engage in this strategy. Indeed, there may be sound logic in simply leaving the fossil fuel sector to its fate without an Evergreen intervention.

Secondly, there is little discussion about regulatory concerns and/or support for *en masse* Evergreen buyout strategies. Yet Investors will need to ensure that buy-outs occur within the parameters of anti-trust law and that tax policy implications are fully considered, as discussed below.

Anti-trust law (also known as competition law) regulates the conduct of business to ensure free and fair competition for the ultimate benefit of consumer welfare. Anti-trust and competition legislation is well-established in the United States,[796] the

---

793. Greenpeace, "Fear of change could precipitate demise of Europe's energy giants," *bank-track.org* (Feb. 27, 2014), http://www.banktrack.org/show/news/fear_of_change_could_pre cipitate_demise_of_europe_s_energy_giants#tab_news_main.
794. Capital Institute, *supra* n. 784, at 7-8.
795. Evergreen Investment Architects, *supra* n. 786, at 16.
796. See: the *Sherman Act of 1890* (26 Stat. 20915 USC.§§ 1–7); the *Clayton Antitrust Act of 1914* (Pub.L. 63–212, 38 Stat.730, enacted Oct. 15, 1914, codified at 15 USC. §§ 12–27, 29 USC. §§ 52–53); and the *Federal Trade Commission Act of 1914* (15 USC §§ 41-58).

UK/Europe[797] and Australia[798] and comprises three broad pillars: to prohibit anti-competitive practices that lead to creation of monopolies and monopolistic abuse of power; to restrict corporate mergers and acquisitions that would have the effect of substantially lessening domestic competition; and to prohibit collusion amongst competitors and the formation of cartels that restrain trade.

The latter two pillars of anti-trust law – substantial lessening of competition and collusive/cartel behavior – raise concerns for Evergreen buy-out strategies. The front-end acquisition stage may be seen as a substantial lessening of competition in the coal (or other fossil fuel) market. As pension funds acquire more and more fossil fuel companies that are responsible for a significant proportion of coal production within a jurisdiction then they are, by definition, decreasing competition in the coal market. To this end, and in the current "stranded assets" environment, Investors may have a "failing firm" defense arguing that, absent the acquisition, the fossil fuel firm would have failed and exited the market, which would have represented a higher risk of substantially lessening competition than the acquisition. In such a case, the competition regulator may be disposed to view the buy-out as relatively benign.

Nonetheless, the back-end transformation stage may be seen as a substantial lessening of competition in the RE market if the majority of "new green" companies are owned by the same pension funds. To mitigate this concern, it would be prudent for multiple pension funds to engage in evergreen buy-out strategies, thus spreading ownership throughout the market.

However, in so doing, pension funds would also need to ensure that they are not breaching the third pillar of anti-trust law by colluding or engaging in cartel-like behavior. A potential vehicle to immunize against this risk is a joint venture between multiple Investors. Nonetheless, Investors would need to seek expert legal advice on the formation of that joint venture in order to satisfy a competition regulator that the joint venture was not created with the purpose of engaging in cartel conduct.

Finally, a competition regulator may explicitly approve Evergreen buy-out strategies, or at least declare regulatory forbearance with regard to them, by issuing new sector-specific M&A guidelines. These guidelines may specify that potential investors offer undertakings or assurances as part of the Evergreen M&A process relating to competitive and independent conduct.

Regarding tax implications, it is important to note that any asset acquisition will attract tax liabilities and benefits. Depending on the jurisdiction within which the acquisition occurs, tax implications for the buyer/investor may include:[799] acquiring the tax liabilities of the acquired business (particularly if the transfer is deemed to constitute a transfer of an on-going concern); paying value-added tax (VAT) on asset sales (usually if the enterprise is not an on-going concern); tax deductions on stock

---

797. *Competition Act 1998* (c.41) and the *Enterprise Act 2002* (c.40). The European Union has competence to deal with anti-competitive conduct and, pursuant to section 60 of the *Competition Act 1998*, UK rules are to be applied consistent with European jurisprudence.
798. *Competition and Consumer Act 2010* (Cth), which replaced the *Trade Practices Act 1974* (Cth).
799. Taxand, *Global Guide to M&A Tax* (Taxand, 2013), 22, http://www.taxand.com/sites/defaul t/files/taxand/documents/MA_Tax_Guide_2013.pdf.

assets that qualify for capital allowances and goodwill; and/or a carry-over of the target enterprise's existing tax attributes, such as tax credits or net operating losses (NOLs).

A key question for the Evergreen buyer of a sinking fossil fuel company is whether the target company's NOLs are available after the acquisition is made. If so, then the buyer/Investor can write-down the losses of the acquired enterprise and cache them for a time when the enterprise becomes profitable. The availability of NOLs post-acquisition would increase investor attraction of an Evergreen buy-out and thus encourage take up and multiplication. Accordingly, Investors would need to seek expert advice regarding the tax implications of each Evergreen buy-out prior to embarking upon the acquisition process.

§7.06   CONCLUDING REMARKS

Broadening the regulatory sights beyond conservative or low risk-tolerant financial actors (such as banks and pension funds) to include capital providers with greater risk tolerances (such as venture capital funds and private equity firms) enables consideration of entrepreneurial green investment modalities to augment renewable energy and clean tech. Innovative modalities of equity investment (such as tax equity partnership structures and Evergreen buy-out strategies) and debt financing (through climate and green bonds) engage a range of private finance actors both conservative and entrepreneurial. In so doing, these modalities can facilitate the financial and regulatory learning-by-doing that is required for timely transition to a low-carbon and climate resilient economy.

One jurisdiction that understands the financial/regulatory green nexus is China. Accordingly, the next and final chapter is devoted to China's policy framework for green growth and the pivotal role of its corporate champions in the global transition to a low-carbon future.

CHAPTER 8

Looking to the Future: Chinese-Global Green Growth

§8.01   OVERVIEW

In order to transition to a truly *global* low-carbon economy, we need to broaden the focus once more; this time going beyond climate-related activity in market economies such as the United States to the investing activities from and within emerging economies, particularly China. Indeed, this book could not be complete without mentioning the role of Chinese enterprises in renewable energy and clean tech investments and a low-carbon transition on a global scale.

Why focus on China in particular? China is transitioning to a consumer economy and its energy demands continue to grow. It became "the biggest energy consumer on the planet" when it consumed over one fifth of the world's energy in 2010, effectively doubling its energy consumption in just ten years.[800] In order to satisfy its domestic energy demands, China is burning significant amounts of fossil fuels especially coal. For example, in 2013 China accounted for 47 per cent of global coal consumption and became the world's largest net importer of crude oil.[801] Concomitantly however, China is also making significant investments in clean energy solutions. It is now the world's largest producer of wind power, installing nearly 30 per cent of new global capacity in

---

800. The Economist, "Fossil Fuels Usage Hits Global Record," *The Weekend Australian Financial Review* (Jun. 11-12, 2011) 11, 12.
801. "China consumes nearly as much coal as the rest of the world combined," *US Energy Information Administration* (Jan. 29, 2013), http://www.eia.gov/todayinenergy/detail.cfm?id = 9751; "China is not the world's largest net importer of petroleum and other liquid fuels," *U.S Energy Information Administration* (Mar. 24, 2014), http://www.eia.gov/todayinenergy/detail.cfm?id = 15531.

2012 alone; and three of its state-owned power generators are the world's largest owners of solar assets.[802]

The dominant Chinese investors in renewable sectors are state-owned enterprises (particularly in wind) and state-supported private enterprises (particularly in solar). Moreover, the Chinese banking sector is dominated by state-owned banks. Accordingly, we need to better understand the patterns and motivations of Chinese green "state capital" both domestically and abroad in order to look to the future of a global energy transition.

This final chapter begins with some background on China's "going global" policy which has spurred overseas direct investment (ODI) particularly by corporate champions, and hones in on the increasing role of sustainability and green strategy to China's Five Years Plans. The chapter then moves to a discussion of China's "green" policy mechanisms coupled with the rise and role of state-owned and state-supported enterprises in facilitating China's green growth. By using publicly available data, the chapter pinpoints where and how investment is occurring in terms of sectors, geography and key financing mechanisms. In so doing, this chapter also offers insights as to why Chinese enterprises are motivated to make green investments on foreign soil and how this might play out in the future.

§8.02   CHINA GOES GLOBAL

[A]   Policy Background

In 2001, China initiated its global strategy under the label *zouchuqu* which literally means "go out" but can also be interpreted as "go global."[803] This "going global" strategy was unveiled in China's Tenth Five Year plan in the same year that China acceded to the World Trade Organization (WTO). Three major initiatives were instigated as part of this strategy to motivate China's investment reach and impact on a global scale. First, China started issuing shares in its major state-owned enterprises (SOEs) on foreign exchanges. For example, Sinopec and CNOOC, two central government owned energy SOEs, were listed on the New York, Hong Kong and London exchanges in 2001. Listings created huge capital gains, which enabled SOEs to become "powerful in terms of their ability to generate capital and to augment their value in terms of market capitalization."[804] Secondly, in 2003 the Ministry of Commerce (MOFCOM) was established as responsible for domestic and international trade and

---

802. International Renewable Energy Agency (IRENA), "30 Years of Policies for Wind Energy: Lessons from 12 Wind Energy Markets" (IRENA, 2012), https://www.irena.org/DocumentDownloads/Publications/GWEC_China.pdf; X. Tan, Y. Zhao, C. Polycarp and K. Bai, "China's Overseas investments in the Wind and Solar Industries: Trends and Drivers," *World Resources Institute* (April 2013), http://www.wri.org/sites/default/files/pdf/chinas_overseas_investments_in_wind_and_solar_trends_and_drivers.pdf.
803. See L.K. Cheng and Z. Ma, "China's Outward Foreign Direct Investment" in Robert C. Feenstra and Shang-Jin Wei (eds.), *China's Growing Role in World Trade* (University of Chicago Press: Chicago, 2010) 547, 550.
804. K.E. Brødsgaard, "Politics and business group formation in China: the Party in control?" (2012) 211 *The China Quarterly* 624, 630.

international economic cooperation, and the central Chinese State-owned Assets Supervision and Administration Commission (SASAC) was created to exercise owner-ship over China's central SOEs. Thirdly, President Jiang Zemin announced that the "going out" policy included: increased ODI; undertaking construction and engineering projects abroad; and exporting labor services.[805] Cheng and Ma note that it is difficult to catalogue exact policy measures due to a lack of publicly available information; however initial measures certainly included increased financial support for national corporate champions.[806]

Perhaps most importantly, the going global strategy motivated Chinese SOEs to actively seek out and *acquire* foreign assets and equity interests as opposed to merely trading in global commodities and raw materials.[807] Indeed, Liao and Zhang assert that it was after the GFC in particular that China's ODI experienced "explosive growth"[808] due in part to increased acquisition of foreign entities that had become bankrupt or required significant capital injections which could not be provided by their home-state. The aversion to risky lending in Europe and North America post-GFC led to a higher cost of capital for investors from these regions. This shift allowed Chinese investors with access to relatively low-cost capital from their home financial institutions to enter those markets. For example, in 2012, the State Grid Corporation of China acquired electricity transmission assets in Brazil from the debt-laden Spanish construction firm ACS, and acquired a 25 percent stake in Portugal's national power company REN, which was forced into privatization under the EU bailout plan.[809]

The phenomenon of "state capitalism" embodies a form of hybrid capitalism pursuant to which a government actively promotes economic growth by picking and/or backing national champions while also using capitalist tools to this end, such as stock market listing, external financing and subjecting those champions to global competition.[810] State capitalism contrasts markedly with the low interventionist or "light touch" government regulation of corporate actors in liberal (market) economies, which has been the dominant focus of this book. State capitalism is certainly not a new phenomenon; however, since the GFC it has taken the form of unprecedented patterns of investment within the borders of developed nations by corporate champions from emerging economies, particularly China.[811]

---

805. Cheng and Ma, *supra* n. 803, at 550.
806. *Ibid.*
807. N.C. Howson, "China's Acquisitions Abroad—Global Ambitions, Domestic Effects" (2006) *Law Quadrangle Notes* 73; E. Hong and L. Sun, *Go overseas via direct investment: internationalization strategy of Chinese corporations in a comparative prism*, Discussion Paper 40 (London: University of London, Department of Financial and Management Studies, School of Oriental and African Studies, 2004). See generally G. Gilligan and M. Bowman, "State Capital: Global and Australian Perspectives," (2014) 37 *Seattle University Law Review* 597.
808. S. Liao and Y. Zhang, "A New Context for Managing Overseas Direct Investment by Chinese State-Owned Enterprises" (2014) 7(1) *China Economic Journal* 126, 128.
809. P. Domingues and P. Ho, "China buys up Spain's assets," *Wall Street Journal* (May 29, 2012), http://online.wsj.com/news/articles/SB10001424052702303807404577433934077603506.
810. A. Musacchio and S.G. Lazzarini, *Leviathan in Business: Varieties of State Capitalism and Their Implications for Economic Performance* (Harvard Business School Working Paper 12-108, Jun. 4, 2012).
811. See Gilligan and Bowman, *supra* n. 807.

[B]   The Rise and Role of State-Owned Enterprises to Going Global

Despite the fact that China is a relatively new outward investor it now has a market presence in virtually every country and every market in the world.[812] Key to this expansion has been the rise and role of its SOEs. For example, in 2010 ODI flows by Chinese SOEs accounted for nearly 62 percent of China's total ODI flows.[813] As such, some understanding of SOEs is required at this point.

Oi describes the evolution of the Chinese "corporate state" from the initial local town and village enterprises (TVE) to show that Chinese SOEs were traditionally an organizational form not a legal one.[814] Specifically, Oi describes the county as similar to "a large multi-level corporation…at the top of a corporate hierarchy as the corporate headquarters, the township as the regional headquarters, and the villages as companies within the larger corporation."[815] Importantly, TVEs were forms of government not private ownership. Traditionally, therefore, an SOE did not have separate legal personality nor issue stock or equity ("ownership") in itself; instead it was administratively controlled by the state, which had the right to appoint management and appropriate revenues or profits.

This model has fed into current day. Indeed, China's state-led growth has been termed *local state corporatism* whereby local governments in China are economic actors and not merely administrative ones. Due to the historical-cultural beginnings of TVEs, Chinese local governments are very entrepreneurial: they are heavily involved with business and they want companies to succeed,[816] whether SOE or private. Moreover, Li et al. describe how, since the 1978 economic reform, local governments give priority to economic growth above all else:

> This is because central government evaluates the performance of local governments according to economic output; and also, as the decentralization of power is not fully guaranteed or stipulated by law, local protectionism emerges and local authorities tend to maximize their short-term benefits.[817]

In this way, local governments have developed their own economic identity and have become economically self-reliant.[818] Some commentators have even asserted that "the

---

812. S. Husted and S. Nishioka, "China's fare share? The growth of Chinese exports in world trade" (2013) 149 *Review of World Economics* 565, 567.
813. Liao and Zhang, *supra* n. 808, at 127.
814. J.C. Oi, "The role of the local state in China's transitional economy" (1995) 144 *The China Quarterly* 1132-1149, 1138-1139. See also Howson, *supra* n. 807.
815. Oi, *supra* n. 814, at 1138.
816. T. Ruskola, *Legal Orientalism: China, the US and Modern Law* (Harvard University Press: Cambridge MA, 2013).
817. W. Li, M. Beresford and G. Song, "Market Failure or Governmental Failure? A Study of China's Water Abstraction Policies" (2011) 208 *The China Quarterly* 951, 962. See also: D. Yang, "Patterns of China's Regional Deelopment Strategy" (1990) 122 *The China Quarterly* 250; G. Long and M.k.Ng, "The Political Economy of Intra-Provincial Disparities in Post-Reform China: A Case Study of Jiangsu Province" (2001) 32 *Geoforum* 215.
818. Li et al., *supra* n. 817, at 962-3.

entrepreneurial interests of local governments have compromised their role as agents of the central state."[819]

As such, a radically different government/business model exists in China compared with that of Anglo-American nations in which sub-national governments are inextricably subsumed within a federalist structure; and a sharp divide tends to exist between business and government with ministerial intervention or favoritism for particular companies usually not encouraged. Apprehending these cultural-historical differences is essential to understanding Chinese investment generally.

Although China commenced a corporatization program via the 1994 Company Law and 2006 PRC Company law, this did not result in mass privatization.[820] Indeed, top executives at 53 of the most important central (or federal level) SOEs have ministerial-level status and equal rank to provincial governors.[821] Perhaps unsurprisingly, government policies have almost exclusively preferred central SOEs in order to facilitate their going out,[822] which has included significant ODI finance administered and guaranteed by state-owned banks, the China Development Bank and the Export-Import Bank of China. This institutional support for SOEs has garnered increasing criticism in the context of macroeconomic policy and internal competitiveness.[823]

Nonetheless, there is complexity to the SOE discussion, particularly in the context of green investment, which prompts a definitional question: what exactly is an SOE? Does that term only embody state ownership or does it extend to state control? The central SASAC represents and manages central SOEs (except for state-owned banks, which are managed by the Ministry of Finance); the central SASAC also owns most of those SOEs, and publishes a list accordingly.[824] This is where the simplicity ends. There are five levels of government in China – central, provincial, prefecture/city, county, and township – and all levels bar townships have their own SASACs. So for provincial and city-level companies, the process to determine SOE status is much more complicated than simply checking the central SASAC register. One needs to know where a company is headquartered and then check the relevant local SASAC listings to find it by a process of elimination. If one does not find it, then the natural assumption is that it is a privately owned enterprise (POE). However, even when a company is not state-owned it can still be state-controlled or state-supported. For example, a company might have a majority shareholder that is a private individual with government status, in which case the government will undoubtedly have some influence even though the company is ostensibly a POE. Moreover, while central SOEs are privileged in their access to finance, local SOEs are less so due to the profit-maximizing nature of local governments. If a local SOE is performing poorly then it can be chilled from financial

---

819. Oi, *supra* n. 814, at 1139.
820. N.C. Howson, "'Quack Corporate Governance' as Traditional Chinese Medicine: The Securities Regulation Cannibalization of China's Corporate Law and a State Regulator's Battle Against State Political Economic Power" (2014) 37(2) *Seattle University Law Review* 667.
821. Brødsgaard, *supra* n. 804, at 634.
822. C. Li, "The end of the CCP's resilient authoritarianism? A tripartite assessment of shifting power in China" (2012) 211 *The China Quarterly* 595, 608.
823. See e.g. *ibid.*; Oi, *supra* n. 814, at 1138-1139; Liao and Zhang, *supra* n. 808.
824. See http://www.sasac.gov.cn/n2963340/n2971121/n4956567/4956583.html.

access. Conversely if a POE is doing well, for example by creating employment and economic success, it is likely to get local state support.

As such, there is a valid distinction between an SOE and a *state-supported enterprise* in China. This is particularly important to grasp in the green financing and investment space. Whilst some active corporate players in clean tech and renewable energy investments are central SOEs, many others are local SOEs or POEs that have access to local financing through local governments and/or association with central SOEs and state banks. In short, the SOE/POE distinction is not always useful in this space. As such, this chapter includes the role of "state-supported enterprises" (SSEs) in green investment alongside SOEs and POEs, which delivers a more accurate and nuanced understanding of Chinese state capitalism in this context.

It is also worth noting the vigorous debate about the extent to which ODI decisions by SOEs are being exercised independently of their sovereign sponsor, that is, for commercial not political reasons. The traditional view of SOE investment activity abroad is that capital allocation is government-directed. However, some commentators have noted that multiple external parties are involved in SOE investment decision-making abroad, including domestic consultants, corporate partners and financiers, such that decisions cannot be made solely by a government entity.[825] With regard to China in particular, Nicholas Calcina Howson muses whether the going global strategy is being directed by the corporations themselves, citing the action of CNOOC in bidding for Unocal in 2005 despite opposition by central government.[826] Similarly, KPMG and the University of Sydney have reported that "Chinese SOEs abroad have shown strong commercial motivations, similar to those of multinational corporations from developed countries".[827] It is outside the purpose of this chapter to add to the rich debate on SOE political/commercial motivations. However, it is likely to remain a vibrant source of political-economic discord for the intermediate future between investing and investee nations, particularly in the Asia-Pacific region.

[C]   Going Global Responsibly?

The reliance on state-backing combined with an ever-increasing global reach has raised discussion in investee nations and also within China about the need for China to go global *responsibly*. This notion of responsibility is bifocal. First, there is the argument based in international economic law that China has a global responsibility to maintain

---

825. See e.g. Gilligan and Bowman, *supra* n. 807; Clayton Utz, *Digging Deep: Chinese Investment In Australian Energy And Resources* (Clayton Utz, 2013); B. Kotschwar, T. H Moran and J. Muir, *Chinese Investment in Latin American Resources: The Good, the Bad, and the Ugly* (Petersen Institute for International Economics, Working Paper 12-3, 2012) at http://www.piie.com/pu blications/interstitial.cfm?ResearchID = 2046.
826. N.C. Howson, "China's Acquisitions Abroad – Global Ambitions, Domestic Effects" (2006) 3 *Law Quadrangle Notes* 73.
827. KPMG and the University of Sydney, *Demystifying Chinese Investment* (KPMG, 2012), 13. See also KPMG and the University of Sydney, *Demystifying Chinese Investment in Australia: March 2014 Update* (KPMG, 2014).

free and competitive trade between nation states.[828] Secondly, there is the need for responsible and sustainable behavior by Chinese SOEs and SSEs in foreign jurisdictions.[829] Indeed, prompted by its own concerns regarding international competitiveness and its self-perception as *fuzeren de daguo* or a "responsible great power," China has demonstrated sensitivity in central policy promulgations around the issue of corporate conduct on foreign soil.[830]

Moreover, domestic pressures within China have prompted it to consider its internal sustainability. China is experiencing intractable environmental and energy concerns within its own borders. The nature of these concerns were starkly revealed in the OECD/FAO *Agricultural Outlook 2013-2023* report, which detailed unprecedented levels of air pollution with attendant health risks; decreasing availability of arable land due to droughts, diminishing soil quality and encroaching urban development; and a rapidly growing middle class with matching appetite for energy, water and food consumption beyond current supplies.[831] Concomitantly, China's reliance on fossil fuels has been domestically and globally significant. In 2013, coal comprised more than 60 percent of the country's primary energy resources[832] and China accounted for 47 percent of global coal consumption, almost as much as the rest of the world combined.[833] Of the 2.9 billion tons of global coal demand growth since 2000, China accounted for 2.3 billion tons (82 percent). Moreover, in September 2013, China's net imports of petroleum and other liquids exceeded those of the United States, which made China the largest global net importer of such products and heavily-reliant on Saudi Arabia.[834]

The upshot is that China has had to reconsider its development from a qualitative and not just quantitative perspective via a strategy of "green growth":

> At home, China needs to achieve the transformation of its development mode – from an unsustainable model with high pollution, high-resources use, high-carbon emissions, export-dependence, and labor intensiveness – to a greener, more balanced, and innovative-driven growth model.[835]

---

828. Liao and Zhang, *supra* n. 808, at 126.
829. C. Maurin and P. Yeophantong, "Going Global Responsibly? China's Strategies Towards "Sustainable" Overseas Investments" (2013) 86(2) *Public Affairs* 281.
830. E.g. SASAC of the PRC, *CSR Guidelines for State-Owned Enterprises* (Jan. 4, 2008, Beijing), available online at http://www.sasac.gov.cn/n2963340/n2964712/4891623.htm; Green Credit Directive *infra*. See generally Maurin and Yeophantong, *supra* n. 829.
831. OECD/ FAO, *OECD/FAO Agricultural Outlook 2013-2023 (OECD Publishing, 2013)*, 65-105, *available at* http://www.keepeek.com/Digital-Asset-Management/oecd/agriculture-and-food /oecd-fao-agricultural-outlook-2013_agr_outlook-2013-en#page4. See also Li et al., *supra* n. 817, at 956.
832. J.C.K. Daly, "Chinese Coal Use to Hit 4.8 billion metric tons annually by 2020," *The Diplomat* (Dec. 10, 2013), http://thediplomat.com/2013/12/china-coal-use/.
833. "China consumes nearly as much coal as the rest of the world combined," *US Energy Information Administration* (Jan. 29, 2013), http://www.eia.gov/todayinenergy/detail.cfm?i d=9751.
834. "China is not the world's largest net importer of petroleum and other liquid fuels," *US Energy Information Administration* (Mar. 24, 2014), http://www.eia.gov/todayinenergy/detail.cfm?i d=15531.
835. Liao and Zhang, *supra* n. 808, at 127.

China's Twelfth Five Year Plan (2011–2015) initiated a model of "sustainable" growth. Specifically, the central government committed to investing US$468billion in green sectors, more than double the previous five-year period, with a focus on clean technologies, sustainable energy and resource conservation.[836] The levers and limits to achieving these goals are explored in detail throughout the remainder of this chapter.

§8.03   CHINA AND GREEN GROWTH: POLICY FRAMEWORK

[A]   Key Targets

The Twelfth Five Year Plan demonstrated China's ambition to make the difficult transition towards a more sustainable model. It emphasized renewable energy and energy efficiency as integral elements of China's wider strategic and economic objectives. In this respect the key targets are:

- Energy Use – a 16 percent reduction in energy intensity (energy consumed per unit of GDP) and an increase of 11.4 percent in the share of energy produced by non-fossil fuel sources.
- Pollution – an 8 percent reduction target for sulfur dioxide and chemical oxygen demand and a 10 percent reduction target for ammonia nitrogen and nitrogen oxides, the latter of which come mainly from China's dominant coal sector. There is also a focus on cutting heavy-metal pollution for industry.
- Carbon – a 40-45 percent reduction in carbon intensity (carbon emitted per unit of GDP) from 2005 levels by 2020, with an interim target of 17 percent reduction by 2015.
- Water Usage – a water intensity (water consumed per unit of value-added industrial output) reduction of 30 percent.
- Forestry – an increase in forest cover to 21.66 percent, and forest growing stock by 600 million cubic meters.

These high-level targets are supplemented by a range of policy measures including market and taxation mechanisms discussed below, and also corporate disclosure,[837] government "green procurement,"[838] pollution liability insurance for companies with

---

836. http://www.ipc-undp.org/pub/IPCWorkingPaper95.pdf.
837. See: The explanatory statement and guide of the Green Securities Policy, *Instructing Opinions on How to Enhance Environmental Protection Monitoring and Management of Listed Companies*, released by the MEP in February 2008 (SEPA, 2008); SSE's environmental disclosure guidelines, *Guidelines on Environmental Information Disclosure of Listed Companies on Shanghai Stock Exchange*, issued in May 2008 (SSE, 2008); and A Guide on Listed Companies' Environmental Information Disclosure (Draft) (MEP, September 2010).
838. The series of laws include: *Clean Production Promotion Law of the People's Republic of China of 2002; Notification on Resource-saving Activities by the State Council Office, Opinions of Implementing Government Procurement of Energy-saving Products* and *Decision on Carrying Out Scientist Development Concept by Strengthening Environment Protection of 2005*; and *Opinions on Implementation of Government Procurement for Environmental Labeling Products of 2006*.

high environmental risks,[839] an Ecological Compensation Mechanism that rewards conservation measures,[840] and even a "cadre system" for promotions, demotions, and salaries to incentivize the attainment of policy goals by government officials.[841] In total, China's Twelfth Five Year Plan has embodied one of the world's most comprehensive environmental policy regimes.

[B]   Policies That Encourage Green Investment Within China

The key anchor in the regulatory regime is the *Renewable Energy Law*, which provides the framework for securing the development of renewable energy in China. It was first introduced in 2005 and subsequently amended in 2009 to include a RET and co-ordination efforts regarding transmission, grid connection and energy storage. It focuses on the development and utilization of renewable energy to improve the country's energy structure, ensure stable energy supply, and to prevent pollution and ecological damage.[842]

[1]   Incentives

In 2009, China set a RET of 15 percent non-fossil fuel energy consumption by 2020.[843] Alternative energy sources are expected to account for more than 20 percent of the country's total electricity generation by 2020.[844]

Moreover, China has implemented FITs for both solar and wind. In 2009 China introduced a wind power generation FIT, which applies for the entire operation period of a wind farm. Similarly, the *Notice of the National Development and Reform Commission on Perfection of Policy Regarding Feed-in Tariff of Power Generated by Solar PV* was introduced in 2011 and sets out the FIT for solar projects. That policy was revised in 2013 by the *Notice on Improving the Development of Solar PV Industry by Utilizing the Price Leverage Effect* which divided the country into three different regions with a differentiated FIT for each region.

The National Development and Reform Commission (NDRC) is China's national authority for accepting and approving all fixed-asset investment projects including

---

839. The Ministry of Environmental Protection (MEP) and the China Insurance Regulatory Commission (CIRC), *Guiding Opinions on Pilot Scheme for Compulsory Environmental Pollution Liability Insurance 2013.*
840. *Ordinance of Ecological Compensation*; I.S. Chang, Y.X. Yang, J. Wu, M.M. Shi, "Ecological Compensation in China – Progress, Problems and Prospects" (August 2013) *Advanced Materials Research* 726.
841. Farber, *supra* n. 24, at 368.
842. X. Qiu and H. Li, "Energy Regulation and Legislation in China," *Environmental Law Institute* (2012), http://www.epa.gov/ogc/china/Qiu.pdf.
843. E. Martinot and L. Junfeng, "Renewable Energy Policy Update for China," *Renewable Energy World* (Jul. 21, 2010), http://www.renewableenergyworld.com/rea/news/article/2010/07/re newable-energy-policy-update-for-china.
844. "Asia-Pacific Renewable Energy Policy Handbook 2014," *Global Data* (April 2014), http://w ww.whatech.com/market-research-reports/press-release/23754-china-s-rapid-renewable-ene rgy-growth-driven-by-ambitious-government-policies.

renewable projects. It has made clear that it can adjust the tariffs going forward, based on factors such as investment cost changes and technology development. Accordingly, regulators retain some flexibility in order to review and prospectively revise these measures.

Despite these policies however, grid connection and transmission remain key concerns.[845] Along with the expected rapid expansion of solar power generation capacity, the transmission capacity of the national grid must grow to meet the anticipated growth of China's solar PV power stations. The same issue continues to vex China's wind industry: in 2012 only about 70 percent of China's total wind power capacity was connected to the grid.

[2]   *Taxation Measures*

China's green tax policy is balanced between incentives and penalties and focuses on green buildings and resource efficiency in energy, water and materials.[846] In 2013 China ranked sixth globally on *The KPMG Green Tax Index,* which outlines the green tax landscape around the world and ranks a government by its activity "in using its tax system to drive sustainable business and achieve green policy objectives."[847]

Specifically, the Chinese Corporate Income Tax Law and its implementing regulations provide tax incentives for a number of green sectors:[848]

- As a form of tax credit, enterprises that purchase and use qualified energy saving equipment can apply for a tax deduction of 10 percent of the amount invested. The deduction can be carried forward for five years.
- A custom duty and VAT exemption exists for certain imported energy efficient equipment.
- Energy services companies can claim a VAT exemption on the transfer of assets to clients at the end of a project, and assets can be transferred as if fully depreciated for corporate income tax purposes.
- Revenues earned by projects in energy and water conservation, environmental protection and/or the Clean Development Mechanism are eligible for a three year exemption and a 50 percent reduction in corporate income taxes.
- Many enterprises engaging in activities in the green sector are considered high and new tech enterprises, which can be granted a reduction in corporate income tax liability.

---

845. See e.g. J. Yang, "Analysis: China looks at subsidy 'adjustment,'" *Wind Power Monthly* (Mar. 19, 2014), http://www.windpowermonthly.com/article/1285959/analysis-china-looks-subsidy-adjustment; International Renewable Energy Agency, *supra* n. 802.
846. KMPG, "The KPMG Green Tax Index 2013: An exploration of Green Tax Incentives and Penalties" (KPMG International, 2013), 5.
847. *Ibid.,* 2.
848. See "Selected Tax Incentives in China's Renewable Energy Sector," *China Briefing* (Jun. 15, 2011), http://www.china-briefing.com/news/2011/06/15/selected-tax-incentives-in-chinas-renewable-energy-sector.html; KMPG, *supra* n.845, at 12.

[3] National Emissions Trading Scheme

The NDRC intends to implement a national ETS or "total emissions control scheme" in the Thirteenth Five Year Plan (2016-2020).

Prior to launching a national trading scheme, the trialing of schemes is taking place in seven geographically and economically diverse localities being the cities of Beijing, Chongquing, Shanghai, Tianjin and Shenzhen and the provinces of Guangdon and Huei. A practical obstacle for moving to a national ETS is that all of the pilots adopt slightly different approaches regarding measuring, reporting and verification; and regulatory inflexibility on electricity prices has resulted in corporate resistance and low trading.[849] The NDRC is attempting to find common ground between them.[850]

[4] Foreign Investment Catalogues

In 2008, the NDRC and MOFCOM jointly issued the *Catalogue of Foreign Investment Advantageous Industries in Central and Western China* (Central and Western Catalogue), which is a revision of the original policy issued in 2000 and supplements the *Foreign Investment Industrial Guidance Catalogue* (Guidance Catalogue). Both Catalogues provide blueprints for foreign investments in China.[851]

There are three categories of foreign investment into China: "encouraged," "restricted" and "prohibited." Categorization impacts upon tax incentives, approval requirements and ease of market entry for foreign enterprises on Chinese soil. Energy saving and environmental protection is one of the most important objectives embodied in the Guidance Catalogue. The "use of sea water, treatment and recycling of industrial wastewater" and "development technology of bio-energy" are listed under "encouraged" categories. Conversely, "development of the coal and its accompanying resource" have been downgraded from encouraged to "permitted."

For foreign investors, the benefits of engaging in encouraged or permitted industries include simpler approval procedures for establishing foreign investment enterprises within China. Moreover, foreign investment enterprises engaging in "encouraged" industries enjoy exemption from customs duty for imported equipment for self-use, which includes corresponding technology, accessories and spare parts.

---

849. M. Adams, "Trials and Tribulations: China Experiments with Carbon Trading," *The Economist Intelligence Unit* (August 2013), http://www.eiu.com/Handlers/WhitepaperHandler.ashx?fi = China_Carbon_Trading.pdf&mode = wp&campaignid = ChinaCarbon.
850. R. Song and H. Lei, "Emissions Trading in China: First Reports from the Field," *World Resources Institute* (Jan. 24, 2014), http://www.wri.org/blog/2014/01/emissions-trading-china-first-rep orts-field.
851. China Law Update, "Foreign Investment Catalogues and the Investment Environment in China," *China Law Update* (2009), http://lawrwx.fyfz.cn.

[C]   Green Banking and Financing Policies

It is clear that China's green policy framework is comprehensive, even visionary. Yet how and from where does capital and investment flow in order to mobilize that policy framework?

The *Renewable Energy Law* does not specify financing modalities but provides a platform for subsequent regulations on funding initiatives, including preferential bank lending for renewable energy development and projects.[852] This paved the way for China's State Environmental Protection Agency (SEPA), the People's Bank of China, and the China Banking Regulatory Commission (CBRC) to jointly issue their *Comments on the Implementation of Environmental Policies and Regulations to Prevent Credit Risks* in 2007, which became known as the *China Green Credit Policy*.

The green credit policy has a dual purpose. First, it directs credit away from highly polluting and high energy-consuming enterprises and projects; secondly, it directs credit towards projects for energy conservation and GHG emissions reductions on preferential terms.[853]

In March 2012, the CBRC issued a Directive to banks for operationalizing the green credit policy.[854] The Directive is notable for its comprehensiveness in requiring banks to make certain that their clients develop environmental and social risk assessment criteria, take mitigation actions, implement risk response plans, and ensure that international environmental and social standards are upheld – not only during initial assessment stages but also during the life of a project. Failure to comply by mid-level managers can prompt the CBRC to prevent appointments to senior positions within Chinese state-owned banks;[855] and there are specific financing consequences for clients such as suspension or termination of loans if they do not meet their ongoing obligations.

There is no equivalent banking regulation in the United States or Europe. Accordingly, the Directive has been described as "China's most important and useful social and environmental regulation, and one of the most advanced banking regulations at the transnational level."[856]

Moreover, new public corporate-environmental disclosure requirements in China have regulatory consequences for the financial system. Pursuant to the *Measures for*

---

852. "Chapter VI: Economic Incentives and Supervisory Measure," Articles 24- 25, *The Renewable Energy Law of the People's Republic of China* 2005, available at http://www.martinot.info/China_RE_Law_Beijing_Review.pdf.
853. M. Aizawa, "China's Green Credit Policy: Building Sustainability in the Financial Sector", *China Environment Forum* (Feb. 24, 2011), http://www.wilsoncenter.org/sites/default/files/Motoko%20Aizawa.pdf.
854. China Banking Regulatory Commission, *Notice of the CBRC on Issuing the Green Credit Guidelines* (Feb. 24, 2012), http://www.cbrc.gov.cn/EngdocView.do?docID = 3CE646AB629B46B9B533B1D8D9FF8C4A.
855. J. Thomä, "UN highlights China's progress on 'greening' its finance sector," *China Dialogue* (Feb. 4, 2014), https://www.chinadialogue.net/article/show/single/en/6704-UN-highlights-China-s-progress-on-greening-its-finance-sector.
856. Per Paulina Garzon quoted in D. Hill, "What good are China's green policies if its banks don't listen?,"*theguardian.com* (May 17, 2014), http://www.theguardian.com/environment/andes-to-the-amazon/2014/may/16/what-good-chinas-green-policies-banks-dont-listen.

Corporate Environmental Credit Evaluation (Trial), issued by the CBRC, NDRC, Ministry for Environmental Protection and the People's Bank, Chinese companies in heavy polluting industries are assigned an environmental credit rating based on a four-color scale, which is then directly linked to credit approval. Since March 2014, banks have been discouraged from lending to companies with a red (the worst) rating; and insurance companies can restrict businesses by the control of premium rates.[857]

Strikingly, these financial regulations put responsibility on banks to ensure that corporate borrowers comply with Chinese environmental regulations, which are often strong on paper but have been notoriously difficult to enforce.[858] Being enlisted to give teeth to environmental policies transmutes banks into quasi-regulatory bodies akin to an EPA.

In short, China has made clear that its finance sector is in the frontline of going global *responsibly*.

Yet the nascency of these policies means their actual impact is difficult to assess. Specifically, debates have emerged about whether Chinese banks are operationalizing or ignoring the green credit policy. For example, in April 2014 Friends of the Earth alleged that a coal power plant in Bosnia and Herzegovina and a copper project in Ecuador, which had received China Development Bank finance, fell short of appropriate socio-environmental standards.[859] In contradistinction, banks in Hebei Province in 2013 reportedly rejected 20 percent of loan applications to restricted industries such as steel, cement and glassmaking, and increased lending to projects for energy saving and technological upgrades.[860]

Implementation will remain a live issue going forward.

§8.04 CHINESE GREEN INVESTMENT AND FINANCE: PATTERNS, LEVERS AND LIMITS

The aforementioned green policies comprise the principle framework for the development of renewable energy and clean tech sectors via enhanced investment and finance. This comprehensive policy framework is encouraging significant expansion in both Chinese green FDI and ODI. Regarding FDI, China's renewable energy capacity increased six fold from 27.8 Gigawatts (GW) in 2001 to 183 GW in 2013. Similar expansions have occurred regarding ODI. According to Bloomberg New Energy Finance, China was the top investor nation in renewable energy and energy smart tech

---

857. Thomä, *supra* n. 855; Bloomberg News, "China to Assign Some Credit Ratings on Environmental Protection," *Bloomberg.com* (Jan. 3, 2014), http://www.bloomberg.com/news/2014-01-02/china-to-assign-some-credit-ratings-on-environmental-protection.html.
858. R. Edward, "China's Green Credit Policy under Scrutiny," *Eco-Financas* (Feb. 19, 2013) http://ef.amazonia.org.br/2013/02/chinas-green-credit-policy-under-scrutiny/#comments (accessed Aug. 30, 2014).
859. See BankTrack, *China Sustainable Finance Newsletter* (Issue 21, April 2014), 3.
860. *Ibid.*, 2.

in 2013 at US$61.3billion compared to investments from the United States of $48.4billion.[861] UNEP further reported that 2013 was the first year that Chinese green investment exceeded that for the whole of Europe.[862] The reasons for these developments have been not only a sharp reduction in the cost of PV systems but also reduced investor confidence due to regulatory uncertainty around renewable energy policy in Europe and the United States.[863]

By contrast, Chinese investor enthusiasm has been encouraged by Chinese investment policy incentives. This, in turn, has facilitated major advances in wind and solar power, which are the staple renewable energy portfolios for Chinese green FDI and ODI. Specifically, from 2006 to 2012, wind investments comprised nearly 60 percent and solar comprised nearly 30 percent of Chinese outbound clean tech investment.[864]

[A]   Wind and Solar

China installed nearly 30 percent of new global wind capacity in 2012.[865] Already the world's largest producer of wind power, China aims to move from its current installed capacity of 75GW to 200GW by 2020. By contrast, the EU has a current installed capacity of 90GW.[866]

The national wind power generation market is mainly shared among the "Big Five" power producers and a few more large SOEs. The big five producers are China Datang Corporation, China Guodian Corporation, China Huadian Group, China Huangeng Group and China Power Investment Corporation. They are all central SOEs and administered by central SASAC. In 2012, these firms accounted for more than 80 percent of the total wind power market.[867]

In 2013, China's solar developers installed a record 12GW of Solar PV projects; no country had added more than 8GW in a single year previously.[868] In the same year, China's state-owned power generators China Power Investment Corporation, China Three Gorges and China Huadian Corporation became the world's largest owners of solar assets.

According to data from the World Resources Institute (WRI), nearly half of the investments from 2002-2012 were made in new solar PV-based electricity generation

---

861. Bloomberg New Energy Finance, "Clean Energy Investment Falls for Second Year," *bnef.com* (Jan. 15, 2014), http://about.bnef.com/press-releases/clean-energy-investment-falls-for-second-year/.
862. United Nations Environmental Program (UNEP), *"Global Trends in Renewable Energy Investment 2014*, http://fs-unep-centre.org/sites/default/files/attachments/14008nef_visual_14_key_findings.pdf.
863. *Ibid.*
864. Azure International, *China Cleantech Outbound Investment In Transition: Key Drivers, Important Trends, Major Players* (Azure International: Beijing, June 2014).
865. International Renewable Energy Agency, *supra* n. 802.
866. *Ibid.*
867. http://www.censere.com/articles/238-development-and-trend-of-china-wind-power-sector.
868. International Renewable Energy Agency, *supra* n. 802.

either as green-field investments or through joint ventures.[869] Importantly, China's leading solar manufacturers are also its leading ODI champions. Yingli Solar, Suntech Power and Trina Solar are among the top ten global manufacturers and are ranked first, third, and fourth respectively in the WRI data for number of overseas investments. In contrast, relatively small companies like LDK Solar, Jiangsu Zongyi, Sunlan Solar, and Hareon Solar have put most of their overseas investment into building solar plants.

[B]   Financing Arrangements

Access to abundant and relatively low-cost capital provided by Chinese financial institutions (such as China Development Bank) has enabled Chinese wind and solar SOEs, and to a lesser extent privately-held companies, to invest overseas. Financing has been available both as lines of credit to corporate entities and as project finance in order to acquire and develop overseas power plants.[870] For example, state-owned Chinese banks reportedly agreed to loan nearly US$41billion to Chinese solar companies from January 2010 to September 2011.[871]

Other sources of capital that have incentivized ODI expansion in solar and wind include the Chinese green stimulus package of 2009 (US$46billion), IPO raisings from the capital markets in 2009 (US$5.9billion), and Chinese government loan guarantees in 2010 (US$36billion).[872]

Nonetheless, like Europe and the United States, China experienced a supply glut in the solar market against weak demand partly due to the success of its green tax breaks and subsidies. Some big players like Suntech and LDK Solar were declared insolvent with the latter escaping liquidation due to a consortium loan of US$321million led by China Development Bank.[873]

Chinese financial institutions have also indirectly supported wind and solar ODI by providing finance to engineering, procurement and construction (EPC) companies, and by offering export credit. Accordingly, many wind and solar manufacturers that are POEs or local SOEs and even some foreign enterprises, have bundled their products and investments with state-owned developers and EPC companies to expand overseas. This bundling has allowed them to benefit from interest rates of 1.6 percent paid by SOEs when borrowing from state banks, which is lower than the rate of 4.7 percent for

---

869. Tan et al., *supra* n. 802.
870. W. Pentland, "China's Coming Solyndra Crisis," *Forbes* (Sep. 27, 2011), http://www.forbes. com; Mercom Capital Group Website, http://mercomcapital.com/loans-and-credit-agreement s-involving-chinese-banks-to-chinese-solar-companies-since-jan-2010; A. McCrone, E. Usher, V. Sontag-O'Brian, U. Moslener, J.G. Andreas and C. Gruning (ed.), "Global Tends in renewable energy Investment 2011," (UNEP, Bloomberg and Frankfurt School, 2011); Goldw ind, *2011 Annual Report*, 11, http.goldwindglobal.com.
871. Mercom Capital Group Website, http://mercomcapital.com/loans-and-credit-agreements-inv olving-chinese-banks-to-chinese-solar-companies-since-jan-2010.
872. Bloomberg New Energy Finance, "Joined at the Hip: the US-China Clean Energy Relationship" (2010), http://bnef.com/PressReleases/view/116.
873. I. Clover, "LDK Solar receives $321 million state-backed loan," *PV Magazine* (May 27, 2014), http://www.pv-magazine.com/news/details/beitrag/ldk-solar-receives-321-million-state-bac ked-loan_100015205/#axzz38LBw3SY4.

private companies.[874] For example, the China Development Bank lent US$44million directly to the American-based company SPI to build a solar power project in New Jersey due to its close working relationship with LDK Solar.[875]

Exemplars of state financing are Goldwind (wind) and Hanergy (solar). Goldwind is a SOE and China's second largest wind turbine manufacturer. It is headquartered in China with foreign offices situated in Chicago, Germany and Australia. In 2010, the China Development Bank offered Goldwind a US$6billion credit facility to help foster international expansion. Subsequent to a further $1.5billion credit facility from the state-owned Industrial & Commercial Bank of China Ltd in 2011, Goldwind signed another financing agreement with China Development Bank in 2012 for 35billion-yuan (US$5.5billion).[876] The aim of the latter arrangement was to provide new capital to facilitate Goldwind's development ambitions in the United States.[877]

Hanergy Group is an example of a private company receiving significant state support due to its potential market significance as a major solar PV energy provider. It has secured lines of credit from several Chinese banks, including China Minsheng Banking Corp. and the Asia Financial Cooperation Association, totaling approximately 20billion yuan (US$3.3billion).[878] Concomitantly, Hanergy is also receiving financial support from foreign sources. The American renewable energy investor Greenback Renewable Energy Corp and Hanergy Holdings America Inc. entered into a memorandum of understanding (MOU) in 2014 pursuant to which Hanergy will develop commercial solar power to be bought and financed by Greenback. This joint venture is set to provide Hanergy with a long-term source of capital.[879]

[C]   Foreign Regulatory Levers and Limits

According to WRI data, Chinese green ODI has concentrated in the United States, Germany, Italy, Australia, South Africa, Pakistan and Bulgaria.[880] From 2002 to 2012, the United States was the leading destination for China's solar investments and, to a lesser extent, wind investment. European nations have attracted mostly solar investments from China.

These investment statistics do more than reflect undervalued assets in OECD country markets resulting from the financial crisis. They also correlate with the

---

874. The Economist, "The Rise of State Capitalism" (January 2012), http://www.economist.com.
875. Solar New Jersey, "China Development Bank Finances LDK-SPI Solar International Expansion Plans" (2012), http://www.solar-new-jersey.org.
876. D. Cusick, "China's No. 2 turbine maker secures $5.5B from state bank," *Governors' Wind and Energy Coalition* (Feb. 2, 2012), http://www.governorswindenergycoalition.org/?p = 1095.
877. S. Bakewell, "Goldwind Signs $5.5 Billion China Development Bank Wind Pact," *Bloomberg-.com* (Feb. 1, 2012), http://www.bloomberg.com/news/2012-01-31/goldwind-signs-5-5-billio n-china-development-bank-pact-for-wind.html.
878. M. Osborne, "Banks provide Hanergy with US$3.3 billion line of credit," *PV Tech* (Jan. 8, 2014), http://www.pv-tech.org/news/banks_provide_hanergy_with_us3.3_billion_line_of_credit.
879. Media release from Greenbacker, http://greenbackerrenewableenergy.com/documents/GB_H anergy.pdf; A. Colthorpe, "Greenbacker to Finance up to 126MW of Hanergy projects in the US," *PV-Tech* (Apr. 17, 2014), http://www.pv-tech.org/news/greenbacker_to_finance_up_to _126mw_of_hanergy_projects_in_the_us.
880. Tan et al., *supra* n. 802.

domestic incentivizing regulation in investee nations as documented throughout this book, namely tax credits in the United States, FITs in Europe, and the RET in Australia. These policy and market factors have attracted or "pulled" Chinese investments into foreign solar and wind industries.

Accordingly, supportive policies aimed at domestic participants, such as preferential taxes, FITs and RETs or RPSs have encouraged Chinese investors even while not exclusively targeting them. For example:

- The RPS in Illinois attracted Goldwind to secure a 20-year electricity supply contract with Commonwealth Edison Company starting in 2012, which enabled investment in the Shady Oaks Wind Farm.[881]
- Arizona's 2009 renewable energy tax credit enabled Suntech Power to open a US$30million factory for the final assembly of solar panels in Arizona in 2010.[882]
- In South Africa in 2003, the government announced a 10,000 GWh RET to be achieved by 2013, which was supported by a FIT.[883] Subsequently, Chinese companies Yingli Solar and Suntech Power invested in solar PV plants, and Goldwind and Guodian Longyuan invested in wind farm developments in South Africa.[884]

At the international level, bilateral cooperation agreements have been used to formally guarantee support and attract investments from China. For example, the trade promotion agency of Kazakhastan signed an MOU with Goldwind to encourage it to develop wind power projects in that country.[885] Similarly, China's premier Wen Jiabao listed climate change as one of eight measures for cooperation between China and Africa, and committed to help countries in Africa build 100 new clean energy projects.[886]

Yet just as domestic incentivizing regulation encourages foreign investment, restrictive forms of regulation may repel it. Notably, the imposition of import barriers to incentivize local production over and above foreign investment has created barriers to clean tech market expansions globally. Policies that discourage imports, such as local content requirements and anti-dumping duties have had the side effect of shifting China's production activities directly to destination markets or third countries not subject to these restrictions in order to maintain their market share.

---

881. R. Davidson, "Goldwind Purchases 106.5 MW Illinois Project," *Wind Power Monthly* (Dec. 11, 2010), http://www.windpowermonthly.com/article/1047296/goldwind-purchases-1065mw-illinois-project.
882. E. Trevizo, "Suntech Opens in Goodyear, Brings Jobs to West Valley," *The Arizona Republic* (Oct. 12, 2010), http://www.azcentral.com.
883. Department of Minerals and Energy, Republic of South Africa, *White Paper of Renewable Energy* (2003), https://unfccc.int/files/meetings/seminar/application/pdf/sem_sup1_south_africa.pdf.
884. Tan et al., *supra* n. 802, at 17.
885. *Ibid.*
886. People's Daily, "Chinese Premier Announces Eight New Measures to Enhance Cooperation with Africa," *peopledaily.com* (2009), http://english.peopledaily.com.cn/90001/90776/90883/6807055.html.

Local content requirements have been present in some of China's major markets including Brazil, Canada, and India:

- In Brazil, the Proinfa wind FIT program, which was replaced in 2010 by a tendering system, stipulated 70 percent local content. The requirement was later dropped for 1.5 MW-plus turbines when it became clear that there was insufficient local factory capacity to meet the program target.[887]
- In 2009, the Canadian province of Ontario introduced a rule that its FIT for solar PV projects of more than 10kW would only be available for developers using modules with at least 50 percent of their cost based on local goods and services, a benchmark that rose to 60 percent in 2011.[888]
- In 2010, India announced that solar PV project developers participating in the first 150 MW phase of its Solar Mission initiative would only be eligible for support if they used locally assembled modules.[889]

Yet these policies have not necessarily deterred Chinese investors or exporters. For example, Sinovel, a central SOE and the largest wind turbine manufacturer in China, won a bid to supply turbines to a wind power project in Brazil in 2011 under the condition that it produce and procure turbines locally in order to fulfil local content requirements.[890]

In contrast however, anti-dumping duties by the United States on China's solar and wind products have caused Chinese companies to shift their production to jurisdictions not covered by the duties, such as Taiwan and Thailand, and to export their products to the United States from those locations. Anti-dumping measures implemented by the United States and Europe in 2012 resulted in significant tariffs on solar cell imports from China and a plunge in Chinese solar exports by 31 percent in the first half of 2013.[891] Specifically, the US Department of Commerce had raised import tariffs to protect domestic solar manufacturing which, it claimed, was unfairly disadvantaged by the influx of cheap solar panels and equipment into the American market by Chinese SOEs.[892]

In response to China's challenge of those anti-dumping duties, the WTO Panel ruled on July 14, 2014 that the United States had violated international trade agreements by imposing the 2012 anti-dumping tariffs on imports of, amongst other things, Chinese-made solar products.[893] Importantly, the Panel found that the US Department of Commerce had acted inconsistently with obligations set forth in the *WTO Agreement*

---

887. McCrone et al., *supra* n. 870.
888. *Ibid.*
889. *Ibid.*
890. H. Yu, "Sinovel Signed Wind Turbines Supply Contract to Brazil," *The Economic Observer* (2011), http://www.eeo.com.cn/2011/0927/212591.shtml.
891. China Daily, "Exports of Chinese PV products plummet in H1," *chinadaily.com* (Aug. 12, 2013).
892. Energy Matters, "WTO Rules US Solar Panel Anti-Dumping Tariffs Illegal," *energymatters.com* (Jul. 17, 2014), http://www.energymatters.com.au/index.php?main_page = news_article&art icle_id = 4396.
893. *United States — Countervailing Duty Measures on Certain Products from China* (WT/DS437/R), Panel Report, Jul. 14, 2014.

on Subsidies and Countervailing Measures when it initiated those measures against Chinese SOEs solely on the grounds that they were majority owned and controlled by the Government of China and therefore public bodies or "authorities."[894] The Panel found that "ownership and control in and of themselves are not sufficient for determining that an entity is a public body" that exercises "governmental functions" for the purposes of the Agreement.[895]

As international contestation in this space emerges, international law and domestic regulation will similarly need to evolve.

§8.05   FUTURE DIRECTIONS

Looking to the future of Chinese green investment, there are several emerging themes.

The first theme is the role of SOEs in investment generally and green growth specifically. SOE hegemony has crowded out private investors, a phenomenon described as *guo jin min tui*, which has been criticized as unconducive to a competitive domestic market.[896] In its 2013 *Decision on Major Issues Concerning Comprehensively Deepening Reforms*, China set a course for reform that revolves around creating a more market-based and competitive economy and, in so doing, signaled a lessening of state involvement in corporate activity through the implementation of market-oriented SOE reforms.[897] Yet the reality is that such a transition will take time: "It was the case then and continues to be the case today, that SOEs possess unique traits that are products of China's developmental path, and which cannot be quickly changed."[898]

The economic transition that China seeks to make is particularly challenging in light of the historical-cultural context of the corporation in China. As mentioned previously, state-owned banks continue to financially support even unprofitable solar SOEs in a way that Anglo-American nations simply would and could not.

This leads to the second emerging theme, which is the increasing substantiveness of green ODI by private enterprises. ODI generally is increasingly coming from POEs, with some Chinese commentators predicting that POEs will eventually overtake SOEs as the dominant foreign investment modality.[899] Indeed, this changing of the guard has already occurred in the United States. According to the Rhodium Group, POE investment has steadily increased in the United States since 2009, reaching more than half of total transaction value (59 percent) of Chinese investments in 2012. Due to a small

---

894. *United States — Countervailing Duty Measures on Certain Products from China* (WT/DS437/R), Panel Report, Jul. 14, 2014, para. 7.75.
895. *United States — Countervailing Duty Measures on Certain Products from China* (WT/DS437/R), Panel Report, Jul. 14, 2014, paras. 7.70, 7.60-7.74.
896. Liao and Zhang, *supra* n. 808; L. Song, J. Yang and Y. Zhang, "State-Owned Enterprises' Outward Investment and the Structural Reform in China" (2011) 19(4) *China and World Economy* 38.
897. N. Borst, "Economic Reform in the Third Plenum: Balancing State and Market" (2013) XIII(23) *China Brief* 7, 7.
898. R. Chen, "Market Solutions to the Information Challenge of China's Legal System: Overseas Listing of State-owned Enterprises" (2013)1(1) *Peking University Law Journal*, 19.
899. See e.g. Y. Huang, "The Changing Face of Chinese Investment" (2012) 4(2) *East Asia Forum Quarterly* 13; Liao and Zhang, *supra* n. 808.

number of large deals in the real estate and food sectors in 2013, private firms accounted for 87 percent of transactions and 76 percent of total Chinese ODI value in the United States that year.[900] By 2014, POEs were dominating Chinese capital inflows into the United States, accounting for more than 80 percent of transactions.[901]

These general investment patterns mirror changes in green investment and financing whereby POEs that are also SSEs may be gaining an increasingly dominant role in Chinese outbound investing in the clean tech sector. For example, in 2012, Wanxiang Group Co., China's biggest auto-parts maker and a private company with strong state financial backing, won approval from the Committee on Foreign Investment in the United States (CFIUS) to buy most of the assets of A123 Systems Inc., the bankrupt electric-car battery manufacturer backed with US government funds.[902] Wanxiang acquired substantially all of A123's automotive, grid and commercial assets for approximately US$256.6million.

Similarly, in 2012, Hanergy Holding Group, another private company with strong state finance backing, bought Miasole (a Silicon Valley cleantech start up) for US$30million.[903] The following year it also acquired the American firm Global Solar Energy Inc. and bought Solibro, a unit of insolvent German solar group Q-Cells SE.[904]

A third emerging theme is that China may well be breaking new ground by developing a new model for green growth internationally:

> In the past, all the models of economic organization, industrial structure, and business that were formulated by industrialized countries have been copied in China. Now, all of a sudden green growth has become a global trend that is unfamiliar to everyone, including developed countries. Indeed, China is now at the same frontier of green growth as the United States and Europe.[905]

And an important ingredient of this new green model is government intervention in the form of support for corporate and commercial enterprises that are active in the investment space. The US National Intelligence Estimate in November 2008 noted that:

> Today wealth is moving not just from West to East but is concentrating more under state control. In the wake of the 2008-2010 global financial crisis, the State's role in the economy may be gaining more appeal throughout the world. These States

---

900. T. Hanemann and C. Gao, "Chinese FDI in the US: 2013 Recap and 2014 Outlook," *The Rhodium Group* (Jan. 7, 2014), rhg.com. See also M. Bowman, "One More Time: The Ongoing Investment Review of Smithfield-Shuanghui," *CLMR web portal* (Aug. 16, 2013), http://www.clmr.unsw.edu.au/article/risk/one-more-time-ongoing-investment-review-smithfield-shuanghui.
901. Hanemann and Gao, *supra* n. 900; Azure International, *supra* n. 864, at 6.
902. M. Bathon, "Wanxiang Wins U.S approval to Buy battery maker A123," *Bloomberg*.com (Jan. 30, 2013), http://www.bloomberg.com/news/2013-01-29/wanxiang-wins-cfius-approval-to-buy-bankrupt-battery-maker-a123.html.
903. N. Groom, "Hanergy to buy US Solar Miasole: source," *Reuters* (Oct. 1, 2012), http://www.reuters.com/article/2012/10/01/us-solar-miasole-idUSBRE8901G720121001.
904. C. Zhu, "China's Hanergy Buys U.S solar panel maker in technology push," *Reuters* (Jul. 25, 2013), http://www.reuters.com/article/2013/07/25/us-hanergy-gse-idUSBRE96O0MZ20130725.
905. Liao and Zhang, *supra* n. 808, at 134.

are not following the Western liberal model for self-development but are using a different model – "state capitalism."[906]

This is not to say that market economies ought to adopt a state capitalist model. Rather, it reaffirms the case study findings in this book that the role of government support for green growth – whether through amenable policy initiatives or public/co-financing – is inescapable. It is a nexus that China has adopted and exploited to its benefit.

Strikingly, the preceding discussion has revealed that incentivizing regulation within national borders will also attract foreign capital to augment domestic green markets. In so doing, these markets are augmenting transnationally, provided that import and foreign investment restrictions investee nations are minimal and non-discriminatory.

This leads into the final observation and a fecund area for future "green" state capital research. On the one hand, we know that domestic financial incentivizing regulation for RE (that is, tax credits, FITs, grants, RETs) in investee nations aimed at *domestic* private finance actors has the secondary effect of pulling in Chinese ODI. Yet on the other hand, foreign investment law and policy in investee nations invariably sets thresholds for FDI, especially regarding proposed investments by state-backed investors, which can have the effect of deterring Chinese ODI. Certainly there is a view that Chinese enterprises (both SOEs and POEs) have suffered some prejudice in the course of their overseas investment generally.[907] How and to what extent do these two seemingly unrelated spheres of regulation interact with each other and impact upon green investment proposals and activities? Will any regulatory interplay evolve or regress if China augments its sustainability strategies to become an international green growth leader?

§8.06 CONCLUDING REMARKS

China is trying to secure energy supply for its growing population. For these reasons, it has invested heavily in fossil fuels such as coal as well as renewable energy and clean tech, specifically solar and wind. In this way, China appears to be behaving akin to the United States with an "all of the above" energy policy. However, unlike the United States, intractable environmental problems such as air pollution and poor soil quality mitigate against China's continued reliance on fossil fuels.

Three steps are intrinsic to any government regulatory process. First, promulgation of policy; secondly, implementation of policy; and thirdly, achievement of policy objectives. The machinations of this regulatory process can be most clearly witnessed

---

906. Quoted in R. Hormats, "Ensuring a Sound Basis for Global Competition: Competitive Neutrality," *Dipnote: US Department of State Official Blog* (May 6, 2011), http://blogs.state.gov/stori es/2011/05/06/ensuring-sound-basis-global-competition-competitive-neutrality#sthash.i0s6S Hlg.dpuf.

907. See e.g. Liao and Zhang, *supra* n. 808; G. Golding, "Australia's Experience with Foreign Direct Investment by State Controlled Entities: A Move Towards Xenophobia or Greater Openness?" (2014) 37(2) *Seattle University Law Review* 533; M. Bowman, G. Gilligan and J. O'Brien, "Foreign Investment Law and Policy in Australia: A Critical Analysis" (2014) *Law and Financial Markets Review* 65.

within China due to its five level governance structure, which directly shapes resource allocation and political will for implementation of central policies by local (provincial, prefectural and county-level) governments. The second and third steps – implementation and efficacy – are important areas for further research in this space, particularly given that they have impact on the rollout of market-based policies such as the national ETS and non-market policies such as water conservation. Moreover, at a transnational level, implementation and efficacy will dictate the extent to which truly innovative policies like the Green Credit Directive have beneficial impact on the ground in foreign jurisdictions. This in turn will decide whether China's high-level socio-environmental aspirations are achieved. Accordingly, China needs to get all three steps of the regulatory process right. It needs to do this not only for reasons of credibility as an international green growth leader but also for reasons of its self-perceived role as a "responsible great power."

Green investment and market growth is required for the move to a low-carbon economy. In 2013 the World Economic Forum unequivocally urged the G20 to grasp that "Greening investment, and thereby the economy...is the only route to sustained growth and development."[908] China is emerging globally as both an economic leader and potential green growth role model. This chapter has demonstrated that by pursuing green growth at home, China concomitantly facilitates green growth in other jurisdictions. Thus, for China to "go green" as it "goes out" is a fundamental component of a timely global energy as well as economic transition.

---

908. World Economic Forum, *The Green Investment Report: The Ways and Means to Unlock Private Finance for Green Growth* (WEF: Geneva, 2013), 8.

CHAPTER 9

Conclusion

---

§9.01 THE PRECIPICE AND THE BRIDGE

We stand upon a precipice with the 2°C guardrail of dangerous climate change pressed against our collective chest. Looking over the edge into the chasm below it is clear that a more radical and multisectoral approach to GHG emissions reductions is urgently required to avert a cataclysm. Addressing dangerous climate change requires an economic transformation on the scale of the industrial revolution. Moving from a fossil fuel-based existence to a low-carbon one will involve one of the largest market and economic transitions in modern society. Such a massive transition requires financial input and facilitation on an equally grand scale. This, in turn, requires the input and facilitation of financial intermediary actors, both public and private.

Yet little is understood about the relationship between private finance actors and climate change and the interplay with regulatory context. Questions arise regarding the risks and opportunities of climate change for private finance actors; the intrinsic capabilities by which they can assist mitigation; what motivates them to do so; what repels them from doing more. Ironically, the very entities that caused global upheaval due to the GFC could facilitate global salvation in the face of the climate crisis. Indeed, without a realistic understanding of the climate-related motivations and limitations of private finance actors, we cannot hope to make the timely transition to a low-carbon or more resilient global economy.

Banking on Climate Change has sought to advance understanding of these questions through theoretical and empirical investigation. It has conceptualized extant literature using micro (intra-organizational), meso (inter-organizational) and macro (socio-cultural) lenses of focus to shed some light on why finance firms voluntarily adopt climate-related practices. In so doing, the heart of this book comprised data and findings from one of the first qualitative case-studies of early-moving transnational banks in order to provide "real life" learnings regarding the broader finance sector and climate change.

A critical finding from the case study was that banks are driven by business case logic, which comprises profit increase (directly via fee generation and indirectly via competitive edge) and risk mitigation (financial, regulatory, and reputational). Crucial to these findings was a deeper understanding of "corporate reputation" in business practice: it comprises not only the well-established "social reputation" or social license of a firm but also a reputation for good business sense and delivering excellent service that helps large corporate clients to flourish, which I termed "client service reputation." Client service reputation was a prime motivator for climate-related products, services and new market entry. Under the impetus of client service reputation, banks could be agnostic about climate change; their "green" driver was the greenback not a desire to save the world. Importantly, CSR proved to be an extremely limited explanatory tool for what drives actual corporate environmental-social uptake.

Moreover, interplay between reputation and regulation became apparent when examining banks' perspectives of climate change as a risk or an opportunity. In so doing, the case study elucidated the importance of regulatory contexts in shaping bank decision-making. In large part, their perspective was jurisdiction-specific and shaped by government interventions – namely a carbon price, financial incentives for renewables, a GHG reduction target, or even direct coercive social regulation such as the US Community Reinvestment Act – and also by social pressure from NGO campaigns, mass media coverage and civil society responses. The more sophisticated and stable the state interventions, the more that banks saw climate change as an opportunity, and leveraged regulatory incentives to enhance their client service reputation, social reputation, and profits. The weaker or less certain the state interventions, the more important that NGO activity and voluntary industry standards became to mobilize better corporate behavior; and the more likely that banks saw climate change as a risk. In such a regulatory context banks focused on strategies for downside prevention, and their aim was to keep BAU running as smoothly as possible. They were less likely to be proactive and innovative in addressing climate-related issues. Thus, through cross-jurisdictional comparison, the case study helps to predict whether finance actors will adopt reactive or modest strategies motivated primarily by risk mitigation, what I call "BAU with a lemon twist," or proactive strategies that assist climate change mitigation.

The conclusion of the study is that the business case drives voluntary corporate change but, simultaneously, impedes change that is far-reaching and expeditious. It is both a lever *and* a limitation. Central to this conclusion was the discovery that client service reputation – a critical ingredient of the business case – is a double-edged sword. It drove banks to adopt green practices to enhance and protect client service in order to enhance and protect their own profits. Yet, concomitantly and ironically, this meant that banks only made rational and conservative changes that would not compromise their client base or potential for profit maximization.

Importantly, these findings regarding the levers and limits of voluntary "green" action are generalizable to other private finance actors such as institutional investors and insurers.

In short, business case logic is seductive but unsatisfying as an assured modality of corporate change that can facilitate mitigation of climate change and the timely transition to a low-carbon economy.

In this way, the case study showed that the question of why firms *actually* go green can be separated from a normative discussion about why they *ought* to go green. I have argued that we need to appreciate these "real life" aspects in order to make effective policy that can augment best practice. Specifically, I have argued that the business case can provide an available and actionable modality of change if it is harnessed by clever government regulation. Thus, the regulatory suggestions in this book are based on the empirical findings and supported by theories of new governance and Nudge economics. In this way *Banking on Climate Change* has provided empirically-based regulatory recommendations for harnessing private finance actors in ways that are realistic, accurate and fruitful.

Given the heterogeneity of projects and investors, there is no one-size-fits-all prescription for designing effective climate finance policy for mitigation or adaptation. Nonetheless, broadly speaking, federal- and regional-level government intervention can take two forms in order to motivate increased private finance flows to address climate change. The first is direct and coercive (or "command-and-control") regulation, which is considered a high intervention option. The second form is indirect or steering ("nudging") regulation, which is low interventionist.

The advantage of coercive regulation is that it ensures BAU with a lemon twist is not *de rigueur*. The disadvantage is that, on its own, it is ill-equipped to deal with an emerging and dynamic area like climate finance. Climate finance regulation needs flexibility and credibility given the complexity and newness of this area. As such, it needs to be responsive to new information gained through learning-by-doing, and will require input from experts in the field including private finance actors in order to have traction with them. Moreover, coercive regulation forces private finance actors to become quasi-government instruments of climate-related societal benefit. This approach has been adopted by China in the Green Credit Directive. Yet it is not surprising that private finance actors in market economies prefer to not be so coerced.

Indeed, China is emerging globally as both an economic leader and potential green growth role model. An investigation of Chinese green investment and finance modalities showed that by pursuing green growth at home China is concomitantly facilitating the augmentation of green markets globally. For China to "go green" as it "goes out" is a fundamental step toward a timely global energy transition. Moreover, incentivizing regulation in investee nations mobilizes not only their own private sector actors to channel domestic capital into low-carbon initiatives but it also motivates Chinese firms in green ODI. This realization broadens the implications of the empirical case study findings beyond domestic/national landscapes to a truly transnational one.

§9.02   VIEWING CHANGE THROUGH AN INSTITUTIONAL LENS

This book has self-consciously advocated for working with what we've got in order to make beneficial change. In accepting the empirical findings that financial firms privilege business case logic I have recommended that, instead of lamenting that reality, we harness it. This is not due to any lack of imagination on my part. It is due to a practical desire to provide policy recommendations that can be adopted (relatively) quickly and easily in order to mobilize a full range of actors to assist climate change mitigation and adaptation in a timely way.

Nonetheless, I acknowledge that the concept of reality is nebulous and multi-perspectival. As such, investigation of how it is constructed – and can be reconstructed – around ecological norms instead of economic pressures merits further attention. Accordingly, a fecund area for future research is investigation of institutional (not corporate) change as an alternative way of facilitating climate change mitigation and the shift to a low-carbon global economy.

Hoffman and Ventresca propose two approaches to "making change."[909] First, work within the current values framework such that environmental issues are re-framed as synonymous with economic imperatives; that is, focus on win-win or "it depends" modalities. The regulatory recommendations in this book fit within that approach. The second approach is to restructure the current values framework. This would require recalibrating the systemic norms or institutions that influence the framing of the economics versus environment debate in the first place.[910] Hoffman has specifically argued that in order to solve the climate problem, we need to change the institutions not the organizations.[911]

In pursuing an institutional-level approach, the field of reference broadens out from organizational logic (meso level) to a discussion of the influences that *shape* organizational logic (macro level). The line of questioning would change from asking whether a corporation will act if the business case is made out to asking how items get on the business case agenda in the first place, how company activities shape prevailing ideas, and how certain behaviors become normal or acceptable. That is, a macro level approach recalibrates the lens of analysis to focus on socio-cultural influences rather than their organizational effects.

Scholars of institutional theory assert that industry or organizational field norms are a product of their macro political-economic context. Organizations allocate resources based on prevailing macro norms, which in turn reinforces those norms and the symbolic framework. The prevailing norm in capitalist societies is "economic efficiency" whereby organizations primarily allocate resources on the bases of "profit maximization and economic wealth accumulation" in a way that purports to be "value-free."[912] Indeed, data in this study show how banks refuse to choose clients on moral/value bases but are happy to do so on a rational risk/benefit basis. Importantly, some bankers *did* want to influence normative change; however, they chose the mode of influence based on what was most palatable and appropriate to them. For example, enhanced due diligence does not belie a bank's intention to pick clients on a moral basis and thus does not compromise a bank's client service reputation and fee generation. Indeed, choosing clients and projects on an economic basis is simply good business sense, which is appropriate and rational.

---

909. Hoffman and Ventresca, *supra* n. 652, at 1374.
910. *Ibid.*, 1384-1387.
911. Hoffman, *supra* n. 413, at 367-368.
912. Dillard et al., *supra* n. 411, at 528.

As such, we see in practice that the organizational and macro levels influence and inform each other and reinforce prevailing (economic efficiency) norms through recursivity and reinforcing actions such as (financial) resource allocation. This is evidenced, for example, through continued financial support for fossil fuel and high GHG-polluting corporate entities and the reasoning by some bankers of the social legitimacy of doing so. As Lundgren and Catasús note: "By combining what different stakeholders' demand from banks and the banks' perceptions of their role we largely arrive at what banks currently consider is the right thing to do."[913]

Yet systemic norms can change. Specifically, societal and legal norms can be changed by the conscious action of a host of actors including governments, academics, NGOs, media, the scientific community, and corporations themselves.DiMaggio and Powell note that interaction between these actors can resemble institutional "war"[914] and it is at times of conflict that radical (non-isomorphic) institutional change can occur.[915]

Currently there is an institutional void where government leadership on climate change should be; and a number of players are jostling to ensure that their preferences shape the field. As such, there are potential conflicts and tensions between not only actors in the climate field but between institutions. Approaches to climate change are determined by "which actors are engaged, what kinds of problems are debated, how those problems are defined, and what kinds of solutions are considered appropriate."[916] Further research might focus on the political dynamics between government, non-government and corporate actors in the climate field, including the extent to which corporations such as private finance actors are influencing and changing their institutional environment.

Yet is it possible for radical systemic change to occur when the players and the goals remain the same? How can capitalist values and beliefs be changed without changing the capitalist frame within which they are situated and reinforced? If business case logic dominates CSR discourse and if there is corporate capture of the meaning of CSR then what else is to be expected in a liberal capitalist context? Advocates of systemic change suggest the dismantling of capitalist ideals. Certainly, "deep green ecology" scholars are keen to point out that if we want a different outcome from business then we need a different paradigm for them to operate in.[917] Interestingly, in the wake of the GFC, a questioning of liberal capitalism occurred in media outlets as unlikely as the *Financial Times*[918] and *The Economist*.[919]

---

913. Lundgren and Catasús, *supra* n. 164, at 188.
914. DiMaggio and Powell, *supra* n. 431, at 30-31, citing H.C. White, *Identity and Control: A Structural Theory of Social Interaction* (Princeton University Press: Princeton NJ, 1992). See also Hoffman, *supra* n. 418, at 202.
915. Dillard et al., *supra* n. 411, at 514-515.c
916. Hoffman and Ventresca, *supra* n. 652, at 1369.
917. See e.g. R.A. Bucholz, "Corporate Responsibility and the Good Society: From Economics to Ecology" (1991) 34(4) *Business Horizons* 19; Kovel, *supra* n. 353.
918. G. Rachman, "The West Re-Examines the Rat Race," *The Financial Times* (Jun. 1, 2010), 11.
919. A. Wooldridge, "The Visible Hand," *The Economist* (Jan. 21, 2012), http://www.economist.com/node/21542931.

To be sure, a change-the-world approach is quite exciting from a theoretical perspective, especially given the multiple and murky causes of global warming and its systemic (as opposed to atomistic) nature. However, the praxis of *actually* changing the world in this way is complex and challenging.[920]

Part of the challenge is in answering the question "what is it that society wants?" This book did not attempt to address that question. But the empirical findings hint at multiple and potentially conflicting messages. In particular, there appeared to be two dominant normative pressures operating on banks during the period that the research for the case study was conducted. First, there was an entrenched expectation that banks fulfill their traditional role in society by providing excellent customer service and increasing shareholder value. That is, society expects banks to be profitable and not to take charge of environmental protection in the way an EPA should. A second expectation was just emerging in the form of societal pressure regarding *how* banks make their money with a call for more ethical and transparent business practices in the wake of the GFC. Are these pressures conflicting on the subject of climate change? It depends. Perhaps yes, if banks continue to heavily fund fossil fuels and other unsustainable industries. Perhaps not, if they support lucrative low-carbon markets and technology. Overall, however, what is "acceptable" or "responsible" bank practice is inextricably linked with how society views the role of a bank, how banks view their role in society, and whether and to what extent those views can change.

If further research on institutional change or CSR is to occur, then surely the "nature of the obligations and/or expectations the wider society wishes to impose on business" must be co-examined.[921] As Hoffman stated in 2001:

> To profess, as many do today, that industry is finally seeing the light is to argue that the light has always been there to see. In fact it has not. How companies define their responsibility toward the environment is a direct reflection of how we, as a society, view the environmental issue and the role of business in responding to it.[922]

In other words, do *we* care enough to change?

§9.03   FROM DREAM TO REALITY

There is a Hollywood movie called *Field of Dreams*. In this movie the main character, played by actor Kevin Kostner, builds a giant baseball stadium on a farm so that ghosts of legendary baseballers-past have a venue to play at night after night while avid fans can enjoy the spectacle. The catchphrase of this movie is "build it and they will come," referring to the deliberate transition of a cornfield into a baseball stadium. Without the architecture, there is no game.

A curious analogy exists between that movie and the findings in this book. If national/regional governments supply the basic architecture of the climate finance

---

920. See Hoffman, *supra* n. 413.
921. O'Dwyer, *supra* n. 291, at 550-551.
922. Hoffman, *supra* n. 418, 22.

game then the skilled financial experts can play within it, doing what they do best but for the benefit of the rest of us.

An external expert said this about mitigating climate change: "It is not a blame or guilt game; let's just focus on solving the problem. We need people with lesser morals on board to fix the problem" (CCC-1). He was referring to private finance actors. These actors may not be driven by ethical imperatives and they may not be acting from a (purely) better-the-world place; but they are economic gatekeepers with access to money and the innate ability to move it around. They are critical entities to get on board to help address climate change which is as much an economic and fiscal issue as an environmental and ethical one.

Yet due to inherent limitations in business case logic and disinterest in CSR as a driver, private finance actors will not necessarily "do the right thing" for society and a low-carbon economy on their own or, if they do, they will not take green action far or fast enough to assist timely mitigation of global warming. Government intervention that harnesses their devotion to the business case, for the good of society, is also required.

We no longer need to assume what motivates private finance actors or to project upon them some artificial conception of corporate care that we think *should* motivate them. We now know from their own mouths what drives them and why. We just need to capitalize on it. This book has made some recommendations for how to help make "the dream" a timely reality. Specifically, regulators in market economies can *harness* the profit motive and expertise of private finance actors to better society. I think this is preferable to hoping those actors (especially banks) will "make good" all on their own. To this end, *Banking on Climate Change* has highlighted which areas can be leveraged and harnessed to maximum effect, which areas will yield little reward, and which areas have as yet untested potential. As such, this is not the end of the story but hopefully just the beginning.

Table of Cases

United Kingdom

Brady (1987) 3 BCC 535, 142
Gaiman v. National Association for Mental Health [1971] Ch 317, 330, 142

Australia

Bell Group Ltd (in liq) v. Westpac Banking Corporation (No 9) (2008) 39 WAR 1, 144
Mills v. Mills (1938) 60 CLR 150, 144
Provident International Corp v. International Leasing Corp Ltd [1969] 1 NSWLR 424, 144
Whitehouse v. Carlton Hotel Pty Ltd (1987) 162 CLR 285, 144

United States of America

American Electric Power Company, Inc., v. Connecticut, 564 U. S._ (2011), 48
Aronson v. Lewis 473 A2d 805, 812 (Del 1984), 147
Brehm v. Eisner, 746 A.2d 244 (Del. 2000), 147
Burks v. Lasker, 441 US 471 (1979), 146
Coalition for Responsible Regulation, Inc v. Environmental Protection Agency, 684 F3d 102, 113 (DC Cir 2012), 10
Comer v. Murphy Oil USA Inc. (case 1:11-cv-00220-LG -RHW), filed on 27 May 2011 in the Southern District Court of Mississippi, 48
Dodge v. Ford Motor Co 170 NW 668, 684 (Mich 1919), 147
Dodge v. Woolsey 59 US 331 (1855), 146
FRIENDS OF THE EARTH, INC., et al., Plaintiffs, v. ROBERT MOSBACHER, JR., et al., Defendants No. C 02-04106 JSW UNITED STATES DISTRICT COURT FOR THE NORTHERN DISTRICT OF CALIFORNIA 488 F. Supp. 2d 889; 2007 US Dist. LEXIS 24268 (March 30, 2007, Decided; March 30, 2007, Filed), 48
Gantler v. Stephens 965 A2d 695, 705-06 (Del 2009), 50, 147
Gray v. President 3 Mass (3 Tyng) 364, 379 (1807), 146
Historic Boardwalk Hall, LLC v. Commissioner, 136 T.C. 1 (2011) (United States), 203

In re Goldman Sachs Grp., Inc. S'holder Litig., No. 5215-VCG, 2011 WL 4826104, at *23 (Del. Ch. Oct. 12, 2011), 148

Lockyer v. General Motors No 3: 06. Civ 05755 (ND California, filed 20 September, 2006), 47

Massachusetts, et al., Petitioners v. Environmental Protection Agency, et al. 549 US 497 (2007), 10

re Citigroup Inc. S'holder Derivative Litig., 964 A.2d 106 (Del. Ch. 2009), 50

re Lehman Bros. Holdings Inc., No. 08-13555 UMP) at 4 (Bankr. S.D.N.Y. Mar. 11, 2010), 148

Table of Statutes

International Conventions, Agreements and Resolutions

Kyoto Protocol to the Framework Convention on Climate Change (1998) 37 ILM 22 (Kyoto Protocol), 1

United Nations Framework Convention on Climate Change, *Report of the Conference of the Parties on its Fifteenth Session, held in Copenhagen from 7 to 19 December 2009,* FCCC/CP/2009/11/Add.1 (30 March 2010), Decision 2/CP.15 (Copenhagen Accord), 5

United Nations Framework Convention on Climate Change *Report of the Conference of the Parties serving as the meeting of the Parties to the Kyoto Protocol on its Sixth Session, held in Cancun from 29 November to 10 December 2010,* FCCC/KP/CMP/2010/12/Add.1 (15 March 2011), Decisions 1/CMP.6 and 2/CMP.6 and *Report of the Conference of the Parties on its Sixteenth Session, held in Cancun from 29 November to 10 December 2010,* FCCC/CP/2010/7/Add.1 (15 March 2011) Decision 1/CP.16 (Cancun agreements), 5

United Nations Framework Convention on Climate Change *Report of the Conference of the Parties serving as the meeting of the Parties to the Kyoto Protocol on its seventh session, held in Durban from 28 November to 11 December 2011,* FCCC/KP/CMP/2011/10/Add.1 (15 March 2012), Decisions 1/-5/CMP.7 (Durban Report), 5

United Nations General Assembly Resolution *Report of the World Commission on Environment and Development* (A/RES/42/187) 11 December 1987 (Brundtland Report), 12

National Legislation

American Clean Energy and Security Act of 2009 (H.R.2454) (United States of America), 7, 114

American Recovery and Reinvestment Act (2009) (Pub Law 111-115) (United States of America), 8

Australian Securities and Investments Commission Act 2001 (Commonwealth of Australia)

Bank of England Act 1998 (United Kingdom), 135

Banking Act 1959 (Commonwealth of Australia), 136

Banking Act 1987 (United Kingdom), 133

Clayton Antitrust Act of 1914 (Pub.L. 63–212, 38 Stat.730, enacted October 15, 1914, codified at 15 USC. §§ 12–27, 29 USC. §§ 52–53) (United States of America), 212

Clean Energy Act 2011 (Commonwealth of Australia), 7, 116, 157

Clean Energy (Consequential Amendments) Act 2011 (Commonwealth of Australia), 7

Clean Energy Legislation Amendment Act 2012 (Commonwealth of Australia), 7

Clean Energy Legislation (Carbon Tax Repeal) Act 2014 (Commonwealth of Australia), 7

Clean Energy Regulator Act 2011 (Commonwealth of Australia), 7

Clean Production Promotion Law of the People's Republic of China of 2002 (People's Republic of China), 222

Climate Change Authority Act 2011 (Commonwealth of Australia), 7

Climate Change Levy (General) Regulations 2001 No 838 (United Kingdom), 164

Code of Federal Regulations Title 12 Banks and Banking (United States of America), 136

Community Reinvestment Act of 1977 (Pub.L. 95-128, title VIII of the Housing and Community Development Act of 1977, 91 Stat. 1147, 12 USC. ch 30) (United States of America), 112

Companies Act 2006 (c.46) (United Kingdom), 142

Company Law 1994 (People's Republic of China), 219

Company Law 2006 (People's Republic of China), 219

Competition Act 1998 (c.41) (United Kingdom), 213

Competition and Consumer Act 2010 (Commonwealth of Australia), 213

Comprehensive Environmental Response, Compensation and Liability Act 1980 (Pub L No 96-510, 94 Stat 2767) (United States of America) (CERCLA), 39

Corporations Act 2001 (Commonwealth of Australia), 135, 144

CRC Energy Efficiency Scheme Order 2013: Climate Change 2013 No. 1119 (United Kingdom), 6

Customs Tariff Amendment (Carbon Tax Repeal) Act 2014 (Commonwealth of Australia), 7

Dodd-Frank Wall Street Reform and Consumer Protection Act (Pub L 111-203, H.R. 4173) (United States of America), 136, 138, 139, 141, 178, 193

Employee Retirement Income Security Act of 1974 (ERISA) (United States of America), 33

Energy Efficiency Opportunities Act 2006 (Commonwealth of Australia), 9

Enterprise Act 2002 (c.40) (United Kingdom), 213

Excise Tariff Amendment (Carbon Tax Repeal) Act 2013 2014 (Commonwealth of Australia), 7

Federal Reserve Act 1913 (P.L. 63-43, 38 STAT. 251, 12 USC 221) (United States of America), 136, 138

Federal Trade Commission Act of 1914 (15 USC §§ 41-58) (United States of America), 212

Financial Sector (Collection of Data) Act 2001 (Commonwealth of Australia), 135

Financial Services Act 1986 (United Kingdom), 133

Financial Services Act 2012 (United Kingdom), 134

Financial Services and Markets Act 2000 (United Kingdom), 133, 134

Internal Revenue Code 26 USC. (United States of America), 8

Legislation Amendment (Financial Claims Scheme and Other Measures) Act 2008 (Commonwealth of Australia), 135

Money Laundering Regulations 1993 (United Kingdom), 134

National Bank Act 1864 (Chapter 106, 13 STAT. 99) (United States of America), 136, 138

Notification on Resource-saving Activities by the State Council Office, Opinions of Implementing Government (People's Republic of China), 222

Ozone Protection and Synthetic Greenhouse Gas (Import Levy) Amendment (Carbon Tax Repeal) Act 2014 (Commonwealth of Australia), 7

Ozone Protection and Synthetic Greenhouse Gas (Import Levy) (Transitional Provisions) Act 2014 (Commonwealth of Australia), 7

Ozone Protection and Synthetic Greenhouse Gas (Manufacture Levy) Amendment (Carbon Tax Repeal) Act 2014 (Commonwealth of Australia), 7

Pensions Act 1995 (United Kingdom), 33

Procurement of Energy-saving Products and *Decision on Carrying Out Scientist Development Concept by Strengthening Environment Protection of 2005* (People's Republic of China), 222

Renewable Energy Sources Act 2000 or *Erneuerbare Energien Gesetz* (Germany) (EEG 2000), 8

Renewable Energy Sources Act 2014 or *Erneuerbare Energien Gesetz* (Germany) (EEG 2014), 7–8

Reserve Bank Act 1959 (Commonwealth of Australia), 135

Sherman Act of 1890 (26 Stat. 20915 USC.§§ 1–7) (United States of America), 212

Superannuation Industry (Supervision) Act 1993 (Commonwealth of Australia), 33

Tax Relief, Unemployment Insurance Reauthorization, and Job Creation Act of 2010 (H.R. 4853) (United States of America), 115

Trade Practices Act 1974 (Commonwealth of Australia), 213

True-up Shortfall Levy (Excise) (Carbon Tax Repeal) Act 2014 (Commonwealth of Australia), 7

True-up Shortfall Levy (General) (Carbon Tax Repeal) Act 2014 (Commonwealth of Australia), 7

United States Code Title (USC) *12 Banks and Banking* (United States of America), 136

Index

A

Annual General Meeting (AGM), 44, 46
Anti-dumping, 231, 232
Anti-trust law, 213
Australian Prudential Regulation
    Authority (APRA), 135, 136
Australian Securities and Investments
    Commission (ASIC), 135, 193
Authorized deposit-taking institution
    (ADI), 135–137

B

Banking industry, 15, 17–20, 27, 37,
    42–43, 58, 60, 62, 94, 96, 98, 109,
    116, 127, 129, 131, 132, 134, 150,
    151, 170, 174, 175
Bank regulation, 133–142, 201
Banks, 2, 21, 80, 93, 131, 173, 201, 220,
    241
BankTrack, 43, 44, 46, 51, 52, 159, 160,
    209, 212, 227
Behavioral economics, 186, 193
Beta, 197–201
Board (of Directors), 44, 59, 60, 93, 95,
    96, 100–101, 118, 119, 121, 123,
    130, 136–138, 142, 143, 146–149,
    162, 190, 203, 210, 243
Bonds (green, climate), 12, 20, 201,
    205–209, 214
Business-as-usual (BAU), 2, 69, 70, 73,

94, 152–158, 162–165, 167, 172,
    174, 179, 187, 238, 239
Business case, 19, 20, 35, 55, 63, 64,
    70–81, 88, 89, 91, 94–97, 101,
    105, 107, 111, 115, 118, 119, 122,
    125, 126, 130–132, 149–153, 156,
    163, 167, 169, 170, 174–177, 182,
    183, 188, 189, 195, 197, 201, 206,
    238–241, 243

C

California Public Employees Retirement
    System (CalPERS), 50
Capital asset pricing model (CAPM),
    198–202, 206
Carbon capture and storage (CCS), 52,
    57, 165
Carbon dioxide (CO2), 1, 9, 10, 44,
    48–49, 51, 52, 58, 159
Carbon Disclosure Project (CDP), 27, 38,
    56, 83, 205
Carbon Principles, 51–53, 83, 108, 109,
    114, 154–156
Case-studies, xii, 17–20, 53, 64, 72,
    91–127, 129–173, 177, 183–188,
    192, 193, 210, 218, 235, 237–239,
    242
China, 2, 5, 18, 43, 207, 211, 214–222,
    239
China Banking Regulatory Commission
    (CBRC), 226, 227

Chinese, 18, 20, 219–235, 239
Chinese Ministry of Commerce
     (MOFCOM), 216–217, 225
Cleantech, 57, 228
Client service reputation, 20, 98–101,
     105, 109–111, 114–118, 120, 123,
     129, 131, 132, 143, 145, 149, 158,
     160162, 163, 167, 170–172, 174,
     177, 179, 183, 185, 186187, 189,
     238, 240
Climate change, xi, xii 1–62, 72, 73, 76,
     79, 83, 86, 91–96, 100, 101, 109,
     111, 113, 117, 118, 122–124, 129,
     130, 143, 150, 151, 154157,
     159–165, 168, 169, 171–175,
     177–152, 184–189, 192, 194,
     205–207, 209, 211, 231, 237–243
Climate finance, xi, 2, 11–21, 37, 40, 47,
     63, 79, 92, 95, 109, 170, 178–183,
     189, 191, 193–195, 201–205, 207,
     239, 242–243
Climate Investment Fund (CIF), 13
Climate Principles, 51–53, 57, 155, 156
Climate risk, 3, 11, 16, 26–57, 83, 114,
     153, 154, 157, 159, 169, 179, 177,
     187, 188
Code of Federal Regulations (CFR), 41,
     136
Company law, 20, 50, 142–149, 210, 219
Competition law, 134, 212
Conference of the Parties (COP), 5, 31
Corporate climate finance, xi, 2, 18–21,
     47, 63, 92, 94, 170, 181–183, 189,
     194, 202–205
Corporate environmental responsibility
     (CER), 71, 76, 77, 173
Corporate finance theory, 197–202
Corporate financial performance (CFP),
     73, 75
Corporate governance, 15, 20, 30, 33,
     35, 36, 49, 51, 5960, 62, 77, 82,
     93, 130, 132–149, 161, 211, 219
Corporate law, 30, 82, 132–149, 219
Corporate reputation, 19, 80, 92, 97–99,
     118, 127, 129, 170, 171, 179, 183,

189, 238
Corporate Social Responsibility (CSR),
     19, 38, 66–71, 74–78, 81, 83,
     87–89, 92–94, 98, 99, 102, 103,
     105–110, 118–127, 144, 152, 153,
     161, 166, 167, 170–173, 176177,
     221, 238, 241, 242243
Credit risk, 40, 42, 116, 226

D

Decision making, xii, 18, 19, 29–30, 34,
     26, 42, 59, 80, 82, 87–89, 92, 96,
     109, 118, 119, 125, 132, 142–144,
     147, 148, 161, 170, 173, 177, 178,
     184, 185, 187, 238
Directors' duties, 20, 142, 145–149
Divestment, 35, 44–45, 49, 53, 55
Drivers, 64–81, 92, 94–118, 113, 114,
     116, 118–125, 129–131, 152, 161,
     169–173, 176, 182, 183, 216, 228,
     238, 243

E

Early-moving, 18–21, 57, 63, 84, 118,
     151, 169, 237
Emissions trading scheme (ETS), 1, 6, 7,
     56, 111, 113, 114, 116, 154, 192,
     225, 236
Empirical, xi, xii 17–21, 47, 59, 60, 63,
     66–67, 73–75, 77, 79, 80, 86–87,
     85, 112, 126, 129, 156, 167,
     169–195, 200, 237, 239
Engineering, procurement, construction
     (EPC), 229
Environmental non-government
     organization (ENGO), 71, 104,
     105
Environmental, social, governance
     (ESG) (risks, factors, issues), 3,
     16, 29–31, 33–36, 38, 50, 51, 58,
     107, 113, 156, 210

Environment Protection Authority
(EPA), 10, 117, 160, 227, 242
Equator Principles, 43, 51–53, 79, 83,
98, 109, 113, 151, 154–156
Equity, 2, 3, 15, 17, 20, 22, 27, 35,
37, 40, 47, 55, 56, 59, 60, 93,
104, 114, 115, 123, 124, 176,
197, 200–206, 209, 214, 217,
218
European Union (EU), xi, 1, 6, 7, 56,
113, 178, 181, 213, 217, 228
Evergreen, 20, 197, 209–214

F

Federal Deposit Insurance Corporation
(US) (FDIC), 136, 137,
139–141
Federal Reserve Bank (US) (FRB), 136,
137
Feed-in-tariffs (FITs), 7, 8, 112, 113,
116, 187–188, 190–192, 208, 223,
230–231, 235, 239
Fiduciary (duty, obligation, capitalism),
16, 30, 33–35, 37, 42, 142,
146, 147, 175, 209, 210, 212
Financial activities tax (FAT), 179–180
Financial Conduct Authority (UK)
(FCA), 134, 135, 193
Financial institutions, 2, 11, 13–15, 18,
21–26, 43, 44, 51, 52, 54, 59, 60,
105, 133, 135–137, 155, 179–180,
203–205, 217, 229
Financial intermediaries, 1–3, 14, 22,
25–26, 58, 237
Financial Services Authority (UK) (FSA),
133–135
Financial Stability Oversight Council
(US) (FSOC), 139–141
Financial transactions tax (FTT), 179,
180
Foreign direct investment (FDI), 216,
227, 228, 234, 235

G

Global financial crisis (GFC), xi, 11, 17,
37, 42, 50, 113, 121, 123, 125,
134, 137, 148, 173, 176, 178, 180,
197, 204, 217, 234, 237, 241,
242
Going global, 216–218, 220–222, 227
Green, xii, 1, 2, 4, 6, 7, 9–14, 18–20, 26,
36, 41, 43–45, 48, 51, 56, 57,
63–91, 94, 97, 98, 100, 101, 104,
106, 107, 111, 114, 118–120, 123,
126, 127, 129–133, 142, 143,
146–167, 170–176, 179–181,
188–193, 197, 201, 205–236, 238,
239, 241, 243
Green Climate Fund (GCF), 1, 2, 13–14,
206
Greenhouse gas (GHG), 1, 3, 4, 6, 7, 9,
10, 12–15, 18, 21, 26, 31, 36, 38,
40, 46–49, 51–55, 60, 73, 76, 102,
113, 114, 124, 129–131, 153, 157,
162, 164, 166, 174, 175, 186, 188,
189, 191, 192, 205, 206, 208, 211,
226, 237, 238, 241
Green Investment Bank (UK) (GIB), 9,
190–192, 201
Greenwash, 159, 181

I

Incentivizing (regulation), 7–9, 112, 114,
116, 187–190, 192, 194, 208, 231,
235, 239
Incremental change, 164
Initial public offering (IPO), 22, 161,
201, 229
Institutional investors, 2, 16, 17, 21, 22,
24, 30–37, 39, 40, 42, 44, 49, 50,
175–177, 179, 182, 199, 200, 206,
209, 210, 238
Institutional theory, 64, 81, 83, 85, 89,
150, 240

Institutions, xii, 2, 11, 13–18, 20–27,
    30–37, 39, 41–44, 49–52, 54, 56,
    59, 64, 67, 71, 80–85, 89, 97, 98,
    103, 105, 112, 133–142, 149–151,
    155, 156, 162, 175–177, 179–182,
    184, 199, 200, 205, 206, 209–211,
    217, 219, 229, 238–242
Insurance (companies, industry), 15–18,
    21–23, 25–34, 37, 42, 48, 52, 53,
    55, 79, 115, 134–142, 175, 222,
    223, 227
Insurers, 2, 3, 16, 17, 21, 24, 28–32, 39,
    49, 53, 55, 175, 176, 206, 238
Inter-governmental Panel on Climate
    Change (IPCC), 1, 4, 27, 29
International Finance Corporation (IFC),
    12, 32, 51, 54, 56, 201, 209
Investment, 1–3, 7–9, 13–18, 20–27,
    31–42, 44–56, 58, 60, 74, 75, 79,
    87, 91–93, 95, 96, 98, 101,
    103–106, 109, 111–115, 117, 118,
    121–124, 133, 135, 138, 148, 149,
    153, 157, 159, 161–163, 166, 169,
    171, 176–179, 182, 186, 187,
    189–194, 197, 198, 200–204,
    206–212, 214–217, 219–221,
    223–231, 233, 234
Investment tax credit (ITC), 8, 115, 202

K

Key performance indicator (KPI), 106,
    161, 162, 174
Kohlberg Kravis Roberts (KKR), 104
Kyoto Protocol, 1, 5, 56

L

Liberal (economies), 217
Litigation risk, 3, 17, 38, 47–49
Low-carbon, xi, 2, 3, 9, 13, 14, 17, 18,
    21, 27, 32, 37, 38, 51, 52, 57,
    61–63, 87, 100, 115, 120, 129,

    130, 151, 154, 157, 159, 160,
    162–166, 168, 169, 173, 174, 176,
    177, 183, 187, 189, 190, 194, 200,
    201, 206, 209, 212, 214, 215,
    236–240, 242, 243

M

Macro, 12, 19, 63, 64, 78, 81–85, 89–91,
    126, 149, 166, 198, 204, 219, 237,
    240, 241
Mainstreaming, 20, 58, 63, 127,
    129–131, 149–151, 153, 163, 167,
    174, 206
Market economies, 6, 18–21, 32, 83,
    132, 150, 175, 182, 191, 215, 217,
    235, 239, 243
Mergers and acquisitions (M&A), 56,
    112, 213
Meso, 19, 63–81, 85, 89, 91, 237, 240
Micro, 19, 27, 50, 63, 64, 85–88, 91, 107,
    115, 119, 124, 237
Mountain top removal (MTR), 54, 55,
    108, 153
Multilateral, 12–14, 24, 43, 209

N

National Development and Reform
    Commission (China) (NDRC),
    207, 223, 225, 227
Network change potential, 18, 20, 58,
    62, 92, 124, 129, 174, 189
New governance theory, 180–182, 189,
    192
Non-government organizations (NGOs),
    3, 26, 43–46, 49, 52–55, 61, 71,
    73, 79, 91–94, 98, 99, 102–110,
    117, 121, 123, 150, 153–162, 166,
    167, 169, 171, 172, 176, 181, 189,
    194, 205, 209, 238, 241
Nudge (theory, economics, regulation),
    20, 178, 183–194, 239

O

Office of Comptroller of Currency (OCC), 136–138, 141
Office of Thrift Supervision (OTS), 138
Opportunities, 2, 6, 9, 14, 16–19, 21, 26–57, 61, 63, 76–78, 86, 94–96, 100, 107, 111, 113–118, 124, 126, 149, 150, 153, 155, 157, 162–164, 166, 168, 170–172, 175, 176, 182, 183, 186, 187, 212, 237, 238
Organisation for Economic Co-operation and Development (OECD), 39, 165, 211, 221, 230
Overseas direct investment (ODI), 216–220, 227–230, 233–235, 239

P

Pension funds, 2, 14–16, 18, 20–22, 24, 31–33, 35, 49, 95, 100, 112, 135, 149, 153, 175, 189, 197, 206, 214, 213, 214
Photovoltaic (solar) (PV), 8, 115, 118, 190, 204, 223, 224, 228–232
Power purchase agreement (PPA), 190
Principles for Sustainable Insurance (PSI), 29, 30, 34, 42
Private capital, xi, 37, 125
Private equity (PE) firms, 2, 15, 17, 20, 27, 40, 104, 176, 197, 200, 201, 214
Private finance actors, xi, 2, 3, 14–16, 18–63, 81, 89–92, 125, 127, 129, 133, 168, 169, 172, 174–179, 181–183, 185–189, 197, 198, 200, 206, 209, 214, 235, 237–239, 241, 243
Privately owned enterprise (POE), 219, 220, 229, 233–235
Production tax credits (PTCs), 8, 115, 202, 203
Profit maximization, 16, 33, 35, 163, 167, 188, 219, 238, 240

Project finance, 17, 25, 38, 51, 52, 56, 115, 120, 155, 156, 201, 202, 229
Prudential regulation, 134–136, 179
Prudential Regulation Authority (UK) (PRA), 134
Public finance (actors), xi, 2, 10–12, 14, 49

Q

Qualitative (study, research), xi, 55, 92–95, 237

R

Rainforest Action Network (RAN), 44, 47, 53, 54, 104, 108, 159
Reinsurers, 2, 16, 21, 24, 27–32, 39, 55
Renewable energy (RE), 2, 7–9, 13, 17, 20, 27, 29, 32, 38, 39, 49, 51, 52, 54–57, 104, 106, 107, 113, 115–118, 120, 122, 123, 125, 132, 158, 160, 163, 177, 183, 188, 192, 202–205, 208, 210, 211, 214, 215, 220, 222–224, 226–230, 235
Renewable energy certificates (RECs), 117, 118, 190
Renewable energy targets (RETs), 8, 32, 117, 171, 192, 223, 231, 235
Renewable portfolio standard (RPS), 8, 231
Reputation risk, 3, 42–47, 79, 111, 113, 116, 156
Reserve Bank of Australia (RBA), 135
Responsible investing (socially), xii, 16, 32–36, 55, 74, 176
Risk/return, 2–4, 6, 7, 11, 13, 15–22, 24, 26–58, 61, 63, 64, 68, 73–81, 83, 85–87, 91–95, 97, 98, 101, 103, 105, 107, 111–118, 121, 123, 124, 130–135, 138, 139, 141, 144, 148–150, 156–160, 162–164, 166,

169–172, 175–179, 183, 186–189,
197–214, 217, 221, 223, 226, 234,
237, 238, 240

169, 206, 209, 235–237, 239

U

S

United Kingdom (UK), 4, 6, 7, 9–11, 22,
25, 32, 33, 38, 43, 50, 55, 59, 65,
Security market line (SML), 199, 200
69, 72,–74, 77, 79, 81, 85, 88, 93,
Shareholder wealth maximization, 20,
113–114, 120, 133–135, 142–144,
30, 74, 88, 126, 146, 149
148, 153, 155, 166, 175, 180, 181,
Short-termism, 132, 149
190–193, 203
Social and environmental performance
United Nations Environment Program
(CSP), 51, 61, 73, 75, 192
(UNEP), 11, 14, 15, 30, 31, 42,
Social license, 78, 80, 89, 98, 102,
56, 121, 122, 126, 173, 177, 201,
106–109, 111, 114, 115, 170,
228, 229
238
United Nations Environment Program
Socially responsible investment (SRI),
Finance Initiative (UNEPFI), 3,
16, 33, 74, 176
15, 27, 29–31, 54, 94, 205, 211
Social reputation, 80, 84, 97–99,
United Nations Framework Convention
102–110, 116, 118, 129, 131, 145,
on Climate Change (UNFCCC), 5,
156, 161, 170–172, 185, 186, 189,
13, 14, 31, 211
238
United Nations Principles for
Soft law, 29, 38, 52, 57, 130, 151, 154,
Responsible Investing (UNPRI),
180
34–36,
Solar, 8, 12, 113, 115, 117, 118, 121,
United States (US), xi, 1, 2, 7, 10, 12–14,
156, 188, 190, 204, 211, 216, 223,
18, 19, 28, 29, 32–34, 36, 37, 39,
224, 228–235
41, 43–45, 48–50, 52, 55, 56, 59,
State capitalism, 217, 220, 230, 235
76, 84, 92, 93, 100, 104, 108, 111,
State owned enterprises (SOEs),
112, 114–116, 125, 133, 136–142,
216–221, 228–230, 232, 233,
146–149, 154, 156, 157, 159, 160,
235
166, 169, 172, 178–180, 187, 188,
State supported enterprises (SSEs), 216,
193, 203, 207, 208, 211, 212, 215,
220–222, 234
218, 221, 222, 228–235
Superannuation, 25, 33, 35, 118
United States Code (USC), 8, 10, 112,
Sustainable supply chain networks
136–138, 212
(SSCN), 61
United States Securities and Exchange
Commission (SEC), 41, 42, 45,
46, 141, 142

T

Taxation, 6, 52, 179–180, 202–205, 222,   **V**
224
Texas Utilities (TXU), 44, 104, 108   Value-added tax (VAT), 213, 224
Transformational change, 149–167, 176   Values, 16, 30–31, 33–36, 41, 50, 56, 57,
Transnational, 2–11, 14, 35, 83, 93, 94,   65, 70–80, 82, 84, 87, 96, 97, 106,

119, 124, 126, 127, 130, 131, 143, 145–149, 155, 158, 161, 170, 179, 185, 200, 203, 209, 213, 216, 222, 233, 234, 240–242
Venture capital (VC) funds, 2, 15, 17, 20, 24, 27, 176, 197, 201, 204, 214
Voluntarism, 130, 170, 174, 177, 183

216, 223, 224, 228–232, 235
World Bank, 2, 4, 5, 12, 13, 56, 85, 156, 178
World Business Council for Sustainable Development (WBCSD), 68
World Wildlife Fund (WWF), 43, 77

W

Wind, 8, 9, 12, 29, 31, 55, 113, 115, 117, 118, 156, 177, 208, 211–212, 215,

1. Jan Job de Vries Robbe, *Innovations in Securitisation. Yearbook 2006,* 2006 (ISBN 90-11 - 2533-7).
2. Jim Bartos, *United States Securities Law: A Practical Guide,* Third edition, 2006 (ISBN 90-411-2362-8).
3. Hui Huang, *International Securities Markets: Insider Trading Law in China,* 2006 (ISBN 90-411-2557-4).
4. Barth et al., *Financial Restructuring and Reform in Post-WTO China,* 2007 (ISBN 90-411- 2573-6).
5. Zhongfei Zhou, *Banking Laws in China,* 2007 (ISBN 978-90-411-2519-4).
6. Jan Job de Vries Robbe & Paul Ali, *Expansion and Diversification in Securitization Yearbook, 2007,* 2007 (ISBN 978-90-411-2661-0).
7. Ross Buckley, *International Financial System. Policy and Regulation,* 2008 (ISBN 978- 90-411-2746-4).
8. Jan Job de Vries Robbe, *Securitization Law and Practice: In the Face of the Credit Crunch,* 2008 (ISBN 978-90-411-2715-0).
9. Phoebus Athanassiou, *Hedge Fund Regulation in the European Union: Current Trends and Future Prospects,* 2009 (ISBN 978-90-411-2856-0).
10. Jan Job de Vries Robbe, *Structured Finance: On from the Credit Crunch: The Road to Recovery,* 2009 (ISBN 978-90-411-2787-7).
11. Gaetane Schaeken Willemaers, *The EU Issuer-Disclosure Regime: Objectives and Proposals for Reform,* 2011 (ISBN 978-90-411-3394-6).
12. Panagiotis Delimatsis & Nils Herger (eds), *Financial Regulation at the Crossroads: Implications for Supervision, Institutional Design and Trade,* 2011 (ISBN 978-90-411-3355-7).
13. Raffaele Scalcione, *The Derivatives Revolution: A Trapped Innovation and a Blueprint for Regulatory Reform,* 2011 (ISBN 978-90-411-3430-1).
14. Douglas Arner & Ross Buckley, *From Crisis to Crisis: The Global Financial System and Regulatory Failure,* 2011 (ISBN 978-90-411-3354-0).
15. Anton P. Trichardt, *Letters of Comfort: A Trans-Systemic Analysis,* 2012 (ISBN 978-90- 411-3600-8).
16. Eddy Wymeersch, *Alternative Investment Fund Regulation,* 2012 (ISBN 978-90-411-3690- 9).
17. Asif H. Qureshi & Xuan Gao (eds), *International Economic Organizations and Law: The Perspective and Role of the Legal Counsel,* 2012 (ISBN 97890-411-3427-1).
18. Rhys Bollen, *The Law and Regulation of Payment Services: A Comparative Study,* 2012 (ISBN 978-90-411-3818-7).
19. Ramandeep Kaur Chhina, *Standby Letters of Credit in International Trade,* 2012 (ISBN 978-90-411-4560-4).

20. Dirk A. Zetzsche (ed.), *The Alternative Investment Fund Managers Directive: European Regulation of Alternative Investment Funds*, 2012 (ISBN 978-90-411-4044-9).
21. Lorenzo Sasso, *Capital Structure and Corporate Governance: The Role of Hybrid Financial Instruments*, 2013 (ISBN 978-90-411-4843-8).
22. Pablo Iglesias-Rodríguez, *The Accountability of Financial Regulators: A European and International Perspective*, 2013 (ISBN 978-90-411-3874-3).
23. Seraina Neva Grünewald, *The Resolution of Cross-Border Banking Crises in the European Union: A Legal Study from the Perspective of Burden Sharing*, 2014 (ISBN 978-90-411-4909-1).
24. Megan Bowman, *Banking on Climate Change: How Finance Actors and Transnational Regulatory Regimes are Responding*, 2015 (ISBN 978-90-411-5223-7).